Taxation, Innovation and the Environment

OECD

ORGANISATION FOR ECONOMIC CO-OPERATION AND DEVELOPMENT

The OECD is a unique forum where governments work together to address the economic, social and environmental challenges of globalisation. The OECD is also at the forefront of efforts to understand and to help governments respond to new developments and concerns, such as corporate governance, the information economy and the challenges of an ageing population. The Organisation provides a setting where governments can compare policy experiences, seek answers to common problems, identify good practice and work to co-ordinate domestic and international policies.

The OECD member countries are: Australia, Austria, Belgium, Canada, Chile, the Czech Republic, Denmark, Finland, France, Germany, Greece, Hungary, Iceland, Ireland, Italy, Japan, Korea, Luxembourg, Mexico, the Netherlands, New Zealand, Norway, Poland, Portugal, the Slovak Republic, Slovenia, Spain, Sweden, Switzerland, Turkey, the United Kingdom and the United States. The European Commission takes part in the work of the OECD.

OECD Publishing disseminates widely the results of the Organisation's statistics gathering and research on economic, social and environmental issues, as well as the conventions, guidelines and standards agreed by its members.

This work is published on the responsibility of the Secretary-General of the OECD. The opinions expressed and arguments employed herein do not necessarily reflect the official views of the Organisation or of the governments of its member countries.

ISBN 978-92-64-08762-0 (print)
ISBN 978-92-64-08763-7 (PDF)

Also available in French: *La fiscalité, l'innovation et l'environnement*

The statistical data for Israel are supplied by and under the responsibility of the relevant Israeli authorities. The use of such data by the OECD is without prejudice to the status of the Golan Heights, East Jerusalem and Israeli settlements in the West Bank under the terms of international law.

Corrigenda to OECD publications may be found on line at: *www.oecd.org/publishing/corrigenda*.

Foreword

Today's environmental challenges demand the concerted efforts of citizens, firms and governments to encourage less pollution and environmental degradation and change existing patterns of demand and supply. The OECD's Green Growth Strategy (www.oecd.org/greengrowth) aims to inform debate and assist governments' efforts to develop mutually reinforcing environmental and economic policies – illustrating that "green" and "growth" are compatible.

Environmentally related taxes can effectively achieve many environmental goals and their use is widening within OECD countries. But to meet environmental targets at least-cost, we must move beyond current technologies and know-how: innovation is critical. The project leading to this synthesis report explores the benefits of environmentally related taxes that will accrue when higher pollution costs make it economically inviting to invest in the development of new green technologies. A number of case studies have been prepared, some investigating the role of tax design and others looking at ways in which environmentally related taxes can encourage innovation.

We can see that environmentally related taxation does induce innovation, with firms responding in positive ways to market signals – developing new products, creating novel means to neutralise pollutants and altering production practices to make them cleaner. To bring about the widest range of innovations, environmentally related taxes must be properly designed, and be predictable to give businesses confidence that the clean technologies they develop today will have a market in the future.

Angel Gurría
Secretary-General

Acknowledgements

This book is a product of the Joint Meetings of Tax and Environment Experts, a group under the OECD's Committee on Fiscal Affairs and Environment Policy Committee. Preliminary versions of this publication were presented to this group and participants provided valuable direction, comments and suggestions.

In-depth case studies investigating the effectiveness of environmentally related taxation in inducing different types of innovation provided the basis for this publication. These case studies have been undertaken by a range of external experts, whose work provided illuminating conclusions. Summaries of these case studies are provided in the second half of this book.

This publication has been prepared by Michael Ash, seconded to the OECD from the Government of Canada, in close co-operation with Nils Axel Braathen, Nick Johnstone, Ivan Haščič and Anthony Cox of the OECD's Environment Directorate and with Jens Lundsgaard and Stephen Matthews of the OECD's Centre for Tax Policy and Administration.

Table of Contents

Abbreviations 9

Executive Summary 11

Chapter 1. **Introduction** 17

    1.1. The double market failure: Innovation undersupply and pollution
        oversupply. 18
    1.2. Innovation and low-cost, efficient environmental outcomes 23
    1.3. The intersection of taxation, innovation and the environment 27

    Notes 28

    References 29

Chapter 2. **Current Use of Environmentally Related Taxation** 31

    2.1. Revenues from environmentally related taxation across countries. 32
    2.2. Taxes on specific pollutants 36
    2.3. Exemptions and reductions in environmentally related taxation 51
    2.4. Tradable permits 58
    2.5. Conclusions. 59

    Notes 60

    References 61

Chapter 3. **Effectiveness of Environmentally Related Taxation on Innovation** 63

    3.1. Measuring innovation 64
    3.2. Identifying the benefits and drawbacks of innovation. 70
    3.3. Case studies of environmentally related taxation and the inducement
        to innovate 72
    3.4. Environmentally related taxation and different types of innovation. 79
    3.5. Innovation degree: Incremental *versus* breakthrough technologies. 82
    3.6. Constraints to innovation in response to environmentally related taxation . . 83
    3.7. The adoption and transfer of environmentally related innovation 87
    3.8. Conclusions. 90

    Notes 91

    References 92

Chapter 4. **Tax Design Considerations and other Tax-based Instruments** 95

    4.1. Identifying the appropriate level of the tax 96
    4.2. The extent of the tax base 109
    4.3. Administering the tax 110
    4.4. Tax-based policy instruments 111

4.5. The choice of tax instrument 122

4.6. Creating a policy package: Combinations of environmental
and innovation instruments 125

4.7. Conclusions 129

Notes 130

References 131

Chapter 5. **A Guide to Environmentally Related Taxation for Policy Makers** 135

5.1. Why taxes? 136

5.2. Making effective environmentally related taxation 138

5.3. Using the revenue generated 141

5.4. Overcoming challenges to implementing environmentally related taxes 143

5.5. Environmentally related taxes alone are not the answer 147

5.6. Conclusions 148

Notes 148

References 149

Case Studies 151

Annex A. **Sweden's Charge on NO_x Emissions** 153

Annex B. **Water Pricing in Israel** 167

Annex C. **Cross-country Fuel Taxes and Vehicle Emission Standards** 175

Annex D. **Switzerland's Tax on Volatile Organic Compounds** 187

Annex E. **R&D and Environmental Investments Tax Credits in Spain** 197

Annex F. **Korea's Emission Trading System for NO_x and SO_x** 209

Annex G. **UK Firms' Innovation Responses to Public Incentives:
An Interview-based Approach** 217

Annex H. **The UK's Climate Change Levy and Climate Change Agreements:
An Econometric Approach** 228

Annex I. **Japan's Tax on SO_x Emissions** 239

Tables

2.1. Extent of tax instrument utilisation 45

2.2. Taxes on chlorinated solvents 49

2.3. Pesticide and fertiliser taxes 49

2.4. Full exemptions for agriculture from environmentally related taxes 52

2.5. Tax rates on electricity in OECD countries 55

2.6. Environmental impacts of selected tax reductions/exemptions
in the Netherlands 57

4.1. Inducements for innovation by tax instrument 124

4.2. Welfare effects of taxes and R&D subsidies 127

A.1. Adoption of NO_x mitigation technology in Sweden 155

A.2. NO_x patent applications across countries 156

A.3. Plants subject to the NO_x tax: Descriptive statistics 157

B.1. Agricultural prices for fresh water in Israel 168

B.2. Domestic water prices in Israel 169

C.1. Empirical results: Emission abatement technologies 183

C.2. Empirical results: Input (improved engine design) technologies 184

C.3. Empirical results: Output technologies 185
D.1. Largest VOC reductions by industry 189
E.1. Use of reasoned reports in Spain.. 198
E.2. Sequential impact of tax credits 200
E.3. R&D&I tax credits and tax credit use 201
E.4. Impact of R&D&I tax credit on use of EI credit 201
E.5. Environmental Investments tax credits and tax credit use................. 202
E.6. Impact of environmental investments tax credit in use of R&D&I tax credit..... 202
E.7. Characteristics of tax credit use 203
F.1. Implementation progression of cap-and-trade programme 210
F.2. Pollution impact of low-NO_x burners 213
F.3. NO_x reduction efficiencies by low-NO_x burners 213
F.4. Patents by technical field in Korea 214
G.1. Drivers of innovation and construction of indices 220
G.2. Survey results and energy intensity 221
G.3. Survey results and productivity... 222
G.4. Survey results and innovation... 224
H.1. Rates of the Climate Change Levy.. 229
H.2. Descriptive statistics by CCA participation status 231
H.3. CCA participation and environmental performance 233
H.4. CCA participation and innovation performance............................ 236
 I.1. Annual average rate of change of SO_x reduction 247

Figures

1.1. Estimated effects of innovation.. 23
1.2. Drivers of innovation... 25
1.3. Chain-linked model of innovation 25
2.1. Revenues from environmentally related taxation as percentage of GDP 33
2.2. Revenues from environmentally related taxation as percentage
     of total tax revenues 34
2.3. Composition of environmentally related tax revenues in the OECD 36
2.4. Composition of environmentally related tax revenues by country 37
2.5. Tax rates on motor fuel... 37
2.6. Real changes in tax rates on petrol..................................... 40
2.7. One-off motor vehicle taxes... 41
2.8. CO_2 component of one-off taxes....................................... 42
2.9. Implicit carbon price and motor vehicle taxes 42
2.10. Total CO_2 components of motor vehicle taxes 43
2.11. Tax rates on light fuel oil 46
2.12. Taxes on NO_x emissions to air....................................... 47
2.13. Tax rates on landfill.. 51
3.1. Direct government share of total R&D expenditures 65
3.2. Environmental R&D expenditures in total government R&D allocations 66
3.3. Energy R&D expenditures in total government R&D expenditures 67
3.4. Environmental impacts and economic externalities of innovations........... 72
3.5. Types of environmentally related innovation 80
4.1. Innovation impacts with taxation and tradable permits................... 100

4.2. Categories of tax-based measures. 111

4.3. Tax subsidy for R&D in OECD countries 119

4.4. Determinants of emissions and scope for innovation 123

A.1. Effectiveness of Swedish charge on NO_x emissions 154

A.2. Changes in NO_x emission intensities 158

A.3. NO_x emission intensities at individual plants. 160

A.4. Declining marginal NO_x abatement cost curves 161

B.1. Agricultural output value per unit of irrigation water 170

B.2. Impact of the national water saving campaigns. 171

C.1. Excise tax rates on diesel in select OECD countries 176

C.2. Regulatory tailpipe limits for petrol-driven vehicles 177

C.3. Engine calibration and emission levels 179

C.4. Patent applications for relevant vehicle technologies 181

C.5. Patent applications for the four technological categories 181

E.1. R&D&I and Environmental Investments tax credit use by firm size 199

E.2. Patent applications in Spain and EU15. 204

F.1. Targets for ambient NO_2 and PM_{10} concentrations 210

F.2. NO_x emission trends in Korea 211

F.3. NO_2 concentration trends in Korea. 212

F.4. SO_x emission trends in Korea 212

F.5. SO_2 concentration trends in Korea 212

F.6. SO_x abatement patents in Korea 214

F.7. NO_x abatement patents in Korea. 215

F.8. Budget for environmental R&D 215

H.1. Index of patents in the United Kingdom 235

I.1. Tax rates for current SO_x emissions 241

I.2. Trends in SO_x emissions. 243

I.3. Factors of SO_x emissions 244

I.4. FGD sales and patents. 248

This book has...

StatLinks

**A service that delivers Excel® files
from the printed page!**

Look for the *StatLinks* at the bottom right-hand corner of the tables or graphs in this book.
To download the matching Excel® spreadsheet, just type the link into your Internet browser,
starting with the *http://dx.doi.org* prefix.
If you're reading the PDF e-book edition, and your PC is connected to the Internet, simply
click on the link. You'll find *StatLinks* appearing in more OECD books.

Abbreviations

CCA	Climate Change Agreement (United Kingdom)
CCL	Climate Change Levy (United Kingdom)
CCR	Climate change related
CDM	Clean Development Mechanism of the Kyoto Protocol
CL	Compensation Law (Japan)
CO	Carbon monoxide
CO$_2$	Carbon dioxide
CO$_2$e	Carbon dioxide equivalent (in terms of global warming potential)
ECA	Enhanced Capital Allowance scheme (United Kingdom)
EI	Environmental Investments tax credit (Spain)
EPER	European Pollution and Emissions Register
EU ETS	European Union Emission Trading System
EU15	Austria, Belgium, Denmark, Finland, France, Germany, Greece, Ireland, Italy, Luxembourg, the Netherlands, Portugal, Spain, Sweden and the United Kingdom
FE	Fixed effects
FGD	Flue gas desulphurisation
GDP	Gross domestic product
GHG	Greenhouse gas
GWh	Gigawatt hour
HC	Hydrocarbon
HFC	Hydrofluorocarbon
IEA	International Energy Agency
IP	Intellectual property
IV	Instrumental variable
kcal	Kilocalorie
kWh	Kilowatt hour
LNB	Low-NO$_x$ burner
LNG	Liquefied natural gas
LPG	Liquefied petroleum gas
MAC	Marginal abatement cost
MD	Marginal damage
MOE	Ministry of the Environment (Japan)
MWh	Megawatt hour
NA	Negotiated agreement between industry and government
Nm^3	Normal cubic metre ("normal" in terms of the individual gas)
NO$_x$	Nitrous oxide
OLS	Ordinary least squares regression

PCA	Pollution Control Agreement (Japan)
PM/PM$_{10}$	Particulate matter/particulate matter ≤ 10 μm
Ppm	Parts per million
R&D	Research and development
R&D&I	Research and Development and Technological Innovation tax credit (Spain)
SCR	Selective catalytic reduction
SEPA	Swedish Environmental Protection Agency
SME	Small and medium-sized enterprise
SNCR	Selective non-catalytic reduction
SO$_2$	Sulphur dioxide
SO$_x$	Sulphur oxides
TFP	Total factor productivity
TP	Tradable permit
TWh	Terawatt hour
VAT	Value added tax
VOC	Volatile organic compound

Taxation, Innovation and the Environment
© OECD 2010

Executive Summary

Innovation is critical to achieving environmental
outcomes at a reasonable cost

The world is facing a host of environmental challenges. Some are confined to local areas and may be the result of a few polluters, such as mercury emissions to air or sewage discharges in watercourses; others occur at the global level and are brought about by millions of different actors, such as with the emissions of greenhouse gases. While these environmental issues can be thought of as negative side-effects of countries' economic development, it is important to consider as well that as countries grow richer, more dense, and more technically advanced, the desire and ability to confront these challenges grows as well.

Many of the environmental challenges countries face can seem daunting. The consequences of action can appear high if estimates of the cost of environmental remediation rely on the application of existing technologies and technical know-how. Yet, the ability of firms and consumers to innovate – finding new means and technologies to reduce pollution and its effects – can drastically reduce the costs of future environmental policy. Therefore, as discussed in Chapter 1, the key is finding environmental policy tools which ensure that environmental improvement starts now but which also stimulate innovation and development of cleaner technologies for the future.

The issue of the environment and innovation are of importance to governments because market forces alone do not properly address either issue. There is no price on polluting and therefore firms and consumers pollute too much. Conversely, markets may provide too little innovation. Where innovators are not able to reap the full rewards from their innovations, innovation is generally undersupplied. Hence, for environmentally related innovation, the problem is doubly pronounced: innovation is generally undersupplied but even more so in relation to the environment because, without a price on pollution, there is little incentive to use the innovations at all. These features suggest that there is a role for government to address these externalities.

Environmentally related taxation has many
positive features and its use is widening
in OECD economies

Governments have a range of environmental policy tools at their disposal: regulatory (or "command-and-control") instruments, market-based instruments (such as taxes and tradable permits), negotiated agreements, subsidies, environmental management systems and information campaigns. Although no one instrument can be considered best to address every environmental challenge, there has been a growing movement towards environmentally related taxation (and tradable permits) in OECD economies.

Taxes on pollution provide clear incentives to polluters to reduce emissions and seek out cleaner alternatives. By placing a direct cost on environmental damage, profit-maximising firms have increased incentives to economise on its use, just like other inputs to production. Compared to other environmental instruments, such as regulations concerning emission intensities or technology prescriptions, environmentally related taxation encourages both the lowest cost abatement across polluters and provides incentives for abatement at each unit of pollution. These taxes can also be a highly transparent policy approach, allowing citizens to clearly see if individual sectors or pollution sources are being favoured over others.

The use of environmentally related taxation and emission trading systems is widening in OECD economies, as outlined in Chapter 2. An expanding number of jurisdictions are using taxes and charges in areas like waste disposal and on specific pollutants, such as emissions to air of NO_x and SO_x. Moreover, governments are making their existing environmentally related taxes more efficient, both economically and environmentally.

This widening is coupled with a trend that the amount of revenues from environmentally related taxation has been gradually decreasing over the past decade relative to both GDP and total tax revenues. This trend is driven mainly by motor fuel taxes, which account for the vast majority of environmentally related tax revenues. It partly reflects price increases which have stemmed demand for motor fuels in OECD countries and partly a decline in real rates of excise taxes.

The structure of motor fuel taxes is relatively homogenous across countries, but for other environmentally related taxes, there is large variation between countries. In the case of NO_x emissions, tax rates vary more than one hundred times between countries – and many OECD countries do not levy such taxes at all.

Most environmentally related taxes generate very little revenue. Often, tax bases are quite small, making taxes unlikely to raise much revenue even though the resulting incentives can be quite effective from an environmental perspective. In other cases, tax rates can be quite low. Over the medium term, additional revenues from carbon taxes and from the auctioning of tradable permits may increase the role of environmentally related taxation in government budgets.

Environmentally related taxation stimulates
the development and diffusion of new technologies
and practices

In addition to encouraging the adoption of known pollution abatement measures, environmentally related taxes can provide significant incentives for innovation, as firms and consumers seek new, cleaner solutions in response to the price put on pollution. These incentives also make it commercially attractive to invest in R&D activities to develop technologies and consumer products with a lighter environmental footprint, either by the polluter or by a third-party innovator.

The case studies undertaken for this project shed light on how environmentally related taxation can induce innovation, and some of the key findings are presented in Chapter 3. One of the challenges for such studies is to measure innovation. Common approaches include looking at the intent of firms' innovation efforts revealed by the resources they dedicate to research and development activities or investigating the results of their innovative activities materialising as patents. The case studies examining the innovation impacts of the United Kingdom's Climate Change Levy on fossil fuels and electricity found that firms subject to the full rate of the levy patented more than firms subject to a reduced rate only one-fifth of the full rate. This suggests that the cost burden of environmentally related taxation (i.e. the stringency of the tax) does not adversely affect firms' financial capacity to undertake innovation-related activities.

As innovation occurs in many different forms, such as knowing better how to optimise equipment or experimenting with existing processes, patent data or R&D expenditures are not adequate measures alone, as they cannot capture all aspects of innovation. More informal measures, such as interviews and firm-level analysis, can provide strong supplementary information. In Switzerland, the imposition of a tax on volatile organic compounds (VOCs) – quickly vaporising substances that contribute to smog – affected a wide range of small producers, such as printers, paint makers, and metal cleaners. Most of these firms neither had dedicated R&D units nor developed patentable ideas. Nevertheless, interviews with the firms revealed that the adoption of existing technologies coupled with small, firm-level innovations arising from trial-and-error processes led to significant reductions in VOC use.

Putting a price on pollution creates opportunities for a wide range of types of innovation. This gives taxation an advantage over more prescriptive environmental policy instruments which tend to encourage a focus on end-of-pipe innovations (i.e. innovations reducing the emission of pollution but not the creation of it). A typical example is a "scrubber", a device put on the end of a smokestack to limit emissions. Such innovations are important, but are often less efficient than measures which reduce the pollution in the first place. The wide range of actions that can be induced by taxation encourages a more equal mix between cleaner production process innovation and end-of-pipe abatement measures.

Even for firms that do not have the resources or inclination to undertake formalised R&D activities, the presence of environmentally related taxation provides increased incentives to bring in the latest technologies that have already been developed elsewhere. In Sweden, for example, the introduction of a tax on NO_x emissions led to a dramatic increase in the adoption of existing abatement technology: only 7% of firms had adopted abatement technology in the year that the tax was introduced but the fraction rose to 62% the following year.

The wider context plays a significant role in shaping the innovation outcomes of environmentally related taxation: a country's intellectual property rights regime, the system of higher education and cultural norms towards innovation all contribute to a country's innovation capacity. In the Israeli case study, innovations observed in the water sector may result from an innovative culture spanning several decades, in addition to the presence of high water prices and taxes.

It should be noted that the case studies undertaken as part of this project do not provide unambiguous evidence that environmentally related taxation will always lead to innovation and the adoption of new technologies and processes. For example, a cross-country examination of the innovation impacts of petrol prices and taxes, regulations and standards on motor vehicles found linkages between emission regulations and related patents and between fuel taxes and fuel efficiency patents but the results were not completely robust. The study on the United Kingdom found support for the climate change tax encouraging general innovation but not specifically climate change-related innovation. A few reasons why the links between innovation and environmentally related taxation may not be clearly revealed in empirical analyses include:

- First, the use of environmentally related taxation (other than on motor vehicle fuels) is still relatively new, providing limited scope for wide-ranging analysis.

- Second, investigating the innovation effects of environmentally related taxation is significantly more difficult than for other environmental policy tools. Regulatory approaches to environmental policy are often prescriptive (such as setting maximum emission intensities or mandating specific technologies) and targeted at specific sectors or polluters, making it relatively easy to locate any effects. By contrast, the very advantage of using tax instruments is that they promote many diverse innovations. Locating and identifying potential innovations arising from the incentives created by taxation is therefore far more difficult.

- Third, environmentally related taxes may not have been optimally designed which can dampen abatement activities, investment decisions and innovation efforts.

- Finally, many other factors affect firms' innovation efforts. With limited data availability, it can be difficult to disentangle the isolated effect of taxation.

Tax design issues can have a significant effect
on the resulting innovation

The design of environmentally related taxation plays an important role, and is analysed in Chapter 4. As mentioned above, the level of the tax is a significant factor – the higher the rate, the more significant the incentives for innovation. Taxes levied closer to the actual source of pollution (*e.g.* taxes on CO_2 emissions *versus* taxes on motor vehicles) provide a greater range of possibilities for innovation. However, in some cases, taxes levied directly on the pollutants can be difficult to administer, where it requires monitoring of many dispersed and varied sources.

A conducive environment for innovation, characterised by credible policy commitment and predictability in tax rates, is also a critical ingredient to encourage investment in innovative activities. Unlike market uncertainty (such as oil prices), policy uncertainty is more difficult to hedge against. As seen with Japan's SO_x charge, the uncertainty surrounding the viability of the overall scheme had negative effects on patenting in the long run, despite very high tax rates.

It must be recognised that political economy issues can influence tax design and lead to differential impacts on innovation. The low tax rates provided to some households or to energy-intensive/trade-exposed sectors in the United Kingdom provide significantly less incentives for the development of innovation and its adoption. Instead of lower tax rates, other countries have instituted refunding mechanisms, which recycle the revenues back to affected firms on a base different from the collection base. Such mechanisms maintain the marginal incentive to abate (especially where a higher tax rate can be levied because of the existence of revenue recycling) but can weaken some of the incentives to innovate, especially innovation undertaken at the collective level. They may also be at odds with the polluter-pays principle by not making "dirty" products or activities more expensive.

The international aspects of environmentally related taxation are important to consider as well. Like with many environmental policy instruments, there is always concern over introducing policies that are too stringent and cause emission-intensive activities to relocate to other jurisdictions. International co-operation and co-ordination in setting environmental taxes can significantly reduce this risk. Doing so also provides an additional benefit for innovation: the use of environmentally related taxation maximises the international movement of innovation. For two countries using taxes on the same pollutant, an innovation generated in one can necessarily be used in the other. This is less straight forward for regulatory approaches which are typically more prescriptive, potentially limiting the scope for transferring innovations across countries.

Taxes and other environmental policy instruments can complement each other

---

Well-designed taxes put a clear price on the damage to the environment and therefore should overcome much of the environmental externality problem. However, some barriers may require supplementary policy measures. Consumers may not be aware of the full impact of their purchase over the long term and taxes may not affect the incentives for some agents (*e.g.* tenants) if others (*e.g.* property owners) have to pay the tax. Thus, information campaigns and regulations may help complement environmentally related taxation and increase its impact. Such complementarities can help reinforce each instrument. Meanwhile, an overlap of taxes and tradable permits on the same emissions can be problematic, as the tax can have either no net environmental benefit or even cause inefficient abatement across sectors.[*]

Some countries have sought to use the tax system for environmental policy in a number of alternate ways, such as through accelerated depreciation allowances and reduced rates of taxation on environmentally friendly goods. These measures attempt to reduce the cost of "good" actions instead of penalising "bad" actions and they can act similar to subsidies. As a drawback, however, they also tend to favour capital-intensive approaches over simpler approaches. Moreover, these are not costless initiatives – they necessitate that governments find other sources of funds, putting additional pressures on government budgets. If an adequate price is put on pollution via taxation, these instruments are not very cost-effective at inducing additional abatement and innovation.

Many countries have broad innovation policies, although their forms can be quite different. These include supports to universities and researchers, favourable tax treatment of inputs to R&D and of the returns from innovation, intellectual property protection regimes, etc. If these systems are adequate in addressing the undersupply of innovation generally, then they should also be so for environmentally related innovation. Special R&D tax credits targeted at environmental innovation face many of the same drawbacks as other measures stimulating the "good". Most importantly, it has only limited effects on innovation when used as the sole environmental innovation policy instrument: if no cost is put on polluting, adopting technologies brought about by the R&D tax credits provides no benefit to the adopter. Effectively, there is only a benefit to adoption when these actions also reduce some other cost to the adopter. For example, a firm is unlikely to make an investment with any level of tax credit towards a technology that solely reduces carbon emissions if there is no cost at the outset to emit carbon. Where the technology may also save their firm money

---

[*] Taxes may play a role where they are combined with tradable permits that have been auctioned for free. If they are on exactly the same emissions as those covered by the tradable permit scheme, the taxes will lower the price of the permits but recover some of the windfall gains that firms received by not having to buy their permits at auction, which can be desirable from an equity point of view.

(that is, reduce carbon emissions because it increases energy efficiency), only then may an R&D tax credit provide an additional boost and help mitigate the environmental problem.

Environmentally related taxation provides significant incentives for market-ready innovations, but the high-risk, long-term efforts needed for "breakthrough" advances still face barriers – policy and market uncertainty, access to capital and economies of scale – even if all pollutants were taxed optimally. This suggests that broad innovation policies may not adequately address some of the specific issues related to the environment. Additional R&D tax credits targeted to environmental outcomes would likely induce additional innovation but not of the fundamental nature required. Policies outside of the tax system may be required, such as government funding for basic R&D into the development of breakthrough technologies.

This suggests that the optimal approach is to have a strong environmental policy that addresses the oversupply of environmental damage in society; taxes levied directly on environmentally harmful activities should play a significant role. The tax should seek to address the environmental damage but does not need to go above and beyond to specifically address environmental innovation. Concurrently, broad innovation policies should address the undersupply of innovation (including for the environment).

---

*Best practices for implementing environmentally
related taxation rely on a wide range
of considerations*

Based on the findings in this study and others lessons learned by OECD countries, Chapter 5 offers a best practices guide for policy makers. The scope for the expanded use of environmentally related taxes in OECD countries is great, especially in addressing climate change. Bringing in such taxes requires careful consideration of the coverage and design of the tax. To be most effective, environmentally related taxes should cover all sources and all levels of pollution, and governments should not be afraid to levy a tax that will fully address the environmental challenge. While recognising that tax rates should reflect a wide variety of potentially changing factors, they should nevertheless be relatively predictable to strengthen investment and abatement decisions.

The implementation of environmentally related taxation can involve significant political economy challenges. Concerns about the potentially regressive nature of taxes, particularly regarding taxes on water and energy, can bring about attempts by government to modify the tax design in order to reduce the burden on low-income households. While progressivity is a consideration, it is the progressivity of the entire tax and social security system that is important. Therefore, such concerns should be addressed through other means (lower personal income taxes, in-work tax credits, increased social benefits, etc.) rather than the environmentally related tax itself. Separately, there are some concerns that environmentally related taxation can encourage trade-exposed, pollution-intensive activities to relocate to places where such taxes are lower or non-existent. Reduced rates for such activities are common. Yet, the single most important measure to overcome this risk is international co-operation – building similar environmental policies across markets. Finally, citizens in some countries tend to be sceptical of environmentally related taxation, believing that it may simply be a tax grab or may not fully understand why the tax is being levied. Strong communication and credible proponents of the tax (such as a green tax commission) can help overcome some of these issues.

Chapter 1

Introduction

> This chapter introduces why an unregulated market provides too much pollution and too little innovation, the combination of which makes environmentally related innovation doubly undersupplied. It outlines that such innovation is critical for achieving environmental targets cost-effectively. There is discussion of the process of innovation, its drivers and the role of governments and industry. The chapter finishes with a discussion about the role of taxation in correcting these two market failures.

Environmental challenges are growing in prominence across the globe. With rising populations and growing economies, there are increasing pressures on the natural environment. At the same time, economic development and the associated rise in real incomes over most of the world are also creating a green wealth effect – that is, people are willing to allocate a greater proportion of their wealth towards protecting the environment. This growing interest in – and willingness to pay for – environmental preservation and protection is not without limits: achieving environmental goals efficiently and at low cost remains a top priority. Innovation is a key component of this, as attaining strong environmental goals with today's technology and know-how will be much more costly than using new and novel approaches over the coming years and decades. How new ideas and technologies are developed and applied to today's environmental challenges is critical. In this vein, Jaffe and Stavins (1990) suggest that "the effect of public policies on the process of technological change may, in the long run, be among the most significant determinants of success and failure in environmental protection".

This study looks in particular at one aspect of environmental policy – environmentally related taxation – and investigates how it affects the innovation process. Important to this is not only the development of innovation but so too the adoption of innovation by firms.

1.1. The double market failure: Innovation undersupply and pollution oversupply

Governments have a particular interest in environmental innovation simply because normal market mechanisms do not work perfectly. The fields of the environment and innovation are ones fraught with classical economic problems. Ideally, citizens, who "own" the environment and who want less pollution would charge emitters for spoiling their property. Through agreement among market participants, the problem would be solved. Clearly, this does not happen. In the real world, there is an oversupply of pollution because of the lack of prices and ownership rights for harming the environment.

With respect to innovation, inventors would ideally have perfect foresight about the opportunities ahead and have access to all the necessary funding. In addition, they would be able to fully reap all the monopoly benefits that would come from their invention. Again, the real world does not afford such conditions and therefore there is an undersupply of innovation. These market constraints coupled with knowledge spillover effects reduce potential returns from innovation. When the environment and innovation are taken together, Jaffe *et al.* (2005) contend that environmental innovation or technological change is doubly underprovided by markets.

1.1.1. The undersupply of innovation

Innovation plays a central role in promoting long-term economic growth. New products, more efficient processes and novel management methods can all lead to new business opportunities and greater profitability for innovating firms. In the health field, it can lead to groundbreaking medical breakthroughs; in the transportation sector, it can lead to safer and more reliable cars; and, in retailing, it can help to get more products to consumers at lower prices. Basically, innovation expands the range of possibilities available and leads to a more efficient allocation of existing resources.

Imperfections in the marketplace create conditions where the optimal level of innovation is not attained. But how does one know what an "optimal" level of innovation is? In a perfectly efficient market, firms would invest in processes that (hopefully) lead to innovative outcomes. The expected benefits or rate of return that accrue to the inventor determine the initial level of investment. The higher the expected rate of return, the higher the initial investment. Fully functioning markets and complete property rights would ensure that the firm reaps the full benefit of the innovation. Thus, the rate of return to the firm (that is, the private rate of return) would be the same as the rate of return to the entire economy (that is, the social rate of return, which includes that to the inventor) as the firm was able to capture all the benefits.

However, first there are market imperfections that hamper the ability for innovation to be developed and for the inventor to foresee the value of the innovation:

- *Incomplete information*: Critical to the successful creation and deployment of innovative products and processes is that there is a clear understanding about the potential of such an innovation. Yet, there are numerous instances where information is not perfectly transmitted across economic actors or there is uncertainty about the outcomes of certain endeavours. As such, incomplete information can hamper innovation to a level below the social optimum. The predictability of the policy environment is also critically important. In the case of environmentally related taxation or tradable permit systems, for example, changes in the level of a tax rate or in the quantity of allowances can impact on the expected rate of return of a firm. Market-related uncertainty is also a significant issue for any business decision. Investing in research and development activities or yet-unproved technologies can present unknowns that may require a higher hurdle rate of return to overcome, especially where external financing is being sought.[1]

- *Economies of scale*: There are likely to be economies of scale in the inputs to innovation, primarily being investments in R&D. The purchase of physical infrastructure (much of which is likely indivisible) and the hiring of human resources to undertake this research likely has significantly higher returns with a higher initial level of investment, contributing to an increase in the hurdle rate for investment.

Second, the fundamental nature of innovation – that it is basically an idea – suggests further that the market will not provide the inventor with a full recovery of all the benefits of the innovation. There are a number of reasons why this occurs, including:

- *Knowledge externalities*: Since an inventor cannot perfectly stop others from benefitting, either directly or indirectly, from the invention, the private rate of return is lowered due to these knowledge spillovers. This can be thought of, therefore, as the social rate of return remaining the same, in that the economy as a whole derives value from the innovation, but the private rate of return becomes lower, as some of the benefits cannot be internalised by the firm. As firms decide what projects to undertake, these lower rates of private return suggest that fewer projects are undertaken than would be given the social rate of return. This causes an undersupply of innovation compared to the social optimum. Governments have put in place instruments to help inventors appropriate a larger share of the value of their inventions. Other inventors may generate ideas based on the initial idea for which the patent holder may not be remunerated. In other cases, some ideas simply cannot be patented and, as such, they may be copied by competitors.

- *Externalities related to use:* Many times, the value of an innovative product or process grows as users use it – that is, there are dynamic increasing returns to its use. They become better at using and/or making the item, and this knowledge can leak, providing positive externalities to others. The two main categories are:

  ❖ Learning-by-using: New users of technology must learn how to effectively use the innovation and adapt and integrate it into their routines. In some cases, this learning experience can be a source of information for other users, thereby creating externalities for others.

  ❖ Learning-by-doing: In much the same way but from the production aspect, manufacturers learn efficiencies in reproducing the technology. Inasmuch as these knowledge gains can be seen by other manufacturers, they represent an unrecoverable transfer of knowledge wealth to others.

Other people can also just adopt a technology. While not devising better ways to use the technology, their use alone provides benefits to others and can be thought of as network externalities. That is, others' use of technology increases the utility of one's own use because the value of the product has increased. Telephones and social networking sites are classic examples. These returns cannot generally be captured and therefore provide positive externalities to other users.

These various market imperfections and other constraints clearly suggest that the realised level of innovation will be below that of the social optimum unless public policies are put in place to stimulate innovation. Besides only affecting the level of innovation and technological change, these market failures can influence the type of innovation as well. Along the innovation continuum, there is an infinite range of innovations that can span from, at one extreme, innovations with significant public benefits (such as basic research into nuclear fusion, for example) to those with significantly private benefits (such as a more efficient production technique that can be patented and employed by a monopolist) at the other. Firms will focus more attention on innovations with more private benefits. Issues of appropriability and the uncertainty of some significantly longer-term projects suggest that with market failures, innovations with more public aspects are even more reduced than those with more private aspects.

Innovation is critical and governments have long recognised the issues creating an undersupply of innovation. Numerous government programmes and initiatives have been launched in an attempt to spur greater levels of technological change. Five major efforts typify this response (the first deals with the general innovative environment, the remaining four deal with addressing the externality issue more directly):

- *Creating a conductive business and innovative environment:* Reducing barriers to creating and commercialising innovation as well as ensuring adequate returns from its use create a general business climate that is conducive to innovation. This should be in addition to an environment that is supportive of general innovation activities, such as through a society that is research-driven and open to new technologies.

- *Patent protections:* Intellectual property rights regimes provide some legal protections to creators of intellectual property for a number of years; however, such structures are not perfect and cannot prevent all leakages of information.

- *Direct support of basic research*: Governments directly invest in basic research through government laboratories and research stations or through grant-providing bodies. They can also subsidise private firms' R&D efforts, either directly or through joint ventures with higher education institutions.

- *Supply of researchers*: Governments encourage the supply of researchers through university placements. The goal is to both create a more conducive environment for fostering innovation as well as allow for an expansion of R&D budgets that is not simply consumed through higher wages.

- *R&D tax measures*: Most OECD countries employ tax incentives for research and development activities as a means to encourage innovative activities by overcoming the difficulties mentioned before. These measures typically attempt to reduce the marginal cost of capital for firms[2] by providing tax credits for R&D expenses or providing favourable treatment to capital and/or labour expenses.

To overcome the fact that social and private rates of return are different, patent protection regimes attempt to fully internalise the positive externalities for the inventor by increasing the revenues accruing to the inventor, but not affecting the costs to innovate. By contrast, R&D tax credits/subsidies alone seek to lower the costs of innovation, but do not attempt to increase the revenues for the innovator. Both are likely to have scale effects, as the private rate of return is now closer the social rate of return. The difference between the approaches is that while both mechanisms seek to provide a higher return to innovation efforts (approaching the social rate of return), R&D tax credits do it without internalising the externality and therefore maintain the positive spillovers of innovation, benefitting the economy as a whole. Assessing the proper balance, coupled with other pressures on governments, remains a difficult issue.

The case for governments attempting to provide full internalisation of innovation externalities is not as clear cut. On the one hand, ensuring that innovators can internalise a large share of the returns to their creations is important for providing incentives to innovate. On the other hand, the spillovers from innovations positively benefit the rest of the economy by providing impetus and ideas for future growth and additional innovation. This may be especially true with issues such as the environment. Governments must therefore balance these two objectives and the usage of different tools in innovation policy is likely required.

1.1.2. *The oversupply of pollution and the overuse of resources*

Contrary to the undersupply of innovation, unregulated market forces lead to an oversupply of pollution in the economy. Without effective property rights on the environment, polluters do not have to take account of the damage that they are doing to the environment.[3] The effect of the pollution is not (only) felt by the firm but the effects are realised by society at large, which is not compensated for the damage – the negative externality. Under an optimal scenario, polluting firms would choose a production level where their marginal cost of abating emissions was just equal to society's marginal value of the environment – that is, the value of an additional unit of pollution. Without effective mechanisms to translate society's value of the environment into a market-based constraint for firms, pollution will continue to be emitted until the marginal cost to the firm is zero (that is, the input cost of the environment is effectively zero). That is, they will pollute until it is no longer economically profitable for them to do so, which would be well above the societal optimum.

Governments have a range of policy instruments with which to address environmental challenges. Some traditional approaches have relied on prescriptive regulations that have limited the flexibility of firms and the range of potential mitigation measures but have also provided clear paths to pollution reduction. Governments have shifted in recent years to embracing more market-based approaches.

- *Regulatory approaches:* Also known as "command-and-control" approaches, these have traditionally outlined limits and/or approaches for specific industries. These can take the form of emission intensity limits, technology ordinances, or absolute emission limits. They are typically directed at individual industries or specific product characteristics and with the focus usually being on the larger operators.

- *Voluntary approaches:* Governments can also work co-operatively with industrial partners to arrive at binding or non-binding agreements to address emissions, or establish programmes to which firms voluntarily can adhere, thereby reducing the need for legislation.

- *Market-based instruments:* These instruments rely on allowing price signals to motivate firms to find the lowest-cost means of abatement by placing a value on (or at least near) the activity causing environmental damage. These can either take the form of a tax on the pollution, a tax on a proxy to pollution, or an emissions trading system that auctions or freely distributes permits, effectively giving the holder of a permit the right to emit (or that give "credits" to polluters that reduce emissions below a predefined baseline). These permits and credits can typically be traded and banked across time periods and have very similar features and effects to taxes.

- *Subsidies:* Instead of trying to induce abatement by taxing the bad, governments can also try to subsidise the good. By reducing the cost of environmentally friendly actions or products, the structure of demand and supply can be influenced.

- *Information:* In addition to the approaches above, governments have also typically undertaken information campaigns to raise awareness about environmental issues. These can take the form of public-service type messages encouraging citizens to undertake green acts or provide greater information on making environmental choices in consumption, such as detailing information on energy utilisation and expected lifetime costs of certain appliances. This information, which is typically difficult for consumers to collect and compare across different options, can help overcome informational barriers and reinforce environmentally related taxation on energy, for example.[4]

Evaluating which environmental policy instruments are best is a difficult task given the range of potential criteria and the persistence of potential roadblocks to implementing optimal policy design. One of the most important criteria is looking at the ability of environmental policy instruments to achieve the lowest-cost outcome (which includes ensuring that all means to abate are stimulated at all levels of pollution). Especially at the theoretical level, environmentally related taxation and cap-and-trade systems are considered to be the optimal choice, given their ability to achieve the two efficiencies mentioned above (even more so if the exact location of the polluting activity is of limited significance). However, administrative burdens, information constraints, political economy pressures and other issues create scenarios where alternate policy instruments may perform best. For these reasons, other approaches to (either alone or in combination with) environmentally related taxation are sometimes more effective.

1.2. Innovation and low-cost, efficient environmental outcomes

New developments and breakthroughs can have the ability to dramatically lower the costs of achieving environmental goals or achieving strong environmental goals at the same price. Such innovations can be small, such as a firm learning new ways to calibrate industrial machinery to emit fewer pollutants, or more radical, such as the development of alternative energy sources.

1.2.1. Why innovation needs to be central to environmental policy

Economists have created models of climate change, to use a prominent example, to model the effects of innovation on the economic costs of the policies. While the results differ quite dramatically, the common result is that innovation has a strong impact on reducing the financial impacts of meeting environmental challenges. Popp (2004) creates a model where innovation is brought about because of the new environmental policies. The effect of this innovation is an increase in welfare of 10% under an optimal carbon tax scenario, driven primarily by cost savings rather than additional environmental improvement. Gerlagh and Lise (2005) find that including technological change into a climate change model with a constant carbon tax brings about three times more emissions reductions than without innovation being present. Kemfert and Troung (2007) find that accounting for induced technological change significantly reduces the negative GDP impacts of climate change policies. Similarly, Gerlagh (2008) finds that, accounting for technological change in his model, the optimal carbon tax is less than half of what it is in a scenario without innovation.

In modelling undertaken by the OECD, potential innovation was found to have a large impact on the costs and the effect of climate change mitigation policies, as seen in Figure 1.1. With the assumption of two breakthrough technologies (these are undetermined and not-yet-developed technologies that are assumed in the model to be viable in the future),

Figure 1.1. **Estimated effects of innovation**
Estimated emissions permit prices and GDP costs

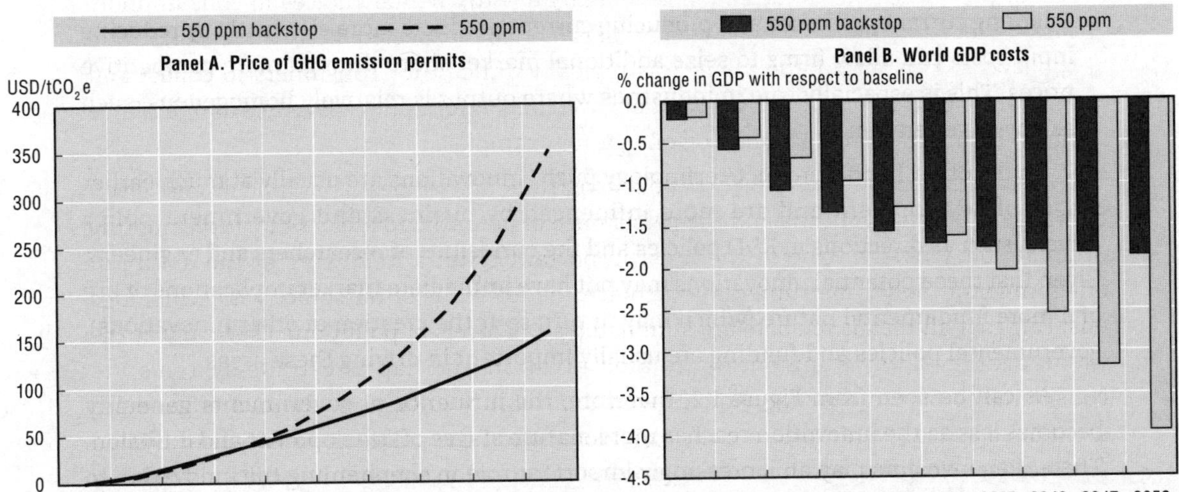

Notes: Emissions of non-CO_2 gases are not covered by the model used in this analysis and are therefore excluded from these simulations. The 550 ppm greenhouse gas concentration stabilisation scenario run here is in fact a 450 ppm CO_2-only scenario and greenhouse gas prices are CO_2 prices. Stabilisation of CO_2 concentration at 450 ppm corresponds to stabilisation of overall greenhouse gas concentration at about 550 ppm.

Source: OECD (2009a).

StatLink ᴍᴤ🌐 http://dx.doi.org/10.1787/888932317160

more expensive abatement efforts through incremental innovation (with higher marginal costs) are now avoided. The result is that the negative impact of climate change policies on GDP in 2050 is reduced by half and significantly lowers the carbon price (through taxes or tradable permits) needed to achieve a 550 ppm stabilisation target for greenhouse gases. Greater effort and greater resources are needed in the short term to bring about these breakthrough technologies compared to relying on incremental innovation alone, resulting in a greater short-term hit to GDP. Over the longer-term, however, these investments provide significant dividends by effectively stabilising GDP losses. Therefore, a central question is what are the factors inducing innovation. Understanding the process is a start.

1.2.2. The innovation process

There are three parts to innovation – the creation and development of the innovation, the adoption (or diffusion) of the innovation within an economy, and, finally, the transfer of the innovation between economies. Looking at how these work and interest is discussed below. In addition, the OECD has developed an innovation strategy (OECD, 2010) to look broadly at the issues of innovation.

The development and drivers for innovation, at a basic level, are well known (see OECD (2009b) for a more extensive discussion of this issue). On the one hand, demand factors create a "market pull" force for innovation. Consumers, reacting to a range of influences and tastes, create demand for new technological advances (and encourage competition to provide existing goods and services at lower cost). Firms react to these forces by investing in R&D and quickly deploying innovations. "Market pull" innovations are typically more developed and market-ready, and firms are more confident in their potential for success in the market. Such "market pull" innovations are typically brought about by two factors.

- Competitive pressures within a well-functioning marketplace are the largest drivers of innovative activity. Developing new products to gain an advantage in the market can provide significant incentives to invest in innovation. The high-tech industry is a prime example, where near-constant product development is critical to a firm's success in the industry.

- Adapting current processes and producing current products more efficiently by reducing input costs can allow firms to seize additional market share through more competitive prices. This is especially true in industries where output is relatively homogenous, such as power generation.

On the other hand, "product/technology push" innovations are usually at much earlier stages of development and are more influenced by business and government policy drivers, such as directions in R&D policies and the curiosities of researchers and engineers. Given that these potential innovations may not have immediate market implications or are of a more fundamental nature (which may in turn spur the creation of other innovations), governmental policies and funding are usually important in driving these areas.

As can be seen from Figure 1.2, therefore, the influence of governments generally becomes less as the innovation reaches more mature stages of development and diffusion. Thereafter, investors play an increasingly important role in shepherding the innovation to market and in reacting to consumer demand.

The actual creation of innovation is not as straightforward as Figure 1.2 suggests, however. Each stage of the innovation process has an impact on other stages, both for the innovation itself and for other innovations. The interrelationships between the knowledge base, the creation process and the development process create a "chain-linked" model of

Figure 1.2. **Drivers of innovation**
Market pull *versus* product/technology push

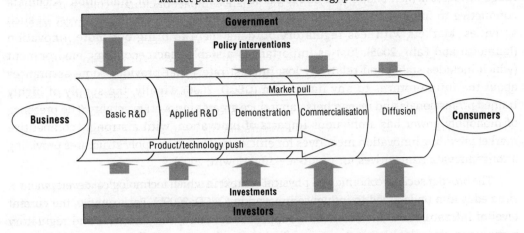

Source: Foxon (2003).

Figure 1.3. **Chain-linked model of innovation**

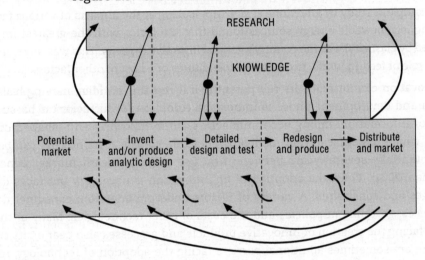

Source: Kline and Rosenberg (1986).

innovation (Kline and Rosenberg, 1986). Figure 1.3 outlines this model, which accounts for the fact that ideas generated by users in the development stage can have impacts on the basic fundamentals of the innovation and can even spur new innovations. This back-and-forth, start-and-stop model reflects well the typically hectic, unscripted and collaborative nature of innovation development.

This model reflects that innovation is much broader than one of its principal means of development, that of R&D activities. Innovation is not only about development of ideas within a firm from a selected group of specialists and the commercialisation thereafter for the firm's use. Innovation is a collaborative and multidisciplinary approach that goes beyond a firm's own walls. It relies on existing expertise elsewhere, including in other fields and by different actors. And the innovations themselves can be used by a wide range of actors, within the firm, within the sector and beyond.

The interrelationships and drivers described in Figures 1.2 and 1.3 interact with a wide range of forces which shape and direct the rate and direction of innovation. A climate conductive to innovation can influence the decisions of individuals to invest in such activities. Markets with less regulatory burdens seem to bring out more innovation (Jaumotte and Pain, 2005). Just as important is a stable macroeconomic environment (which includes stable and relatively low interest rates) that provides some assurances about the future returns to any innovation (OECD, 2006). Finally, the supply of highly trained professionals and researchers can induce greater innovative activity. The presence of monopoly power has ambiguous impacts of innovation, with a strongly competitive market providing innovation incentives for efficiency while monopoly structures providing strong innovation incentives for profit-seeking (Howitt, 2009).

The broader social, economic and physical context in which technological development is situated is also understood to influence innovation (OECD, 2009b). For example, the current level of infrastructure and science to support innovation, the financial and regulatory institutions, the cultural aspects surrounding innovation acceptability and encouragement, and the political drivers differ across countries and exert unique influences. The confluence of these factors can create effective regimes, which can provide powerful support to innovation. As an example, a study by Johnstone et al. (2010) looking at the impacts of various factors on patenting in renewable energy sources found that the factor with the greatest impact (as measured by elasticity) was the country's general inventive capacity. This was more important that the role of feed-in tariffs, taxes, R&D expenditures or a host of other factors.

Innovation cannot reach its full potential if it remains an idea; moving beyond the creation and development phase is important. Adoption (or diffusion) is based on the spread of information among economic actors, often following an S-shaped diffusion pattern, which is similar to that modelling epidemic spreads. The process can be rather slow, potentially several years between first use and significant market penetration (Stoneman, 2001). Thus, the adoption of an innovation is necessary but faces different challenges and constraints. A variety of factors underpin adoption: consumer demand, input prices, government policies and the costs of other technologies. Many of the same barriers facing the creation of innovative products and processes also bear on its creation. Yet, there are sometimes additional barriers facing the adoption of technology, resulting from the fact that new innovations are not always immediately embraced.

Technology lock-in can provide a significant wall to new innovators. Previous innovations, because they were so successful at the time, led to a domination of the market. New innovations face the prospect of having to overcome this inertia. Doing so may require large-scale investments on a number of different fronts. For example, hydrogen-powered motor vehicles would not only have to be accepted by consumers as a smart investment on their own but would also need a network of refilling stations to enable people to effectively make the switch. The technology lock-in of liquid fuels provides a significant barrier to alternative innovations.

Consumers can have very high discount rates,[5] preferring sometimes to purchase lower-cost goods (with higher operating costs) than higher-price goods (with lower operating costs). Also, for some innovations to reach their full potential or usefulness, there must be a network of others users of the same innovation. The first Facebook (or telephone, for that matter) user likely realised this: value of social networking sites, for example, relies on the number of other persons using the technology as well.

One of the largest issues is where the innovation is capital intensive (or where the innovation is imbedded in physical capital). The costs associated with new technology coupled with the likely existence of older, but still useful, capital suggests that innovation adoption will occur as older technology is replaced and new technology is needed. The better an innovation is, the quicker may be the adoption but a complete overhaul of capital is unlikely but for the most significantly advancing innovations. A common measure to look at innovation adoption is to look at the ratio of firms adopting the innovation. However, what is likely more important is how deeply firms adopt the innovation – that is, how quickly they put the innovation in place for all relevant parts of their industrial/creative/service processes. Intra-firm innovation diffusion is typically slower than inter-firm innovation because many firms can undertake some limited innovation adoption with ongoing capital replacement while integrating the innovation throughout the entire firm is more involved (Battisti and Stoneman, 2003; Battisti, 2008), suggesting that for areas of particular interest to governments, changing the direction and speed of adoption may focus on this oft-neglected area.

Finally, the diffusion of innovation is not just limited to firms within the same country – the transfer of innovation across countries (such as in intellectual property) can further spread the reach of the innovation and increase abatement options for foreign polluters. Many of the same issues facing innovation adoption also face innovation transfer. The compatibility and flexibility of countries' environmental policies plays a role in the level of potential transfer. Countries' tax legislation, rules on foreign investment and the stringency of intellectual property regimes also factor into firms' decisions about transferring intellectual property – either in the form of the intellectual property itself or embodied in a product.

1.3. The intersection of taxation, innovation and the environment

Environmentally related taxation is aimed at achieving environmental aims but, by targeting the prices of environmentally harmful consumption, can influence the market pull type of innovation, since firm-level determinants of innovation are centred on prices to firms. In a competitive environment, firms are profit maximising, meaning that the overall mix of input and output prices can greatly determine how and what firms produce. Hicks (1932) first described the impact that this has on technological change through his induced innovation hypothesis:

> [A] change in the relative prices of the factors of production is itself a spur to invention, and to invention of a particular kind – directed to economising the use of a factor which has become relatively expensive.

In order to continue maximising profits, firms will reorient their input/output mixes to maximise revenue while minimising cost. This economising leads firms not only to adjust their production processes but also to adjust their innovation-seeking behaviour, such that it too is reoriented to the new relative prices. When applied to the environmental context, the theoretical example is just as strong. Firms must take into account all factors of production, including their use of the environment (which is consumed when pollution is emitted). The problem is that the environment as an input does not typically have an identifiable (and therefore effective) impact on the firm: there is no price on the use or destruction of the environment.

Clearly, taxation can insert itself here. Taxes, especially excise taxes, can put an explicit price on the environment and therefore should lead to some induced innovation because taxation changes the rate of return to the investor. In the absence of taxation, the theoretical return for inventing a new energy-efficient process should be the future stream of all energy savings. The introduction of taxation creates additional potential returns to the investor: the return on investment is now the future stream of all energy savings plus the reduced tax burden on the energy saved. With a higher expected return, the initial investment (and therefore the resulting level of innovation) should be higher.

Under the induced innovation hypothesis, the firm is still negatively impacted with the environmental tax or regulation that brought about the production change, however. The increased cost of some inputs of production has moved the firm away from their previously optimal point.[6] The additional innovation induced because of this change helps to mitigate, but not fully offset or even exceed the burden on the firm. If there was a net benefit to the firm, a perfectly optimising firm would have done this even in the absence of the new environmental policy. Where there was no incentive before, such as with the emission of many air pollutants, new environmentally related taxation can now offer incentives for abatement. It should be noted that the effects of taxes and tradable permits are generally quite close in this regard (see Box 3.4 for more information).

The tax system can be used in ways other than simply levelling taxes on pollution and taxes on proxies to pollution (such as petrol taxes). Reduced rates of consumption taxes on green products, accelerated depreciation allowances in the corporate income tax code and tax credits for R&D expenditures are also used to encourage environmental protection and innovation. The variety of tax measures can have varying effects not only on the level of innovation but also the type.

Environmentally related taxation is only designed to address the one externality: the oversupply of pollution. By targeting the one externality, it should provide greater incentives for innovation. It does not, however, specifically target the innovation externality. While the incentives to innovate may be greater with environmentally related taxation in place, the barriers to innovation still remain. Therefore, the optimal amount and type of innovation to help solve global environmental challenges will likely not be achieved by environmentally related taxation alone. A strong rationale still exists for other instruments being a part of governments' overall toolkit to specifically address the innovation externality. These policies could include broad-based innovation policies, such as R&D and support to universities (traditional areas of government policy intervention), or more targeted interventions where required.

This report, therefore, attempts to explore a number of key issues, such as whether environmentally related taxation has a positive influence on innovation, what types of taxation are optimal and how taxation affects the range of innovation possibilities. Consideration is also taken of the use of environmentally related taxation in OECD countries. Finally, building upon all of this, a policy maker's guide to environmentally related taxation is provided.

Notes

1. The hurdle rate for known environmental technologies has been well documented. Jaffe and Stavins (1994) outline reasons, such as market failures and managerial constraints, for the apparent paradox where profit-maximising firms do not adopt profitable, energy-saving technologies. Anderson and Newell (2004) find that plants are more influenced by upfront costs than annual cost effects and that adoption hurdle rates are between 50 and 100%.

2. In addition to reducing the marginal cost of capital, governments can also promote innovation by raising the marginal rate of return from innovative activities (for example, learning effects from subsidised R&D can create efficiencies for future R&D projects) but these effects tend not to be as large as those targeting the marginal cost of capital.

3. The scale and complexity of many environmental issues, especially climate change, would imply that the Coase theorem – the somewhat counterintuitive idea that regardless of the allocation of property rights (or lack thereof), economic agents have incentives to resolve issues of externalities to an efficient solution through bargaining as a result of enlightened self-interest – is not practical.

4. On the other hand, combining such information campaigns with a cap-and-trade system will not lead to additional abatement, as long as the total cap of the trading system remains unchanged.

5. These high discount rates may simply reflect the fact that consumers very much prefer consumption in the present period compared to future periods, not that there are necessarily market distortions or failures.

6. This level would not have been optimal for society, given that the environmental damage felt by society was not being taken into account.

References

Anderson, S.T. and R.G. Newell (2004), "Information Programs for Technology Adoption: The Case of Energy-Efficient Audits", *Resource and Energy Economics*, No. 26, pp. 27-50.

Battisti, Giuliana (2008), "Innovations and the Economics of New Technology Spreading within and across Users: Gaps and Way Forward", *Journal of Cleaner Production*, No. 16S1, pp. S22-S31.

Battisti, Giuliana and Paul Stoneman (2003), "Inter- and Intra-firm Effects in the Diffusion of New Process Technology", *Research Policy*, No. 32, pp. 1641-1655.

Foxon, T. (2003), *Inducing Innovation for a Low-Carbon Economy: Drivers, Barriers and Policies*, report prepared for the Carbon Trust, London, available at *www.carbontrust.co.uk/Publications/publicationdetail.htm?productid=CT-2003-07&metaNoCache=1*.

Gerlagh, Reyer (2008), "A Climate-Change Policy Induced Shift from Innovations in Carbon-Energy Production to Carbon-Energy Savings", *Energy Economics*, No. 30, pp. 425-448.

Gerlagh, Reyer and Wietze Lize (2005), "Carbon taxes: A Drop in the Ocean, or a Drop that Erodes the Stone? The Effect of Carbon Taxes on Technological Change", *Ecological Economics*, No. 54, pp. 241-260.

Hicks, John R. (1932), *The Theory of Wages*, Macmillan, London.

Howitt, Peter (2009), "Competition, Innovation and Growth: Theory, Evidence and Policy Challenges", in Chandra, Vandana, Deniz Eröcal, Pier Carlo Padoan and Carlos A. Primo Braga (eds.), *Innovation and Growth: Chasing a Moving Frontier*, OECD, Paris.

Jaffe, Adam B., Richard G. Newell and Robert N. Stavins (2005), "A Tale of Two Market Failures: Technology and Environmental Policy", *Ecological Economics*, No. 54, pp. 164-174.

Jaffe, Adam B. and Robert N. Stavins (1990), "Evaluating the Relative Effectiveness of Economic Incentives and Direct Regulation for Environmental Protection: Impacts on the Diffusion of Technology", paper for the WRI/OECD Symposium *Toward 2000: Environment, Technology and the New Century*, 13-15 June 1990, Annapolis, Maryland.

Jaffe, Adam B. and Robert N. Stavins (1994), "The Energy Paradox and the Diffusion of Conservation Technology", *Resource and Energy Economics*, No. 16, pp. 91-122.

Jaumotte, F. and N. Pain (2005), "Innovation in the Business Sector", *OECD Economics Department Working Papers*, No. 459, OECD, Paris, available at *www.olis.oecd.org/olis/2006doc.nsf/LinkTo/NT000073EA/$FILE/JT03218797.PDF*.

Johnstone, Nick, Ivan Haščič and David Popp (2010), "Renewable Energy Policies and Technological Innovation: Evidence Based on Patent Counts", *Environmental and Resource Economics*, Vol. 45(1), pp. 133-55.

Kemfert, Claudia and Truong Truong (2007), "Impact Assessment of Emissions Stabilization Scenarios with and without Induced Technological Change", *Energy Policy*, Vol. 35, pp. 5337-5345.

Kline, S.J. and N. Rosenberg (1986), "An Overview of Innovation", in R. Landau and N. Rosenberg (eds.), *The Positive Sum Strategy*, National Academic Press, Washington DC.

OECD (2006), *Economic Policy Reforms: Going for Growth*, OECD, Paris, *http://dx.doi.org/10.1787/growth-2006-en.*

OECD (2009a), *The Economics of Climate Change Mitigation: Policies and Options for Global Action Beyond 2012*, OECD, Paris, *http://dx.doi.org/10.1787/9789264073616-en.*

OECD (2009b), *Environmental and Eco-Innovation: Concepts, Evidence and Policies*, OECD, Paris, available at *www.olis.oecd.org/olis/2009doc.nsf/linkto/com-env-epoc-ctpa-cfa(2009)40-final.*

OECD (2010), *The OECD Innovation Strategy: Getting a Head Start on Tomorrow*, OECD, Paris, *http://dx.doi.org/10.1787/9789264083479-en.*

Popp, David (2004), "ENTICE: Endogenous Technological Change in the DICE Model of Global Warming", *Journal of Environmental Economics and Management*, No. 48, pp. 742–768.

Stoneman, Paul (2001), *The Economics of Technology Diffusion*, Blackwell, Oxford.

Chapter 2

Current Use of Environmentally Related Taxation*

> This chapter outlines the usage of environmentally related taxation in OECD countries. It begins by exploring the revenues derived from such taxes, their trends and the role that these taxes play in governments' overall budgets. It goes on to analyse trends in the rates of taxes across countries and how countries are continuing to implement them. The chapter finishes with a discussion on the extent and impact of exemptions and rate reductions within environmentally related taxes.

* The statistical data for Israel are supplied by and under the responsibility of the relevant Israeli authorities. The use of such data by the OECD is without prejudice to the status of the Golan Heights, East Jerusalem and Israeli settlements in the West Bank under the terms of international law.

All OECD countries are seeking to better address environmental challenges. While there are many ways to do this, one of the most interesting is the movement towards a "greening" of government actions. Government fiscal policy – on both the revenue and expenditure side – has a large impact on the economy. Movements towards a more environmentally conscious approach to fiscal policy can translate into changed behaviours within the larger economy.

In particular, the tax system is seen as a medium where governments can have particular influence on the decisions of firms and individuals. Governments have long been conscious of the impact of the tax system on employment, business formation and expansion, and consumption patterns and thus have generally tried to raise revenues without distorting consumption patterns or inhibiting investment decisions. Many of the same ideas can be used in the field of environmentally related taxation; however, a goal of environmentally related taxation is to skew consumption and production patterns and reduce the size of the tax base, which is quite different from the goals of most types of taxation.

2.1. Revenues from environmentally related taxation across countries

While the concept of environmentally related taxes has become more a part of governments' policy dialogue in recent decades, all OECD countries raise revenues through environmentally related taxation and have for many years. Given the definition outlined in Box 2.1, this encompasses a wide range of taxes, such as excise taxes on fossil fuels, motor vehicle registration taxes, taxes on water pollution and waste. Significant differences do exist across countries that reflect historical realties and variances within tax systems. Figure 2.1 shows that environmentally related taxation is a small, but not insignificant, revenue source for governments, averaging around 2% of gross domestic product (GDP).

It is clear that Denmark and the Netherlands lead OECD countries in revenue generated from environmentally related taxation and their shares have been strong over the twelve-year period, contrasted against the general decline for OECD countries. One feature that stands out is the significant geographical differences present. All four countries in the Americas generally have the lowest levels of revenues derived from environmentally related taxation. Three of the four OECD countries in the Pacific have levels below that of the arithmetic average. At the top end, European countries have the most significant revenue levels from environmentally related tax bases. This is consistent with European countries' relatively high overall tax revenue-to-GDP ratios. In the case of Denmark, its tax revenue-to-GDP ratio is the highest in the OECD.

An interesting case is that of Mexico, where the revenues from environmentally related taxation were actually negative in 2008. As with most countries, the vast majority of revenues are normally derived from taxes on motor fuels. The Mexican fuel tax has a unique structure in that it can act inversely to rapid changes to oil prices. In 2002, oil prices were quite low, thereby resulting in relatively high tax rates. By 2008, however, oil prices had increased significantly, resulting in the effective fuel tax rate, and therefore the tax revenue, actually turning negative.

> ## Box 2.1. **Definition of environmentally related taxation**
>
> The OECD, the International Energy Agency (IEA) and the European Commission have agreed to define environmentally related taxes as any compulsory, unrequited payment to general government levied on tax bases deemed to be of particular environmental relevance. The relevant tax bases include energy products, motor vehicles, waste, measured or estimated emissions, natural resources, etc. Taxes are unrequited in the sense that benefits provided by government to taxpayers are not normally in proportion to their payments. Requited compulsory payments to the government that are levied more or less in proportion to services provided (e.g. the amount of wastes collected and treated) can be labelled as fees and charges. The term levy covers both taxes and fees/charges.
>
> Creating any definition of environmentally related taxes is inherently problematic. Taxes may have been implemented for a number of reasons, most likely general revenue-raising, with little to no consideration for the environment. Moreover, some taxes have likely been implemented without stringent assessment of the costs and damages of the pollution, leading to non-optimal rates. Attempting to differentiate taxes based on motivation of the government or exclude some taxes because of their design would, of course, pose significant challenges. Therefore, a broad definition has been used that considers only the type of tax base, not the intention or appropriateness of the instrument.
>
> It should be noted that broad-based taxes, such as value added taxes (VAT), whose tax bases include those which may be environmentally related, are not included as environmentally related taxation in this report. In addition, revenues from the sale of tradable permits and revenues derived from natural resource royalties are not included.

Figure 2.1. **Revenues from environmentally related taxation as percentage of GDP**

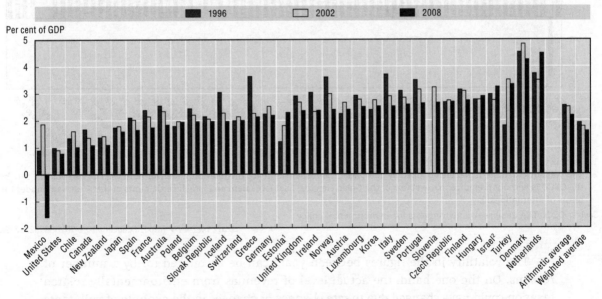

1. Estonia is an accession country to the OECD and has not been included in the averages.
2. The statistical data for Israel are supplied by and under the responsibility of the relevant Israeli authorities. The use of such data by the OECD is without prejudice to the status of the Golan Heights, East Jerusalem and Israeli settlements in the West Bank under the terms of international law.

Source: OECD/EEA database on instruments for environmental policy.

StatLink ⧉ http://dx.doi.org/10.1787/888932317179

Also when analysing the environmentally related tax revenue against total tax revenues, which reflects the importance of the revenues to overall government budgets, much the same trend can be seen in Figure 2.2. The geographic delineations are somewhat less pronounced and the cross-country differences appear less severe. The reliance on environmentally related taxation in some countries, such as Korea, which does not have a high tax-to-GDP ratio, is more pronounced. Turkey stands out for having significantly increased its share of tax revenues from environmentally related bases such that these tax revenues now account for close to 15% of overall tax revenue, well above all other OECD members. This approach is part of a larger tax reform in Turkey to raise additional revenue from consumption and less from other sources, such as income and corporate taxes. Higher fuel taxes have been a deliberate part of their national development plans which seek development in a more sustainable manner, resulting in some of the highest motor fuel prices among OECD countries. On the other hand, the relative level of environmentally related taxation over the ten-year period in countries such as Greece, Mexico and Portugal has been significantly reduced.

Figure 2.2. **Revenues from environmentally related taxation as percentage of total tax revenues**

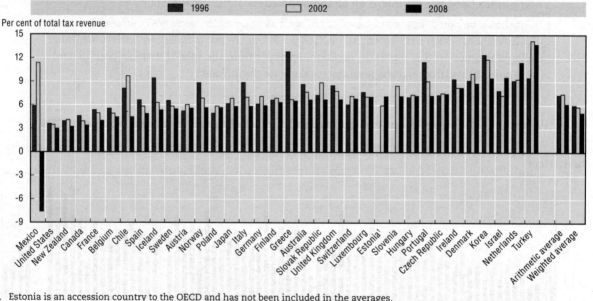

1. Estonia is an accession country to the OECD and has not been included in the averages.
2. The statistical data for Israel are supplied by and under the responsibility of the relevant Israeli authorities. The use of such data by the OECD is without prejudice to the status of the Golan Heights, East Jerusalem and Israeli settlements in the West Bank under the terms of international law.

Source: OECD/EEA database on instruments for environmental policy.

StatLink ᴬᴿᴹᴱ http://dx.doi.org/10.1787/888932317198

The volatility in the figures between years can be accounted for by a number of reasons. On the one hand, the actual level of revenues from environmentally related taxation could have changed due to rate changes or changes in the quantity of pollutants emitted. On the other hand, changes in other government revenues could have occurred, such as income or corporate tax rates, or variations in the bases on which those taxes are levied, such as an economic slowdown that could curtail corporate tax revenues.

Despite this volatility, there is still an overall relative decline in revenues over time, as evidenced by the averages in Figure 2.1 and Figure 2.2. A number of issues contribute to this trend:

- Over this period, oil prices have risen significantly, reducing demand and thereby contributing to downward levels of revenues from these sources compared to other parts of the economy.

- As the construction of environmentally related taxes – typically excise taxes – are levied per unit of product (e.g. EUR 0.10 per litre of petrol), inflation can work to reduce the impact of the tax over the longer term. The nominal value of the tax rate may stay the same but the real value declines, which differs from other tax revenues that are percentage-based (e.g. value added taxes on consumption or income tax rates). Political resistance to tax increases can exacerbate these declines, leading to tax levels that are likely misaligned with the initial rationale for the level of the tax rate. Years of no nominal change in the tax rate can result in significant increases over a short period to compensate.

- The rise of emissions trading systems (which have similar properties to taxes) have meant that some countries are introducing these systems while simultaneously reducing taxes on similar bases. As outlined in Section 2.5, the revenues from auctioning tradable permits are not yet included in the figures of environmentally related taxation. So far, such revenues are modest.

- In the same light, some countries have moved away from taxes in favour of fees (which are similarly not included in the above figures) on the same bases, especially in the transportation area.

- Finally, there is the possibility that some of the impact can be attributed to effectiveness of the taxes themselves in reducing the amount of the pollutants (and therefore tax revenues).

As a counteracting measure, a number of European countries have instituted inflation-adjusted tax rates. These actions remove the political necessity of instituting inflation-related rises and create smoother tax rates across years. Denmark, for example, as part of their 2009 budget process, instituted automatic inflation indexing of energy (including motor fuel) taxes.

It should be noted that the level of revenues raised from environmentally related taxation should not necessarily be taken as a measure of the "environmental friendliness" of a country or of their overall tax system. First, taxes can be non-optimally designed such that they do not necessarily bring about desired behaviour change, and their rates can be set non-optimally, such that they do not necessarily reflect the environmental damage that they cause, despite the fact that they may raise significant revenues. A number of countries have taken to making their environmentally related taxes better designed without necessarily increasing revenues. Additionally, countries may place a greater emphasis on the utilisation of other instruments to address environmental challenges and thereby achieve similar environmental results without the revenues that environmentally related taxation can generate, although often at a higher cost than if well-designed environmentally related taxes had been used. Finally, structural differences across countries' economies can play a role (e.g. some countries may have more emission intensive industries due to location-specific activities).

2.2. Taxes on specific pollutants

Environmentally related taxes have evolved over many decades. In this timeframe, many different events at the local and international level have impacted environmental policy. The result is that, in most OECD countries, there exists a wide range of taxes and charges which may not always be in line with the relevant damages. Different pollutants result in different damages to the environment. High tax rates sometimes occur on some more benign pollutants while more damaging pollution is not taxed. In addition, some pollutants are taxed radically differently based on the source or emitter of the pollution.

The vast majority of environmentally related tax revenues are derived from taxes on energy – of which taxes on motor fuel constitute nearly all of those proceeds. As seen in Figure 2.3, the revenues from these energy taxes account for about two-thirds of the total revenues. In addition, the "other" category, although small, has relatively grown over the period compared to the other categories.

Figure 2.3. **Composition of environmentally related tax revenues in the OECD**

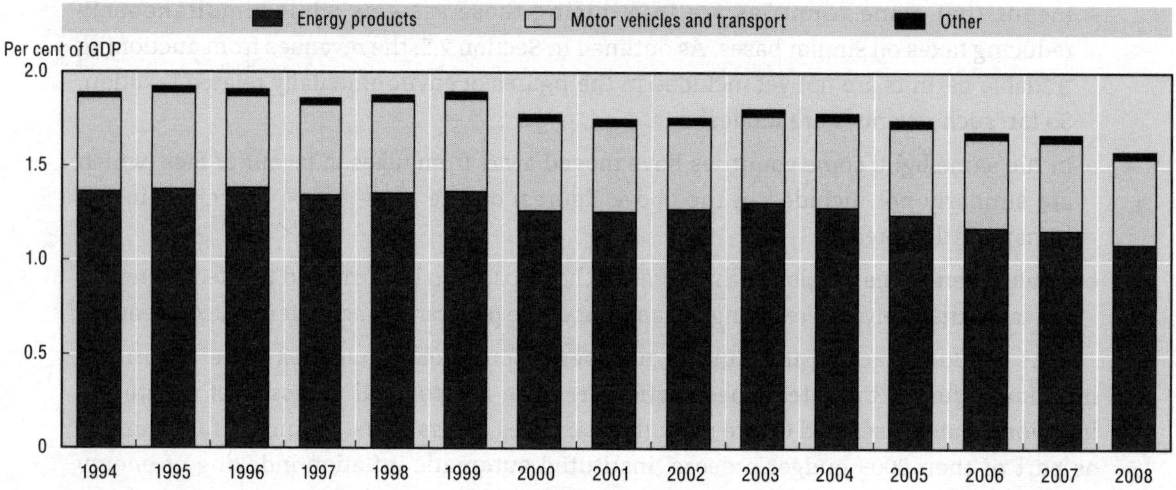

Source: OECD/EEA database on instruments for environmental policy.

StatLink ⟨ms⟩ http://dx.doi.org/10.1787/888932317217

The composition of environmentally related taxation varies across countries as well, as seen in Figure 2.4. Countries such as Poland, the Slovak Republic and Luxembourg[1] rely heavily on energy taxes. Taxes on motor vehicles constitute a significant part for total revenues for Denmark, the Netherlands, Ireland and Norway. Finally, the Netherlands stands out for its relatively large usage of "other" environmentally related taxes.

2.2.1. Motor fuel and motor vehicle taxes

Motor fuel

Excise taxes on fuel have been around for many years, originally being motivated by non-environmental needs alone (such as general revenue generation or sometimes earmarked for specific infrastructure projects). The revenues raised from these taxes are quite high, a result of the significant level of consumption in OECD countries. Figure 2.5 presents the different excise tax rates on petrol and diesel in OECD countries for years 2000 and 2010. Much like the overall analysis above, clear groupings by geographic area are present. North America has the lowest petrol taxes, followed by OECD countries in Asia and

Figure 2.4. **Composition of environmentally related tax revenues by country**

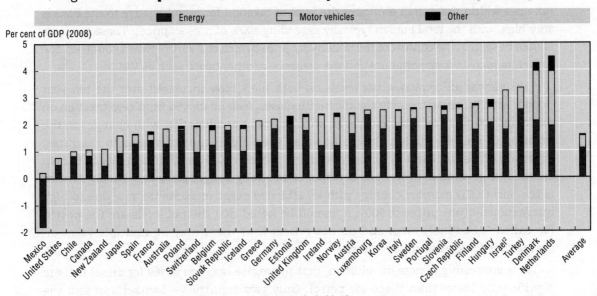

1. Estonia is an accession country to the OECD and has not been included in the average.
2. The statistical data for Israel are supplied by and under the responsibility of the relevant Israeli authorities. The use of such data by the OECD is without prejudice to the status of the Golan Heights, East Jerusalem and Israeli settlements in the West Bank under the terms of international law.

Source: OECD/EEA database on instruments for environmental policy.

StatLink 🔗 http://dx.doi.org/10.1787/888932317236

Figure 2.5. **Tax rates on motor fuel**

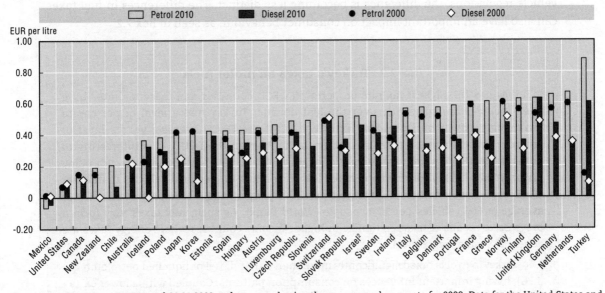

Notes: Rates are as at 01.01.2010 and 01.01.2000 and converted using the average exchange rate for 2009. Data for the United States and Canada include average excise taxes at the state/provincial level. VAT is not included.

1. Estonia is an accession country to the OECD.
2. The statistical data for Israel are supplied by and under the responsibility of the relevant Israeli authorities. The use of such data by the OECD is without prejudice to the status of the Golan Heights, East Jerusalem and Israeli settlements in the West Bank under the terms of international law.

Source: OECD/EEA database on instruments for environmental policy.

StatLink 🔗 http://dx.doi.org/10.1787/888932317255

the Pacific, with European countries having significantly higher tax rates. Compared to other tax rates within the overall economy, the level of taxation for petrol relative to the base is very high, with the total burden typically exceeding 100% of pre-tax prices. These tax rates should be seen in the context of the underlying price (which can vary between countries due to factors such as transportation) and the presence of other taxes, such as VAT.

For almost all countries, tax levels on both fuels have increased over the ten-year period, with Turkey witnessing significant increases. Iceland moved from zero taxation on diesel in 2000 to a tax rate near parity with petrol in combination with changes in the taxation of diesel vehicles. Greece has also seen significant increases in tax rates, especially on petrol. These principally occurred between 2009 and 2010 as a means to help consolidate government revenues in response to strong budget pressures. Finally, Mexico is the only OECD member country with an effective negative excise tax rate due to high international crude prices in 2009. It should be noted that the tax levels are the posted (so-called "headline") rates and do not reflect that there may be multiple rates or exemptions for specific uses or users.

It is interesting to note, in addition, that the excise taxation levels for diesel fuel are significantly lower than those for petrol. Only two countries – Switzerland and the United States – have a higher level of tax for diesel than petrol; Australia and the United Kingdom's rates are the same for both fuel types. The majority of diesel rates are situated within the 70-80% of petrol range, with New Zealand not levying any excise tax on diesel.[2] From an environmental point of view, this is peculiar, as diesel consumption in vehicles has a much larger environmental impact than unleaded petrol, largely due to the significant differences in NO_x and particulate emissions. With more stringent motor vehicle regulations, the difference is becoming less distinct. The differences in fuel taxes can also have an important impact on consumer behaviour, as seen in Box 2.2.

Box 2.2. **Turkey's taxes on motor fuels**

Turkey has the highest tax rate on petrol in OECD countries and these tax rates have been increased significantly over the last decade. Turkey had a level of per capita purchasing power parity in 2007 of only 37% of the OECD average, yet its level of environmentally related taxes is among the highest in the OECD. It is interesting to note that Turkey has increased these taxes on motor fuels alongside tax rates on many luxury goods. Turkey's economy is much less dependent on personal vehicles than other OECD countries, with only 117 vehicles per 1 000 people in 2005, compared to the OECD average of 606 vehicles per 1 000 people (World Bank). As such, fuel taxes may form a progressive means of taxation (unlike in higher-income countries where energy taxes are generally seen as regressive).

In Turkey, petrol is taxed significantly more than either diesel or liquefied petroleum gas (LPG), having important influences on consumption patterns. Coupled with a lower ex-tax price per litre of LPG, lower tax rates have encouraged a significant shift towards LPG-fuelled vehicles. Between 2003 and 2007, the number of cars outfitted to run on LPG more than doubled from 800 000 to over 1.8 million. There was also a significant shift in more standard fuels, with total petrol consumption remaining quite flat while diesel consumption increased significantly. As a percentage of GDP, petrol use declined significantly. The trends suggest that taxes (and underlying prices) can have an important impact on consumer behaviour.

Within these broad categories of petrol and diesel, tax rates also vary based on the fuel characteristics. When leaded petrol was still widely available, governments typically taxed this at a higher rate than unleaded petrol. In the present context, a number of countries differentiate their tax rate based on other criteria related to the characteristics of the fuel, such as the level of sulphur content and the proportion of renewable fuels present.

While Figure 2.5 outlines the actual level of taxes on petrol in OECD, it is difficult to know if these are at the correct level. A multitude of factors go into determining the rate – the environmental damage, the use of roads, the cost of vehicular accidents and the general need for governments to raise revenues. Box 2.3 outlines what one study suggests should be the optimal petrol tax for the American state of California.

Box 2.3. **Multiple externalities and an optimal tax for California**

A prescient example of the multiple externality issue is the calculation of optimal petrol taxes. These taxes obviously have a significant environmental impact, both global and local, although they are levied for a range of reasons. However, determining an "optimal" tax requires looking at all the various impacts that fuel use can have. First, governments need to raise revenue from a variety of sources to fund public services. Since economic theory suggests that changing consumers' preferences leads to some welfare loss, taxes should be focused on those goods that are rather invariable to price changes – that is, those that are price inelastic. Motor fuels meet this criterion. In addition, "optimal" fuels taxes should try to correct negative externalities, which are unwanted effects that accrue to others from an individual's actions. In the case of the environment, one person's combustion of fossil fuel releases pollutants that negatively affect others (without them being compensated). Therefore, taxes should encapsulate the various environmental externalities. Finally, there are other externalities associated with motor fuels. By driving, for example, accidents occur which impose a cost on to taxpayers and congestion reduces the welfare of other drivers. In total, the "optimal" motor fuel tax should incorporate all these various features in the setting of the rate. Therefore, environmentally related taxes go beyond just the environmental considerations, since the focus is on the base.

In their analysis related to the US state of California, Lin and Prince (2009) find an optimal tax rate for petrol of USD 0.36 per litre outside of sales taxes. Most of the tax is based on externalities (USD 0.22 per litre), of which only USD 0.02 is for global pollution (e.g. climate change)* and USD 0.04 is for local pollution. The rest is related to congestion, accidents and oil dependence. The single largest component (USD 0.14 per litre) is due to the attractiveness of taxing petrol because its demand is quite price inelastic – so-called Ramsey taxation.

The climate change component is very small. Even using a different methodology that placed greater emphasis on the damage from climate change would likely not have a significant impact on the overall optimal tax rate. As a means of comparison, the current excise tax is less than a third of this "optimal" level. While the environmental component of the petrol tax is low, it is worth noting that when the overall excise tax rate is lower than the optimal, there will be overconsumption, which induces excess environmental damage. As noted in OECD (2006), the tax rates on petrol in some European countries, however, may be above their optimal level.

* The climate change component is taken from another study. Using different values for the environmental damage would affect the magnitude of this variable but is unlikely to have a large impact on the overall optimal level of the tax, given its small relative contribution.

The general increase in nominal tax rates outlined in Figure 2.5 does not address how these taxes measure against the real impact that they have on influencing consumer behaviour and on government revenues. Figure 2.6 shows the real percentage change of the total tax rate on petrol (excise taxes, no VAT or general sales taxes) across OECD economies over the period 2000-10. While several jurisdictions have seen significant increases in real taxes on petrol, the majority have not. Australia, for example, cut its nominal rate of taxation on petrol. The United States' federal rate remained nearly fixed in nominal terms over the period, while Greece significantly increased petrol taxes as a revenue raiser in at the end of 2009. Nevertheless, these rises did not keep up with inflation, leading to significant declines in the real effect of the tax rate. The average real change in the tax rate on petrol over the period was –8.1% (11.0% decline in the arithmetic average).[3] This, along with rising oil prices bringing about declining consumption, can have an impact on the total environmentally related revenues collected by governments.

Figure 2.6. **Real changes in tax rates on petrol**
Between 2000 and 2010

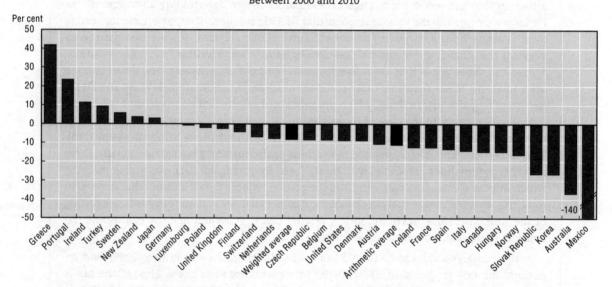

Note: Tax rates constitute all excise taxes levied on petrol as at 01.01.2000 and 01.01.2010. Rates for the United States and Canada include rates at the sub-central level. The weighted average is weighted by revenues from petrol taxes.

Source: OECD/EEA database on instruments for environmental policy.

StatLink ⧉ http://dx.doi.org/10.1787/888932317274

Motor vehicles

Along with motor fuels, taxes on motor vehicles are a major source of revenue for OECD governments. These taxes are generally divided into two categories: those that are one-off (that is, levied on the initial or subsequent sale or import into the country) and those that are recurrent (that is, those that are levied on an annual basis). While theoretically less efficient than taxes on fuel or actual emissions from an environmental point of view, these taxes can nevertheless play a large role in affecting levels of car ownership and the composition of a national fleet of vehicles. In addition, such taxes, particularly those of the one-off kind, can provide a "sticker shock" effect regarding the environmental impact that other taxes may not, as seen in Figure 2.7.

Figure 2.7. **One-off motor vehicle taxes**

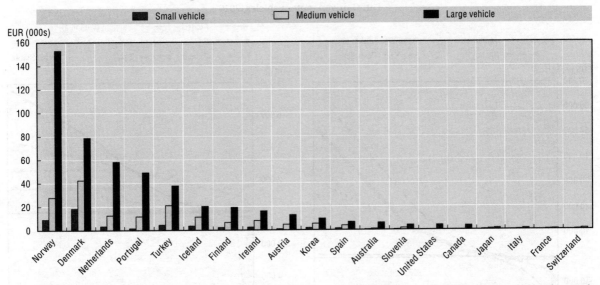

Notes: As at 01.01.2010. One-off taxes on new vehicles only. "Small" refers to a petrol-based car with 53 kW of power, 6.5 l/100 km, 821 kg, 1 000 cc engine, EUR 12 000 pre-tax price; "medium" refers to a petrol-based car with 132 kW of power, 9.4 l/100 km, 1 468 kg, 2 400 cc engine, EUR 25 000 pre-tax price; "large" refers to a petrol-based car/SUV with 300 kW of power, 16.8 l/100 km, 2 587 kg, 6 200 cc engine, EUR 45 000 pre-tax price. Countries with CO_2 components in their taxes on motor vehicles are calculated based on fuel efficiency. For countries with sub-national governments that levy applicable rates, the following jurisdictions are used: New South Wales (Australia), Ontario (Canada), and California (United States). These tax levels do not include non-environmentally related taxes, such as VAT, nor environmentally related tax components that vary significantly between vehicles of a similar size, such as those based on NO_x emissions from each vehicle.

Source: OECD/EEA database on instruments for environmental policy.

StatLink ⧉ http://dx.doi.org/10.1787/888932317293

The manner in which such motor vehicle taxes are being administered has been changing. OECD countries are evermore basing such charges on the characteristics of the vehicle with environmental features being prominently utilised – fuel efficiency, CO_2 emissions per kilometre, engine power, and weight.[4] In a number of cases, more than one of these factors is used to derive a total tax burden per vehicle. This is the case in Norway, where CO_2 emissions, vehicle weight, and engine power are all used to determine the tax level. Coupled with high tax rates for each item, the result is that Norway has substantially higher one-off tax levels on large vehicles than most elsewhere in the OECD, while Denmark has higher taxes for small and medium-sized vehicles, as seen in Figure 2.7. In some cases, the tax burden, especially for larger and more polluting cars, can represent several hundred per cent of the net-of-tax price of the vehicle.

The construction of some of these taxes exacts a significant toll on heavily emitting vehicles. Many countries' taxes involve formulae with many different variables. Figure 2.8 shows only the component of the tax burden (or subsidy level in some cases) on vehicles related to their CO_2 emissions (or fuel efficiency) in OECD countries. OECD (2009b) demonstrates that these taxes can be highly progressive with increasing emission rates, such as in Norway and Portugal. In addition, four countries' CO_2-based component – Austria, Finland, Ireland, and Spain – is dependent on the pre-tax price of the vehicle and in several countries, the tax rates differ between petrol- and diesel-driven vehicles.

Figure 2.9 translates these tax levels solely from the CO_2 component into an equivalent value per tonne of CO_2 emitted over the lifetime of the vehicles, assuming that each vehicle is driven 200 000 km. A uniform rate per tonne of CO_2 would provide a constant tax on emissions, consistent with the damage done to the environment. Fuel

Figure 2.8. **CO_2 component of one-off taxes**

Petrol-driven motor vehicles, 2010

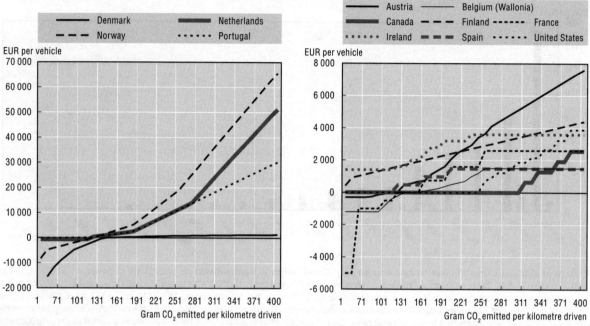

Notes: The CO_2 tax components for Spain, Ireland, Finland and Austria are also dependent on the pre-tax price of the vehicle; for this exercise, a EUR 10 000 vehicle has been used. Note that the axes of the two panels are of a different scale.

Source: Updated data based on OECD (2009b).

StatLink 🔗 http://dx.doi.org/10.1787/888932317312

Figure 2.9. **Implicit carbon price and motor vehicle taxes**

Derived solely from CO_2 component of one-off, petrol-driven motor vehicle taxes

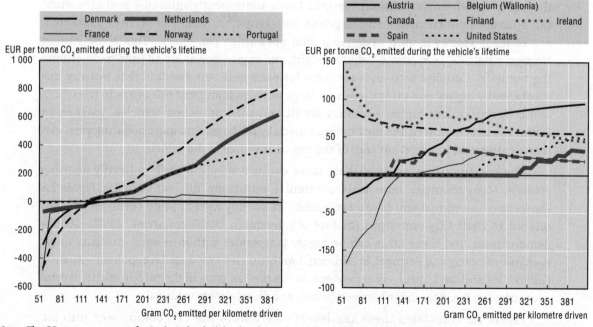

Notes: The CO_2 tax components for Spain, Ireland, Finland and Austria are also dependent on the pre-tax price of the vehicle; for this exercise, a EUR 10 000 vehicle has been used. Vehicle lifetime is assumed to be 200 000 km. Note that the axes of the two panels are of a different scale.

Source: Updated information based on OECD (2009b).

StatLink 🔗 http://dx.doi.org/10.1787/888932317331

taxes work in this way: there is a set rate of tax regardless of the quantity consumed. In about half the OECD countries that impose one-off CO_2 taxes on motor vehicles, there is actually a negative implicit carbon price at certain levels, indicating that society is effectively subsidising carbon emissions via this tax instrument. Through the bonus-malus systems in place, implicit carbon prices rise – sometimes dramatically – as emissions per kilometre rise also. On the other hand, Ireland and Finland have structured their taxes such that the carbon price effectively declines with increasing carbon emission intensity.

At the same time, recurrent (annual) motor vehicle taxes have also been based on CO_2 emissions and fuel efficiency in some OECD economies, providing further incentives for potential abatement. In some cases, such as Ireland and Portugal, both recurrent and one-off taxes are related to the CO_2 emission intensity of the vehicle. Figure 2.10 provides

Figure 2.10. **Total CO_2 components of motor vehicle taxes**

Implicit carbon price from one-off and recurrent taxes related to CO_2 emissions for petrol-driven vehicles

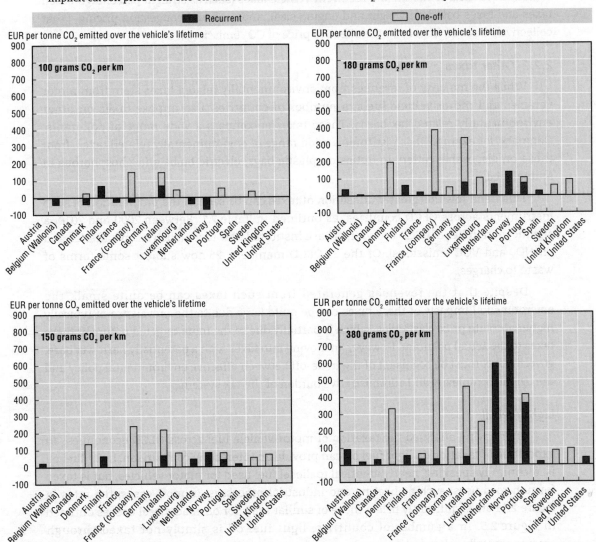

Notes: No discounting has been used for the recurrent taxes and, where the CO_2 component is also related to the price of the vehicle, a net-of-tax value of EUR 10 000 has been used. Vehicle lifetime is assumed to be 200 000 km.

Source: Updated information based on OECD (2009b).

StatLink ⫘ http://dx.doi.org/10.1787/888932317350

an overview of the total effect of one-off and recurrent tax components based on CO_2 emissions (or fuel efficiency), based on selected emission levels. It shows that countries generally favour implicitly progressive rates on carbon emissions from motor vehicles, as vehicles with emission intensities of 380 g CO_2/km have a significantly higher implicit carbon price than those vehicles emitting at a rate of 100 g CO_2/km. This is in spite of the fact that a tonne of CO_2 emitted from a low-emission vehicle causes the same environmental damage as does a tonne from a high-emission vehicle.

These implicit tax rates can reach significant levels. For vehicles emitting 380 g CO_2/km, the implicit carbon price is over EUR 300 per tonne in Denmark, France (for company cars, for which the recurrent charges are significantly different from personal use vehicles), Ireland, the Netherlands, Norway and Portugal, well above the market rate of carbon in the EU ETS. While this issue provides some interesting findings, it is important to remember that one-off and recurrent motor vehicle taxes are part of a range of instruments – including fuel taxes, other components of one-off and recurrent vehicle taxes (such as straight percentage-based taxes and taxes based on weight and engine size), and road and congestion pricing – that collectively have an impact on the effective price of CO_2 emissions.

2.2.2. Other taxes

While the majority of revenues from environmentally related taxes stem from motor vehicles and motor vehicle fuels, a number of countries also impose taxes on other environmentally related tax bases. These taxes encompass a wide range of pollutants. Denmark, for example, has instituted a wide range of environmentally related taxes, from those on disposable cutlery to duties on plastic bags, electric bulbs and phosphorous in animal feed.

This trend towards greater utilisation of taxes can be seen in the proliferation of such instruments across OECD countries, as outlined in Table 2.1. Between 2000 and 2010, a significant number of OECD economies have instituted new taxes (such as on batteries and on NO_x and VOC emissions). Of the 33 OECD members, 25 now subject some forms of waste to charges.

Despite that the revenues generated from such taxes can be quite small, the environmental impacts of these taxes can be quite large due to the prevalence of typically higher absolute price elasticities, which is partially due to the more targeted nature of these taxes to the actual pollutant. However, in moving into taxes with a potentially narrower base, governments must be cognisant of the trade-off between effective environmental policy and the overall complexity and administrative burden of the tax system.

Light fuel oil

After having analysed the taxation of motor vehicle fuel across OECD economies, an exposition of taxation on light fuel oil can provide an interesting comparison. Light fuel oil is, technically speaking, nearly identical to diesel fuel used for motor vehicles, but is taxed for non-road uses, such as heating and industrial processes. In most cases, the rates in Figure 2.11 are significantly below those of similar fuels for on-road diesel use, as outlined in Figure 2.5. In a number of countries, light fuel oil is simply not taxed through environmentally related instruments.

Table 2.1. **Extent of tax instrument utilisation**

Jurisdictions with selected environmentally related tax measures

	2000			2010		
NO$_x$	Czech Republic France Italy Sweden			Australia (ACT, NSW) Canada (BC) Czech Republic Denmark	France Hungary Italy Norway Poland Slovak Republic	Spain (Aragón, Castilla-La Mancha, Galicia) Sweden United States (ME) Estonia[1]
HFCs and ozone-depleting substances	Australia Canada	Czech Republic United States		Australia Canada Czech Republic	Denmark Norway	Poland Slovak Republic United States
VOCs (incl. chlorinated solvents)	Denmark Norway	Poland	Switzerland	Australia (ACT, NSW) Canada (BC) Czech Republic	Denmark Korea Norway Poland	Slovenia Switzerland United States (ME) Estonia[1]
Waste	Belgium Canada (AB, BC, MB, NB, NS, ON, PE, QC, Federal) Czech Republic	Denmark Finland France Germany Greece Hungary Italy	Japan Korea Norway Sweden Switzerland United States (AL, AR, RI, TX, Federal)	Australia (NSW, Federal) Austria (Burgenland, Vienna, Federal) Belgium Canada (AB, BC, MB, NB, NL, NS, ON, PE, QC, SK, Federal) Czech Republic Denmark Finland France	Hungary Iceland Israel (from 2011) Italy Japan Korea Netherlands Norway Poland Portugal	Slovak Republic Spain (Andalusia, Catalonia, Madrid) Sweden Switzerland United Kingdom United States (AL, AK, AR, FL, IN, IA, KS, LA, MD, MS, MO, NE, NJ, NY, OH, RI, SC, TX, VA, WA, Federal)
Batteries	Belgium Canada (BC)	Denmark	Korea	Austria Belgium Canada (BC) Denmark Hungary	Iceland Italy Korea Poland Portugal	Slovak Republic Sweden Switzerland United States (FL, MS, SC, TX)

Notes: The waste category includes charges on landfill or incineration, as well as charges on specific goods that have the potential to cause waste problems (such as paint cans, digital cameras, etc.). Batteries are not included in the waste category, as these are specifically outlined in a separate category.
1. Estonia is an accession country to the OECD.
Source: OECD/EEA database on instruments for environmental policy.

Emissions of nitrogen oxides

Nitrogen oxide (NO$_x$) emissions contribute significantly to local air pollution, as this family of compounds reacts with other substances to create negative environmental and health outcomes. For example, NO$_x$ contributes to ground-level ozone (smog), acid rain, particles in the air, climate change and water quality deterioration, and is generally formed as a result of combustion. Increasingly stringent regulations have significantly reduced NO$_x$ emissions from motor vehicles; in the United States, for example, NO$_x$ emissions from motor vehicles declined by 38% between 1970 and 2008 even as the number of cars and the number of kilometres driven increased significantly. Nevertheless, EPA (2009) estimates indicate that 68% of NO$_x$ emissions in the United States in 2008 were derived from stationary sources.

For many of these reasons, some OECD countries have moved beyond regulation and put in place taxes directly on NO$_x$ emissions to air or implemented tradable permit systems (such as in the United States and Korea). As will be seen with the experience in Sweden in Box 3.2, calculating NO$_x$ emissions can be difficult – a wide range of factors affect its formation in the combustion process. For this reason, sophisticated monitoring systems are typically required to adequately assess emissions. Initial costs may lead to some delays in the implementation of such taxes but, once in place, they provide vast

Figure 2.11. **Tax rates on light fuel oil**

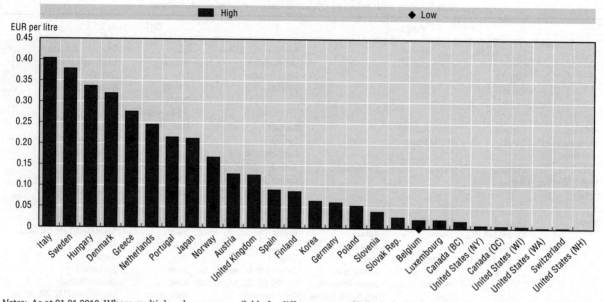

Notes: As at 01.01.2010. Where multiple values were available for different types of light fuel oil (*e.g.* diesel fuel, kerosene), the rate for diesel fuel was used. Reduced rates occur for a range of reasons, such as for different uses, different characteristics, or whether firms have entered into negotiated agreements with governments. Where multiple reduced rates occur, the lowest is indicated for the "low" value.

Source: OECD/EEA database on instruments for environmental policy.

StatLink 🔗 http://dx.doi.org/10.1787/888932317369

quantities of information to both regulators and the regulated that can lead to significant reductions in emissions.

Figure 2.12 outlines the tax rates on NO$_x$ emissions in OECD countries. In general, the rates are quite low, with most under EUR 0.20 per kilogram. On the other hand, Sweden, Norway and in some circumstances New South Wales in Australia all have significantly higher tax levels. Sweden, with the highest rate among OECD countries, implemented the charge with the provision that revenues raised from the tax are recycled back to the energy producers on whom it is levied, based on energy output, while the state of Maine's rate increases with the emission level and is dependent on the total revenue raised from the charge (that is, if a certain revenue threshold is not achieved, the surcharge rate can be doubled). Finally, Australia's charge (see Box 2.4 for more information of New South Wales' Load-Based Licensing System) is likely the most comprehensive, as it varies based on the amount emitted, where the emission occurs, and the time of year, thereby better reflecting the actual damage posed by these emissions.

Chlorinated solvents

A number of countries levy taxes on chlorinated solvents, chemicals which are typically used in certain industrial processes. Some chlorinated solvents contribute to the depletion of the ozone layer, such as chlorofluorocarbons, and have been highly controlled since the Montreal Protocol. Others have continued to be used in selected industries, typically dry cleaning and metal degreasing. These non-ozone depleting substances nevertheless still have significant human health and environmental effects and have typically been subject to some form of environmental control. Starting in the early 1990s, a small number of countries began to address concerns with these chemicals through taxation. Table 2.2 provides an outline.

Figure 2.12. **Taxes on NO$_x$ emissions to air**

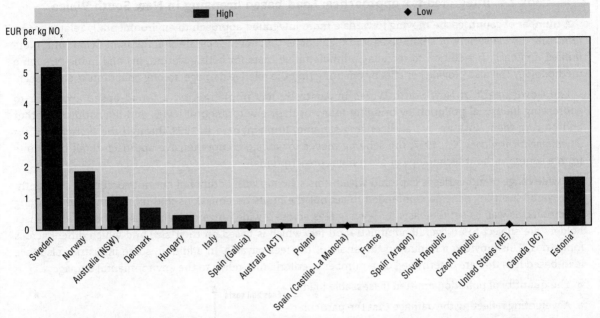

Notes: As at 01.01.2010. High rates represent the highest rate applicable in a country (typically the standard rate) and low rates represent the lowest rate applicable in a jurisdiction (generally based on when, where and how emissions are brought about). For Australia, NSW indicates the state of New South Wales and ACT indicates the Australian Capital Territory; for Spain, Castille-La Mancha indicates the autonomous community of Castille-La Mancha; for the United States, ME indicates the state of Maine; and for Canada, BC indicates the province of British Columbia.
1. Estonia is an accession country to the OECD.
Source: OECD/EEA database on instruments for environmental policy.

StatLink ᄅᆷᆬ http://dx.doi.org/10.1787/888932317388

It is interesting to note that, where data on the effectiveness of the taxation is present, the results are quite striking. Denmark and Norway both had dramatic declines in the level of chlorinated solvents being used within their countries, even though the tax rates were of dissimilar magnitudes. Other countries have used different instruments to reduce the usage of these chemicals, with Sweden outright banning the products,[5] Canada issuing usage permits with maximum values declining over time, and Germany implementing technical standards for their use. For all countries, revenues generated were quite low, reflecting that this was not a prime motivation of the taxation.

Pesticides and fertilisers

Pesticides can be quite harmful to the environment because of their effects on wildlife, biodiversity, and runoff into water systems. They are highly controlled substances in OECD economies with permission for market access only being allowed after stringent regulatory processes by governments. Countries typically rely on a variety of methods to reduce use in the agricultural community, such as education outreach and regulations on usage and mitigation approaches. Yet, only a few countries actually levy taxes on pesticides as a means to reduce their permitted use. While not as potentially toxic as pesticides, fertilisers can also have significant impacts, specifically on water quality due to enriched runoff. Table 2.3 highlights the different approaches that countries have taken in this area. The revenues derived from these taxes also vary significantly, from USD 80 million in Denmark in 2007 for pesticides to USD 11.5 million in Norway, in line with the different tax rates that countries have imposed.

Box 2.4. **Integrated tax approaches: Load-based licensing in New South Wales**

A number of countries are moving towards a more integrated approach to environmentally related taxes, removing a wide variety of small taxes and regulations that cover separate pollutants in favour of a more unified approach. In addition to reducing administrative costs for both governments and industry, such a methodology can also provide benefits by ensuring that the relative damage among pollutants is easily seen.

The Government of New South Wales in Australia has implemented a broad-based approach to addressing industrial pollution by bringing many of their environmental levies and regulations together and pairing them with an overall licensing scheme. Implemented in 1999 through the *Protection of the Environment Operations Act, 1997*, the scheme moves towards a comprehensive approach to all pollutants from a source without technology or abatement prescriptions.

A wide range of industries is captured within this scheme, with additional entrants occurring constantly. The overall licensing scheme is intended to set out the limits on emissions and monitoring and reporting conditions, as well as set the basis for levying fees across a wide range of pollutants. To begin, all license-holders are subject to an overall administration fee, which is based on their size and which differs across industries. This provides a minimum threshold of the fees payable. In addition, some industries also face load-based fees that are determined on a number of criteria that relate to the environmental damage:

- The quantity of pollution emitted (assessable load).

- A weighting reflecting the damage that the particular pollutants cause (pollutant weighting), *e.g.* mercury is weighted at 77 000 while sulphur oxides at 1.5.

- A weighting reflecting the conditions of the local environment (critical zone weighting), *e.g.* areas where VOC emissions are already elevated (such as urban areas) are weighted much higher than other areas (such as rural areas).

- The charge for each unit of pollution (pollution fee units).

- Finally, where the assessable load exceeds a given threshold, the rates are doubled. Above an annual load limit, fees become fines and prosecution can take place.

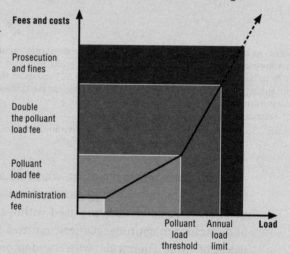

As this process consolidates taxes and fees on a wide range of pollutants, it provides a comprehensive and more efficient system for addressing environmental challenges. For example, firms producing coke are assessed in one system for emissions of: benzene, benzo(a)pyrene, coarse particulates, fine particulates, hydrogen sulphide, nitrogen oxides, sulphur oxides, and volatile organic compounds to air, as well as oil and grease, suspended solids, phenolics, and polycyclic aromatic hydrocarbons to water.

In order to facilitate additional investments in pollution abatement, firms can enter into agreements with the EPA. In return for implementing abating processes, these agreements allow the firm to be assessed on the expected pollution levels after having implemented the abatement measures during the implementation period (up to three years).

This programme provides flexibility to both firms to find additional ways to reduce pollution (and thereby their tax incidence) as well as to governments. Realising that the levy on water pollution was adequate but that those on air pollution were not high enough to meet their intended targets, the government raised these rates without having to relicense affected firms in 2004. In 2001-02, this scheme raised AUD 16 million, rising to AUD 33 million by 2007-08, some of which was due to increasing coverage of the scheme.

Source: NSW EPA (2001) and OECD/EEA database on economic instruments.

Table 2.2. **Taxes on chlorinated solvents**

Location	Name of measure	Type	Rate as at 01.01.2010	Revenue generated	Effectiveness
Denmark	Duty on certain chlorinated solvents.	Tax on emission of dichloromethane, tetrachloroethylene, trichloroethylene.	EUR 0.27/kg	USD 0.1 million (2007).	The law went into effect in January, 1996. It is estimated that consumption of trichloroethylene went from 1 000 tonnes/year before the tax to 356 by 1998. For tetrachloroethylene, consumption went from 720 tonnes/year to 463 by 1998.
Korea	Water effluent charge.	Tax on emission to water of tetrachloroethylene and trichloroethylene.	EUR 186.39/kg		
Norway	Tax on trichloroethane (TRI) and perchloroethlyene (PER).	Tax on harmful input.	EUR 7.10/kg[1]	PER: USD 0.5 million (2008). TRI: USD 0.4 million (2008).	Tax implemented in 2000. It is estimated that TRI usage declined to 139 tonnes in 2001 from 500 in 1999. With respect to PER, usage declined to 32 tonnes in 2001 from 270 in 1999.
Poland	Charge on air pollution.	Tax on emission to air of trichloroethane.	EUR 37.97/kg		

1. For Norway, 41% of the charge is refunded for product that is recycled or properly disposed of.
Source: OECD/EEA database on instruments for environmental policy. Source for effectiveness information for Denmark is Danish Ministry of the Environment (2000) and for Norway is Sterner (2004).

Table 2.3. **Pesticide and fertiliser taxes**

	Description of tax rate as at 01.01.2010	Notes
Canada (British Columbia)	EUR 0.7568 per litre of pesticides.	Earmarked for the residuals stewardship programme.
Denmark	Pesticides: 35 % of retail value for chemical products for disinfection of soil and insecticides; 25% of retail value for chemical deterrents of insects and mammals, chemical products for reduction of plant growth, fungicides, and herbicides; and, 3% of retail value for deterrents of rats, mice, moles and rabbits, and fungicides for wood protection. Fertilisers: EUR 0.67 per kg of nitrogen.	Exports are exempted. Earmarked for the environmental and agricultural sector. Only applies to nitrogen used outside the agricultural sector.
France	Seven pesticide categories with rates ranging from EUR 0.38 per kg to EUR 1.68 per kg.	
Norway	Tax per kg or litre of agricultural pesticides = (base rate * factor)*1 000/standard area dose. Standard area dose is the maximum application rate in kilograms or litres per hectare for the main crop for which the particular pesticide is used. The base rate is set by the government and is the same for all products (was EUR 3.12 per kg or litre in 2005). The factor is a weighting based on the relative risk level of the pesticide according to the following schedule:	

Factor	Products
0.5	Products with low human health risk and low environmental risk.
3	Products with low human health risk and medium environmental risk, or products with medium human health risk and low environmental risk.
5	Products with low human health risk and high environmental risk, or products with medium human health risk and medium environmental risk, or products with high human health risk and low environmental risk.
7	Products with medium human health risk and high environmental risk, or products with high human health risk and medium environmental risk.
9	Products with high human health risk and high environmental risk.
50	Concentrated home garden products.
150	Ready-to-use home garden products.

Sweden	Pesticides: EUR 3.11 per whole kilogram active constituent.	Wood preservatives are exempted.
United States	EUR 0.001-EUR 0.004 per kg.	Earmarked for financing inspection activities.

Source: OECD/EEA database on instruments for environmental policy and OECD (2005).

It is interesting to note the different approaches that countries have taken. Sweden, for example, imposes the same per unit tax on the active ingredient pesticides for all varieties, thereby levying the same rate on rather benign products as those which are more toxic. A percentage tax is used by others, which is dependent on the price of the good. Thus, a heavy user buying pesticides in bulk will pay a smaller amount of tax on each unit of pesticide than a potential hobby gardener next door.

On the other hand, Norway moved away from a system of percentage taxes on imports of pesticides in 1998 in favour of the approach outlined previously. This approach categorises each pesticide based on its negative human health and environmental effects. In doing so, it outlines the specific value of the damage done that is not reliant on the underlying price of the pesticide. Not only does this encourage more conservative use of pesticides in general, it also provides incentives to substitute to less damaging products, as the price among pesticides are differentiated. On the other hand, such a system can present a significant administrative burden – both for regulators and industry. In Norway, this is less of an issue, as only 188 pesticides are approved for use.[6] This level of pesticide registrations differs significantly from that in the United Kingdom by contrast, where 3 075 pesticides are registered for use.[7] The ongoing introduction of the Registration, Evaluation, Assessment, and Restriction of Chemical Substances (REACH) programme within the European Economic Area may present governments with more complete information on making risk-related decisions on environmentally related taxation of pesticides.

Waste

Various programmes exist in OECD countries to reduce the amounts of household and industrial waste. Recycling programmes, composting programmes, manufacturer waste-reduction initiatives, and extended producer responsibility schemes are some of many different types being used. Despite these efforts, residual wastes do exist that require means to dispose of them. These options typically are reduced to two categories: incineration and landfill, both of which have negative environmental effects.

Figure 2.13 outlines the tax rates on landfilling in OECD countries. These rates are dependent upon a wide range of factors that likely vary across countries, such as the composition of the municipal waste, the ability of the landfill to contain environmental damage (such as leakage to groundwater), the availability of other options, and the general availability of land for these purposes. Austria and the Netherlands have the highest rates among OECD countries, while the few US states that actually impose taxes on landfill have the lowest rates.

Unlike many of the other taxes and charges that have thus far been explored, the effectiveness of waste taxes is less direct. In the case of NO_x emissions, for example, firms emitting nitrogen oxides are charged directly on their emissions and, for the most part, have significant ability to control their emissions. On the other hand, there can be a large gulf between the imposition of the tax (when the garbage is dumped into the landfill) and the action that causes the damage (the creation of the waste by the producer). This gulf can present difficulties when attempting to transmit the tax back to the individual households, as households are less able to control issues like packaging which are determined by producers. The effectiveness of waste taxes is somewhat reduced where the presence of recycling and composting options is limited, as well as the availability of illegal dumping.

Figure 2.13. **Tax rates on landfill**

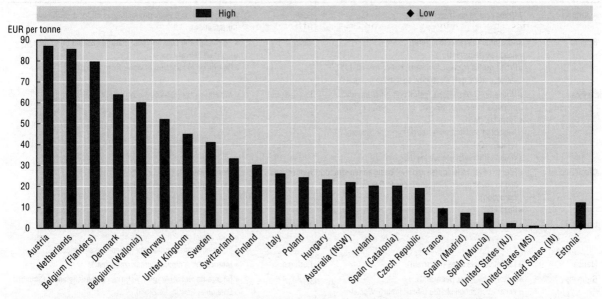

Notes: As at 01.01.2010. Rates on landfilling or transformation of waste (not including incineration) do not include tax bases of hazardous waste or sludge, or taxes/charges on illegal waste disposal. Of note, Israel has a landfill charge to take effect from 2011 at a rate of EUR 9 per tonne.
1. Estonia is an accession country to the OECD.
Source: OECD/EEA database on instruments for environmental policy.

StatLink ᴴᴵᴸᵖ http://dx.doi.org/10.1787/888932317407

2.3. Exemptions and reductions in environmentally related taxation

Although broad-based policies to address environmental challenges are generally most efficient, many times there are features of tax systems that deviate from what may be considered a broad base. These occur for a number of reasons. In some cases, as discussed in Box 2.3, environmentally related taxation may try to address other societal issues beyond the environment. While all pollution sources should face the full cost of the negative externalities to which they contribute, all sources do not contribute to the same externalities.

Tax policy related to agriculture provides an illuminating example of the importance of a full understanding of the role and rationale for deviations from a broad-based tax. Agricultural activities are exempt from a wide range of environmentally related taxes, such as the nitrogen tax in Denmark to the tax on groundwater extraction in the Netherlands and the tyres tax in the Canadian province of Manitoba. By far, however, the most extensive exemptions and reductions for agriculture are those on motor fuels and motor vehicles, as seen in Table 2.4.

In many cases, exemptions for the agricultural community do not make sense from an environmental point of view. The same damage to the aquifer occurs whether a farmer or a non-farmer makes withdrawals. The same is true for the use and disposal of tyres. The opposite may be true in the case of motor vehicle and motor fuel taxation. These environmentally related taxes are sometimes meant to account for both environmentally and non-environmentally related issues, including the usage of publicly funded roads and highways, addressing health concerns related to local air pollution, addressing the costs of motor vehicle collisions, contributing to the general revenue-raising needs of governments and, finally, addressing the contribution to climate change from greenhouse gases. Clearly,

Table 2.4. **Full exemptions for agriculture from environmentally related taxes**

Austria	• Motor vehicle tax (recurrent).	Italy	• Excise duty on energy products (gas oil for greenhouse production).
Belgium	• Additional road tax (recurrent). • Road tax (recurrent). • Excise duties (gas oil, kerosene, heavy fuel oil, LPG, petrol, natural gas, electricity, coal, coke or lignite).	Japan	• Petroleum and coal tax (coal, natural gas and other fuels). • Light oil delivery tax. • Charge on abstraction of water from rivers.
Canada	• Motive fuel taxes (diesel, petrol and other energy products in AB, BC, MB, NB, NL, NS, ON, PE, QC, SK). • Tyres tax (MB). • Additional registration fee large-cylinder capacity vehicles (QC). • Water withdrawal license fee (NS).	Luxembourg	• Mineral oil tax (gas oil, diesel, petrol).
Czech Republic	• Fees to cover watercourse and river basin administration and to cover public interest expenses (tax on water extraction). • Road tax.	Netherlands	• Motor vehicle tax (recurrent). • Tax on groundwater extraction.
Denmark	• Duty on nitrogen.	Norway	• Annual weight-based tax on motor vehicles. • Auto diesel tax. • Electricity consumption tax (greenhouse production).
France	• Tax on vehicle axles (recurrent).	Portugal	• Motor vehicle circulation tax (recurrent).
Germany	• Motor vehicle tax (recurrent). • Water abstraction charge (Mecklenburg-Western Pomerania).	Spain	• Charge on water (Aragón, Asturias, Balearic Islands, Cantabria, Catalonia, Galicia, La Rioja). • Tax on waste (Madrid, Murcia). • Tax on the environmental damage caused by some uses of water from reservoirs (Galicia). • Tax on air pollution (Swine production in Murcia). • Tax on vehicle registration (one-off and recurrent).
Hungary	• Excise tax on diesel. • Tax on motor vehicles (recurrent).	Switzerland	• Distance- and weight-based tax on heavy vehicles. • Motor vehicle tax (Bern).
Iceland	• Excise on motor vehicles (one-off).	United States	• Motor fuel tax (diesel and petrol in CT, IN, MN, NY, SD, WA, WY). • Motor fuel tax (light fuel oil in NY, WA). • Motor fuel tax (natural gas in SD). • Aircraft use tax (MN). • Compressed natural gas tax. • Commercial aviation fuel tax. • Diesel fuel tax. • Gasoline tax. • Non-commercial aviation fuel tax. • Special motor fuels tax. • Heavy truck and trailers tax. • Gas guzzler tax.
Ireland	• Mineral oil tax on coal.	Estonia[1]	• Water abstraction charge.

Note: These are full exemptions specifically for agriculture; there are also reductions in rates of environmentally related taxation for agriculture that have not been included. Broader exemption definitions may indirectly include agriculture and agricultural processes; however, these have not been included.

1. Estonia is an accession country to the OECD.

fuel used in the agricultural sector does not contribute to each of the various effects that fuel and vehicle taxes may be trying to address. On-farm fuel use does not rely on road and highway infrastructure, nor does it likely contribute significantly to urban congestion and any local air pollution might affect fewer people in less densely populated countries. For these reasons, some reductions in the amount of the tax for agriculture may be justified. Since the release into the environment of greenhouse gases for whatever use contributes equally to global warming, for example, full exemptions from environmentally related taxes seem out of place. In all, exemptions in environmentally related taxes need to be carefully assessed, given the variety of considerations that are part of the rate-setting process.

Reductions, exemptions and other features can also result because the introduction of environmentally related taxation typically presents governments with two significant issues: sectoral competitiveness concerns and concerns about distributional impacts (OECD, 2006). These issues generally have impacts on other important government agendas, such as social policy and economic/industrial policy. In order to facilitate the implementation of these taxes, governments typically balance the various impacts and try to find measures to reduce negative consequences. Where some polluters are exempted or impacted to a lesser degree, these measures can result in significantly different values of environmental damage being placed on similar activities.

With respect to the first issue, it should be clear that environmentally related taxation is intended to have competitiveness impacts. A per-unit tax on a particular pollutant should have a much greater impact on a heavily polluting firm producing a given output than one that has found a less polluting method of creating the same thing. The polluting firm is at a competitive disadvantage exactly because is creating more of what society is placing a negative value on. It should even have competitive impacts across substitutes and complements. It is these competitiveness impacts that create the incentives to find less environmentally harmful methods of production and products with fewer negative impacts in their use.

Competitiveness concerns also arise, however, in open economies where domestic policies may be more stringent than elsewhere. Industry and governments are cognisant that policies which set a high rate on domestic firms may encourage production to move across borders, negatively impacting domestic economic performance and having little environmental impact. Clearly, the optimal solution is a large coalition of countries that adopt similar measures that reflect their environmental priorities. Where this is not achievable or where political pressures are significant, remediation measures typically take the form of either exemptions or reductions in the tax rate for heavily polluting industries or tax revenue recycling back to the affected industry (on a different basis than that on which it is collected).

In order to address competitive issues of this nature, however, industries and firms need not necessarily be exempted from environmentally related taxation. Exemptions negate the incentive for firms to undertake pollution abatement measures. By contrast, in this second-best world, targeted revenue recycling can maintain the incentive to reduce pollution while helping to minimise the competitive impact. As will be seen in Box 3.2, for example, a charge on NO_x emissions in Sweden has been successfully implemented with revenues returned to firms based on their energy production.

Border tax adjustments could also address some of these same issues in principle. However, concerns over their administrative burden, a potential escalation in trade disputes, and the fact that OECD analyses show that such mechanisms would have negative overall economic effects concurrent with little pollution abatement in producing countries all suggest that these mechanisms should not be in countries' toolkit (OECD, 2009a).

Regarding the second political economy issue, numerous analyses have shown that environmentally related taxation can have distributional impacts. Lower-income earners typically spend a greater portion of their incomes on goods likely to be impacted by environmentally related taxation, such as motor fuel, home heating, and electricity, although how the revenues are used and the tax's overall impact on the general economy can mitigate some of this impact (Ghersi et al., 2009). Sensitive to these concerns, many times such goods have been exempted from taxation. For example, the UK's Climate Change Levy, intended to help the country reduce carbon emissions through a tax on electricity, gas, coal and liquefied petroleum gas used for energy, exempts the entire household sector. Not wanting to reduce the incentive for all economic agents to reduce their carbon emissions, counteracting measures have been instituted in other jurisdictions. The Canadian province of British Columbia, for example, instituted a carbon tax in 2008. Rising from CAD 10 per tonne to CAD 30 per tonne by 2012, the tax covers all forms of carbon, regardless of source or emitter. To offset the effects, revenues raised by the tax are funnelled into corporate and personal income tax reductions, which include reductions in the bottom two personal tax brackets as well as a refundable tax credit for low-income taxfilers.

In recognition of some of the potential competitiveness and distributional concerns outlined above, governments have taken differential tax approaches to a wide range of potential users, with the differences between households and industry typically being most pronounced. Electricity provides an interesting example for comparison, given its homogenous final state.

About half of the OECD's members have electricity taxes in place.[8] The other half of countries may nevertheless address some of the environmental issues of electricity production by directly taxing the environmentally harmful inputs (e.g. fuels) or the pollutants from electricity production (e.g. carbon dioxide emissions), rather than the final product itself. At the same time, a number of jurisdictions, such as Czech Republic, Sweden and some autonomous communities in Spain, impose additional taxes on nuclear energy to address its unique issues (such as waste, for example).

As can been seen in Table 2.5, there exists a range of tax rates on electricity. Japan has a broadly based tax with a standard rate and no exemptions. The United Kingdom and the Slovak Republic are exceptions in that households are totally exempted from environmentally related taxes on electricity. In some other countries, low-income households face a lower tax burden. By contrast, most jurisdictions provide tax expenditures for businesses compared to the rate for households or the general rate. This can be sector-specific, such as for greenhouse use or for mineralogical procedures, or it can be based on the level of energy consumed. Many countries provide tax relief to energy-intensive firms (which may need to have entered into a negotiated agreement) through exemptions and reductions. The Netherlands' tax structure is the most explicit example of relief provided to increasingly large consumers of energy.

It should be considered that ex-tax prices of electricity in OECD countries are likely to vary across users as well, with large purchasers of power potentially being able to negotiate lower prices than faced elsewhere in the economy. Such structures may compound the differential tax rates outlined in Table 2.5.

It is worth noting that, since electricity generation is covered under the EU ETS, new electricity taxes in Europe would have no impact on the overall level of CO_2 emissions in the European Union, as long as the cap is fixed. Additional taxes on electricity will likely only bring about four effects:

● a higher burden on the electricity sector compared to other sectors (which are covered under the European Union ETS and not subject to tax), as there are now two instruments imposed upon them;

● lower permit prices within the system, as the electricity sector undertakes additional abatement, freeing up permits;

Table 2.5. **Tax rates on electricity in OECD countries**

Tax rates for 2010 in EUR cents per kWh, unless otherwise specified

Austria	
Energy tax	
General rate	1.50
Refunds equivalent to the difference between the full rate and the higher of: *i)* 0.5% of value added; or *ii)* the minimum EU tax rates for enterprises where the sum of taxes on electricity, natural gas, coal and on mineral oils (used as heating fuels) exceed 0.5% of value added	R
Belgium	
Federal tax	0.21
Excise duties	
General rate	0.19
Businesses with NAs or in TP system	0.10
Energy-intensive businesses with NAs or in TP system	0.00
Low-income residents, used in mineralogical processes, agriculture, fishing or forestry	E
Czech Republic	
Electricity tax	
General rate	0.11
Environmentally friendly production, production from already taxed products, track transportation, use in metallurgic processes or mineralogical procedures	E
Additional tax on nuclear energy	0.19
Denmark	
CO_2 tax	
General rate	0.83
Electricity used in public transportation	R
Businesses can obtain a partial reimbursement of 13/18 of the tax on products used in energy intensive processing, and an additional reimbursement of 11/45 of the tax with an NA	R
Electricity tax	
General rate	8.85
Home heating	7.32
Electricity from small plants or derived from wind or water power	E
VAT-registered businesses can obtain reimbursement of the duty paid on electricity except that used for heating. Lawyers, accountants, advertising agencies, etc. are not eligible	R
Finland	
Excise duties	
General rate	0.87
Manufacturing	0.25
Rail use	E
Strategic stockpile fee	0.01
Germany	
Excise duties	
General rate	2.05
Manufacturing, agriculture, forestry	1.23
Buses and railways	1.14
Electricity from wind, solar, geothermal, small hydroelectric, biomass, landfill or sewage gas	E
Ireland	
Electricity tax	
Non-business	0.10
Business	0.05
Households, metallurgical processes, electricity from renewable sources	E
Italy	
State tax on electricity	
Household	0.47
Industrial	0.31
Electrical energy for: heating for industrial processes, factories beyond 1 200 MWh per month, the first 150 kWh per month for households, an input to industrial processes, for public lighting and public transportation, for scientific purposes or for radio and phone communications. Electricity from small renewable sources or methane	E

Table 2.5. **Tax rates on electricity in OECD countries** (cont.)

Tax rates for 2010 in EUR cents per kWh, unless otherwise specified

Sub-national tax on electricity	
Household (but not dwelling)	2.04
Household	1.86
Industrial	0.93
Same exemptions as state tax	E
Japan	
Power resources development promotion tax	0.29
Netherlands	
Energy tax	
< 10 MWh/year	11.14
10-50 MWh/year	4.06
50-10 000 MWh/year	1.08
> 10 000 MWh/year non-business	0.10
> 10 000 MWh/year business	0.05
Electricity for chemical reduction and metallurgical and electrolytical processes, users > 10 000 MWh/year who have an NA	E
Rebate of 50% for non-profits, or places of public worship/philospohical reflection	R
Rebate of EUR 318.62 per connection per year	R
Norway	
Electricity tax	
General rate	1.26
Reduced rate	0.05
Electricity used in chemical reduction or electrolysis, metallurgic and mineralogical processes, the greenhouse industry and in railways. Electricity supplied to energy-intensive enterprises in pulp and paper industry that have an NA and to households and public administration in northern areas	E
Slovak Republic	
Excise duties	
General rate	0.13
Electricity from renewable sources, for energy-intensive industries, mineralogical and metallurgical processes, households, and public transport	E
Spain	
Electricity tax (%)	4.90
Castilla-La Mancha – tax on nuclear electricity	0.15
Extremadura	
Nuclear-derived	0.13
Non-nuclear-derived	0.09
Wind- or solar-derived	E
Sweden	
Electricity tax	
General rate	2.80
Remote areas	1.85
Manufacturing and greenhouses	0.05
Electricity from wind, electricity used in the production of other fuels, electricity for heating	E
Electricity used for the production of heat that is delivered for use in manufacturing industries and for commercial greenhouse cultivation	R
Additional tax on nuclear energy, up to	0.13
United Kingdom	
Climate change levy	
General rate – business	0.53
Businesses with NAs	0.10
Households, electricity for some forms of transportation, and electricity from some renewable sources	E

Note: As at 01.01.2010. E = Exemption, R = Refund, NA = Negotiated Agreement. Common exemptions not mentioned include: production for own use, electricity used in the production or transportation of electricity, diplomatic use, exports, electricity from small generators, electricity from some combined heat and power processes.
Source: OECD/EEA database on instruments for environmental policy.

StatLink ⧉ http://dx.doi.org/10.1787/888932318072

- suboptimal allocation of abatement across firms, raising the overall economic cost of meeting the emissions target; and
- where permits are distributed for free instead of being auctioned, governments can recover some revenues that would otherwise have accrued to them.

In addition to reducing revenues for governments, reductions and exemptions in environmentally related taxation also have non-negligible environmental impacts, resulting in additional environmental damage than if the tax had been levelled uniformly at the posted rate. Beers *et al.* (2007), as outlined in Table 2.6, investigated the environmental and revenue implications from a wide range of tax reductions and exemptions in the Netherlands, some of which may not initially appear to have significant environmental implications.

Table 2.6. **Environmental impacts of selected tax reductions/exemptions in the Netherlands**

	Value of tax reduction/exemption (EUR millions per year)	Greenhouse gas effect (kilotonnes of CO_2 equivalent)	Acidification (tonnes of SO_2 equivalent)	Photochemical ozone creation (tonnes of ethylene equivalent)
Reduced VAT rate on meat	336	116	1 703	18
Energy tax reduction/exemption for large users	1 568	811	19 728	n.a.
Tax deduction for use of public transport in commuter traffic	147	29	5	5
Exemption from excise tax for aviation fuels	1 200	1 272	208	2 433

Source: Beers *et al.* (2007).

StatLink ⟶ http://dx.doi.org/10.1787/888932321701

The value of the Dutch tax reductions and exemptions is generally correlated with the magnitude of the environmental impact. The largest are the reduction/exemption from the energy tax for large users and the exemption from excise tax for aviation fuels. If aggregated together, these four tax exemptions represent EUR 3.3 billion in revenue foregone each year and account for 2.2 million tonnes of additional annual CO_2 emissions. As a means of comparison, the Netherlands' emissions in 1990 (base year for the Kyoto Protocol) were 213 million tonnes, with a target for the 2008-12 period of 200.3 million tonnes.

On a broader level, systems of tax regulation are very large and complex in most countries and understanding their overall impact on specific goods and services can be difficult. In addition to specific exemptions and reductions for goods and services, tax systems can have features which provide indirect preferences for certain items, goods and services with environmental impacts among them. These can take the form of preferred treatment of corporate income for certain firms and sectors, additional deductions for individuals for particular activities, or certain goods and services that might be provided free by governments. On the surface, these treatments may not necessarily appear to have a large impact on the environment but, taken together, they can amount to a significant subsidy that can alter behaviour. In a number of cases, these measures are meant to achieve broader governmental objectives, such as spurring economic activity and helping to overcome barriers that might retard new development. For example, tax concessions are sometimes granted to new fossil fuel extraction operations through tax rate reductions, credits, or the availability of different depreciation schedules. This area of work, known as tax expenditures, focuses on looking at existing tax rates and systems and comparing them against a "baseline" system to determine whether some goods and activities are treated more favourably than others in the economy. This larger field of work is beyond the scope of this publication, but it is important for considering the overall impact of the tax system on the environment.

2.4. Tradable permits

Tradable permits are being utilised more often by governments to address environmental issues, in the form of either "cap-and-trade" systems or "baseline-and-credit" systems.[9] Unlike taxes, which set the price of the pollutant and then allow the market to determine the optimal rate of pollution, cap-and-trade systems fix a set quantity of pollution that can be emitted and allow the market to derive the price (see Box 3.4 for a more complete discussion of taxes and tradable permits). These differences should lead to the same outcome but can reflect preferences for risk tolerance. For example, regarding climate change, a government strongly risk-intolerant to erring in favour of too large a level of emissions would likely prefer a cap-and-trade system, where the amount of carbon emissions is fixed. On the other hand, a government strongly risk-intolerant to potentially high and uncertain carbon prices for industry would likely favour a carbon tax, where the carbon price is fixed and the emission level adjusts accordingly.

In practice, tradable permit schemes have been around for a number of years, with some initial schemes operating in the 1970s. One of earliest, well-known schemes involved permits to control acid rain in the eastern United States. The 1990 *Clean Air Act* Amendment partly replaced existing regulations addressing SO_2 emissions with a programme of tradable permits among polluters with high penalties for non-compliance. Despite initial concerns, the programme proved highly successful. Burtraw (2000) contends that the programme has significantly reduced emissions among participants while also doing so at a cost that was about half of the initial estimate.[10]

This instrument has been evermore used by governments in a variety of new and innovative ways. On the one hand, tradable permits schemes have been used to address relatively small environmental challenges, such as to control salinity in the Hunter River in Australia. Permit holders in this scheme have the right to discharge salty water into the river during times of low flow. The programme has generated less than AUD 0.2 million for 20% of the permits, which last for ten years. On the other end of the spectrum, the issue of climate change is propelling tradable permits to address large-scale, cross-boundary emission problems. The European Union has been leading this effort with its European Union Emissions Trading System (EU ETS), a common approach across member states.

One of the main differences across systems of tradable permits is the initial allocation of permits. The economic efficiency of the trading scheme is not (directly) affected by government's decision to auction or grandfather (that is, distribute freely) permits, as the price of the permits in either scenario will still equate to the marginal abatement cost. However, grandfathering of permits represents a windfall wealth transfer from society to polluting firms and the forgone tax revenue represent an economic inefficiency, in the sense that it cannot be used to compensate society for the pollution damage (such as by, for example, reducing debt, increasing expenditure or reducing distortive taxes).

In the coming years, the impact on government revenues of these systems could be large, as governments move towards more ambitious tradable permit systems where auctioning is a central component. For example, the United Kingdom has identified 7% of their overall Phase II (2008-12) permits within the European Union Emissions Trading Scheme (EU ETS) for auctioning. These 17 million permits per year (85 million over the five-year period for Phase II) should raise significant revenues, even with carbon prices resting around EUR 15 per tonne.

The introduction of tradable permits also presents issues with pre-existing taxes on the same tax bases. Australia, for example, has proposed a comprehensive cap-and-trade system for CO_2-equivalent emissions that will include the transportation sector. Upstream producers of road fuel would be within the trading scheme, leading to higher fuel costs. As a result, the government has proposed that petrol and diesel excise taxes be reduced cent-for-cent against price increases related to higher permit-related costs over the first few years of the scheme (Australian Government, 2008).

It is important to note that government revenue figures regarding environmentally related taxation in this publication do not include revenues derived from the auctioning of permits. Such revenues have, however, in any case been relatively small to date. Accounting experts at the international level are working to determine how to incorporate these revenues into individual countries' national accounts frameworks in the future.

2.5. Conclusions

The revenues from environmentally related taxation form an important component of OECD countries' overall tax revenues, although there is significant variation among countries. These revenues are derived principally from taxes on motor fuels and motor vehicles, with taxes on all other environmentally harmful activities accounting for only a small fraction of the total revenues. Nevertheless, countries are expanding their use of taxes on other environmentally harmful bases, such as with specific emissions to air and water and taxes on waste disposal. It is important to note, however, that the level of revenues raised from environmentally bases is only one potential indicator of how "green" an economy is.

In looking towards the horizon, there are three significant trends that will likely continue to drive development in environmentally related taxation. The first is the greater utilisation of environmentally related taxation to address a wider range of pollutants. Extending the role of taxation to new pollutants beyond taxes on motor fuel and motor vehicles will be driven by the desire to more effectively address environmentally harmful activities that have generally been controlled by regulations or not at all. This will be aided by new technologies and innovations that should make monitoring easier and more cost-effective. Such taxes will likely be on much smaller bases, thereby not leading to significant revenue increases for governments.

The second trend is countries looking to reform existing taxes to make them more effective at meeting environmental targets. This entails ensuring that taxes are not simply levied on bases to raise funds but are structured to more adequately address environmental challenges. Such reforms can be structured to be revenue neutral to governments while bringing about significant environmental benefits.

Finally, the third trend is the significantly larger role of climate change in governments' environmental policies. Even if emissions of greenhouse gases, notably carbon dioxide, were significantly reduced from current levels, the world faces increasing global temperatures and higher climate variability, with the possibility of significant negative repercussions because of the presence of historical emissions in the atmosphere. In addition, the global nature of the problem means that a unit emission of carbon dioxide has the same impact on climate change regardless of from where it comes. As such, responses must be co-ordinated across governments to achieve an effective global mitigation strategy. Mechanisms such as the Kyoto Protocol and its envisaged successors testify to this intention on behalf of governments; implementing comprehensive plans has proved difficult.

Taxes on, or tradable permit systems for, carbon emissions will play a significant role. Some countries have already moved to implement carbon taxes unilaterally. Nevertheless, co-ordination across jurisdictions will be important to achieving global targets. Actions to address climate change also need to address emissions that are currently beyond countries' own jurisdiction. A prime example is the existence of long-established international treaties, such as that for aviation, which has made it difficult for signatories to apply fuel taxes to international trips. Undertaken when climate change concerns were non-existent, these provisions sought to reduce distortions that differential tax rates could bring about. However, recent analysis has shown that aviation accounts for 4.9% of the anthropogenic effects on climate change (Lee *et al.*, 2009), a too significant source to be excluded.

In addition, the use of market-based instruments to tackle climate change could also have important effects on the composition of governments' overall tax revenues, as CO_2 taxes could raise significant sums. Estimates suggest that a tax of USD 25 per tonne CO_2 levied on top of existing price and tax structures would generate annual revenues in 2020 equivalent to 1.9% of GDP in Australia/New Zealand, 1.2% of GDP in Canada, 0.7% of GDP in Europe, 0.5% of GDP in Japan and Korea and 1.0 % of GDP in the United States (calculations based on OECD, 2009a). Therefore, there is a growing role for environmentally related taxation within the OECD (and beyond) both to broaden the bases on which taxes are levied as well as making existing taxes more environmentally effective.

Notes

1. Given the country's small size and relatively low tax rates on motor fuels, Luxembourg realises a significant amount of revenue from the purchase of motor fuel by non-residents (so-called "fuel tourism").

2. In lieu of diesel excise taxes, New Zealand levies a tax on diesel-driven vehicles, being EUR 7.87 per 1 000 km for vehicles less than two tonnes. Such an approach does not give drivers of diesel vehicles any economic incentive to reduce diesel use per kilometre.

3. This value includes the effect of Mexico, where the tax rate fluctuates significantly. However, excluding Mexico over this period does not significantly change the overall impact. The weighted average is calculated with weights based on total revenues derived from motor fuel taxation.

4. See, for example, OECD (2009c) for a discussion of the benefits of using different points of tax incidence to address the varied externalities related to motor vehicle use.

5. In Sweden, the government banned the use of some chlorinated solvents. Due to the lack of acceptable substitutes for these products in some industries, significant public opposition lead to the creation of a number of exemptions. The result is that consumption has decreased but is not near zero. See Sterner (2004) for more information.

6. For the complete list, please see: *http://landbrukstilsynet.mattilsynet.no/plantevernmidler/egodk.cfm*, accessed 14/06/10.

7. For the complete list, please see: *https://secure.pesticides.gov.uk/pestreg/ProdSearch.asp*, accessed 14/06/10.

8. There are some additional taxes on electricity generators, based on revenues, which have not been included in this analysis.

9. A "cap-and-trade" system sets an absolute programme-wide level on emissions. The initial means of distributing permits can vary, with auctioning or based on historical production. By contrast, a "baseline-and-credit" system sets a baseline for individual entities based on historical production and usually as an intensity measure. Under both systems, firms with excess permits can sell these to others in the open market. The result is that, given a baseline-and-credit system does not set an absolute cap on emissions, growth in output of the underlying commodity can lead to increases in the level of emissions compared to a cap-and-trade system. Baseline-and-credit systems also present the difficult task for administrators in deciding what constitutes an acceptable baseline.

10. A wide range of factors can be credited for this (such as a concurrent liberalisation of the railway market), not only the presence of the trading system.

References

Australian Government (2008), *Carbon Pollution Reduction Scheme: Australia's Low Pollution Future*, White Paper, Vol. 1 and 2, Commonwealth of Australia, available at *www.climatechange.gov.au/whitepaper/report/index.html*.

Beers, Cees van *et al.* (2007), "Determining the Environmental Effects of Indirect Subsidies: Integrated Method and Application to the Netherlands", *Applied Economics*, Vol. 39(19), pp. 2465–2482.

Burtraw, Dallas (2000), "Innovation under the Tradable Sulfur Dioxide Emission Permits Program in the US Electricity Sector", Discussion Paper, No. 00-38, Resources for the Future, Washington DC.

Danish Ministry of the Environment (2000), *Economic Instruments in Environmental Protection in Denmark*, available at *www2.mst.dk/udgiv/publications/2000/87-7909-568-2/html/kap05_eng.htm*.

EPA (US Environmental Protection Agency) (2009), *National Emissions Inventory Air Pollutant Emissions Trends Database*, available at *www.epa.gov/ttn/chief/trends/index.html*.

Ghersi, F. *et al.* (2009), "Carbon Tax and Equity: The Importance of Policy Design", paper presented to the 10th Global Conference on Environmental Taxation, in Lisbon, Portugal, 23-26 September 2009.

Lee, David S. *et al.* (2009), "Aviation and Global Climate Change in the 21st Century", *Atmospheric Environment*, No. 43, pp. 22-23.

Lin, C.Y. Cynthia and Lea Prince (2009), "The Optimal Gas Tax for California", *Energy Policy*, Vol. 37, pp. 5173-5183.

New South Wales Environmental Protection Agency (NSW EPA) (2001), *Load-Based Licensing: A Fairer System that Rewards Cleaner Industry*, NSW EPA, Sydney, available at *www.environment.nsw.gov.au/resources/licensing/lbl/lblbooklet.pdf*.

OECD (2005), *Evaluating Agri-Environmental Policies: Design, Practice and Results*, OECD, Paris, *http://dx.doi.org/10.1787/9789264010116-en*.

OECD (2006), *The Political Economy of Environmentally Related Taxes*, OECD, Paris, *http://dx.doi.org/10.1787/9789264025530-en*.

OECD (2009a), *The Economics of Climate Change Mitigation: Policies and Options for Global Action Beyond 2012*, OECD, Paris, *http://dx.doi.org/10.1787/9789264073616-en*.

OECD (2009b), *Incentives for CO_2 emission reductions in current motor vehicle taxes*, OECD, Paris, available at *www.olis.oecd.org/olis/2009doc.nsf/linkto/env-epoc-wpnep-t(2009)2-final*.

OECD (2009c), *The Scope for CO_2-based Differentiation in Motor Vehicle Taxes*, OECD, Paris.

Sterner, Thomas (2004), "Trichloroethylene in Europe", in Winston Harrington, Richard D. Morgenstern and Thomas Sterner (eds.), *Choosing Environmental Policy: Comparing Instruments and Outcomes in the United States and Europe*, Resources for the Future, Washington DC.

Chapter 3

Effectiveness of Environmentally Related Taxation on Innovation

This chapter analyses the effectiveness of environmentally related taxation to bring about innovation. It begins by discussing the challenges to measure innovation empirically and outlines potential metrics. The chapter then delves into a number of case studies to look for potential linkages, finding mixed evidence. It highlights the different types of innovation that environmentally related taxation does (and does not) induce. Constraints to the effectiveness of taxation to induce innovation are also investigated.

The imposition of environmentally related taxation puts an identifiable cost on pollution, providing incentives for profit-maximising firms to reduce their tax burden. They can do this by scaling down operations, abating given current technologies, or inventing/adopting new innovations. The literature is clear that innovation is critical to achieve desirable and lower-cost environmental policy. As governments further adopt market-based approaches for attaining environmental policy outcomes, the question is what effect environmentally related taxation actually has on innovation. This chapter will investigate how to measure innovation, the effectiveness of environmentally related taxation to induce innovation, as well as the presence of constraints to innovation.

3.1. Measuring innovation

Analysing the effectiveness of environmentally related taxation to induce innovation requires metrics to identify and measure innovation (or approximations thereof) in the first place. Yet, the fluid nature of innovation makes measurement – finding applicable data and metrics – difficult. Measuring innovation fundamentally requires specifying what part of the innovation stage is being investigated. On the one hand, inputs to innovation can be measured, such as R&D expenditures. On the other hand, one can measure direct outputs of innovation, such as patents. Given that these are imperfect and sometimes unavailable or not useable, indirect measures of innovation outputs are needed to infer innovation. All of these potential solutions have their benefits and their drawbacks, as outlined below and in Box 3.1.

3.1.1. Input measures of environmentally related innovation

Inputs are only one factor in the overall innovative process but they provide a good source of information on the resources allocated to invention activities. Two main sources of this indicator are expenditures on research and development activities and the number of researchers. The former provides a richer data set, through divisions between public and private expenditures, and potential categorisations among research foci. Inputs are theoretically an important indicator, as they identify the intention of the firm or research institution (given the resources devoted towards the goal). These measures are independent of the outcomes of the R&D process, which does have some factor of luck associated with it. The presence of R&D activities does not necessarily translate into an innovative firm, however. In a survey of a number of countries, the percentage of firms having introduced a product or process innovation was significantly higher than the percentage of firms having performed R&D (OECD, 2009h).

One of the most used – and most widely available – figures is the level of government funds directly allocated to innovation. Direct spending by governments (which does not include that delivered through higher education) typically provides less than half of the total expenditures on R&D in the economy, as seen in Figure 3.1. Moreover, the role of direct government expenditures on R&D has been decreasing in recent years, as funding by the private sector and higher education facilities has relatively increased.

Box 3.1. **Measuring innovation: Is the search different with environmentally related taxes?**

The choice of policy instrument to address environmental issues can likely influence the impact on innovation. More prescriptive approaches, such as technology-based regulatory standards, effectively set a boundary around the range of innovations that can be induced and profitably adopted by firms. Innovations will be limited to the narrow range of the regulations; for example, a regulation requiring scrubbers on coal-fired power plants to reduce airborne pollution will provide incentives over only a very limited range of activities. On the other hand, an emissions tax on the same pollutants vastly increases the type of innovations that a firm can undertake to reduce its tax payment. Thus, one may expect to see in studies a significant difference in favour of the innovative potential of a tax compared to a technology-based standard.

Yet, the practical implications surrounding measurement sometimes lead to empirical work that is not as strong. With patent data, for example, exploring the relationship between patent growth in a specifically defined area (*e.g.* advancements in scrubber design) and the introduction of standards can provide for robust results, as isolating the patent classifications that contain such innovations is clear-cut. The wide-ranging scope of innovation under a well-designed tax, on the other hand, makes the process much more difficult. Taxes can bring about more efficient production, new remediation measures, and even completely new products which are levied across typically larger sectors of the economy. Identifying all the possible areas in which innovation could take place and then looking for potential relationships with tax regimes can prove very difficult for researchers and can therefore lead to less statistically robust results from tax-induced innovation. The case study on the cross-country effects of taxes and standards (see Box 3.6) will highlight this issue in practice.

Figure 3.1. **Direct government share of total R&D expenditures**

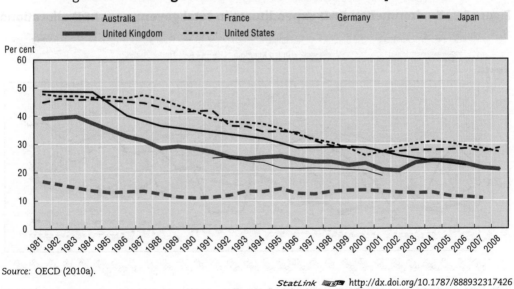

Source: OECD (2010a).

StatLink ⟮⟯ http://dx.doi.org/10.1787/888932317426

There are challenges when attempting to ascertain sub-categories of innovation from the data. Identifying a sole purpose to a set of research can be fraught with issues, for example innovation for environmental aims (see the discussion in Box 3.1). This becomes more apparent as the research becomes more basic in nature. For example, innovations relating to the production of pollution during combustion could be considered innovation relating to everything from the environment to business performance to energy. In recognition of these issues, significant work has been by OECD governments to categorise their expenditures along research focus lines. Figure 3.2 and Figure 3.3 below outline the fraction of government research and development expenditures allocated to the environment and energy, respectively. Since 1981, relative government spending on environmental R&D has increased slightly with France standing out with sustained increases throughout the period. The United States and the United Kingdom have maintained low levels compared to other OECD countries. Large fluctuations can be seen in the levels of Denmark, with significant rises in the mid-1990s.

On the other hand, government R&D expenses for energy purposes present a much different trend: that of long-term decline. Even in recent years, when levels are quite low, they are still above those levels for the environment. While data only goes back to 1981, it is likely that the oil price shocks of the 1970s brought about significant increases in energy R&D on behalf of governments. As real prices of oil returned to less elevated levels, limited R&D funds were slowly redirected to other priorities. The small uptick in 2007 and 2008 suggest that the oil price spike around this period also played a role in changing R&D priorities. It is likely that the smaller scale of the effect compared to the 1970s is a combination of the lag of government response to these price movements and the short-lived nature of the spike. In all, the trend of energy R&D suggests that increased prices can have significant impacts on the direction of R&D trends.

The main issue is that data on private sector R&D is generally not available and the issue is more pronounced for private R&D that is disaggregated by general intention. Environmentally related taxation will stimulate exactly this type of activity, making broad-based linkages between R&D data and environmentally related taxation difficult.

Figure 3.2. **Environmental R&D expenditures in total government R&D allocations**

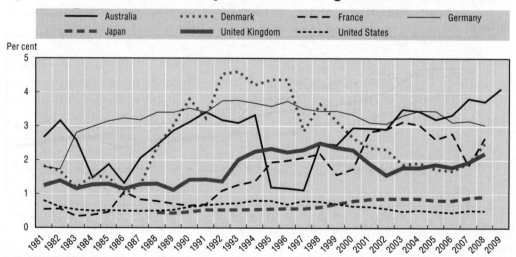

Note: Data is defined by socio-economic objective (in this case, control and care of the environment) through Eurostat's "Nomenclature for the analysis and comparison of scientific programmes and budgets".
Source: OECD (2010b).

StatLink ⟶ http://dx.doi.org/10.1787/888932317445

Figure 3.3. **Energy R&D expenditures in total government R&D expenditures**

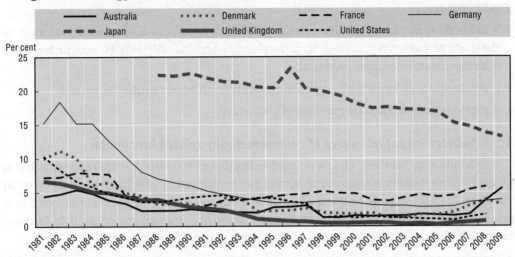

Note: Data is defined by socio-economic objective (in this case, production, distribution and rational utilisation of energy) through Eurostat's "Nomenclature for the analysis and comparison of scientific programmes and budgets".
Source: OECD (2010b).

StatLink ⌘⌘ http://dx.doi.org/10.1787/888932317464

3.1.2. Direct output measures of environmentally related innovation

With increasing digitalisation of data, particularly with respect to patents, more and more information on the outputs to innovation are becoming available. Patents are a valuable measure to researchers because they specifically identify the production of an innovation, when it was created, and by whom. The patents provide valuable information about their inherent nature and the patent system provides clues about an individual patent's value, through information on citations and international transfer. Clearly, patents are a highly useful source of information on innovation.

Although OECD (2009i) finds that most major innovations have been patented, evaluating patent data necessarily excludes some types of innovation. Rule-of-thumb innovations and organisational innovations are difficult, if not impossible, to patent. In addition, patents necessarily reflect the innovative capacity of a country which can be characterised by the productivity of researchers, education policies and other policy tools (Rassenfosse and Pottelsberghe, 2009). Therefore, patent levels can be influenced by the propensity of a country to patent, reflected in their legal, cultural and administrative traditions. Moreover, the actual patent system itself can impact greatly on the level of patents, with administrative fees and the degree of protection a system provides its patent holders. As such, caution must be taken when drawing conclusions from simple cross-country comparisons of patent data.

To overcome some of these issues, the European Patent Office and the OECD have developed a unique database (PATSTAT) that provides detailed information on worldwide patents (OECD, 2004). This database brings together patents from major patenting countries and categorises them according to a number of different standards. The database is updated regularly, containing over 70 million patents with significant information about their history and their intended purpose. This database provides an invaluable resource of researchers and has been used in a number of the case studies undertaken for this project.

Even with excellent databases, search strategies are still critical to obtaining all relevant and useful patents in a given area. Therefore, focusing on "claimed priorities" (those patent applications that have been claimed as priority in an additional patent body beyond the initial body) can provide significant advantages over simple patent searches (OECD, 2009d):

- it helps filter out lower-quality patents that likely have little-to-no economic value, as the costs of patent registration in multiple jurisdictions will only be sought for those with significant economic potential;
- it avoids double counting when pooling patents from across jurisdictions; and
- it provides truly worldwide coverage of patents.

3.1.3. *Indirect output measures of environmentally related innovation*

In addition to relatively clearly defined indicators of innovation – R&D expenditures or patents – more indirect measures can sometimes be utilised to infer innovation when other measures are not available and/or useful. These measures look to the effects of innovation in areas where it would be expected for the firm, instead of at the innovation itself. In terms of taxes on pollution, indirect measures of innovation can include the following:

- *Declining marginal abatement costs*: Environmentally related innovations that are profitable for the firm to implement will help the firm reduce the marginal cost of abatement. Declining (or inward-shifting) marginal abatement costs can therefore be indicative of the integration of innovations into the firm's *modus operandi*.
- *Decoupling of pollution from output*: Decoupling the trends of pollution and outputs can be indicative of innovations being taken up by economic actors, although the means by which decoupling occurs are likely diffuse.
- *Pollution reduction given technology adoption*: Reductions in emissions, accounting for adoption of existing technologies, can provide insight to innovations used by the firms that go beyond standard means of abatement.

It is important to consider that seemingly strong indirect measures of innovation may be occurring because of the influence of non-innovation factors. Efficiency gains, productivity increases or input substitution may be resulting in less pollution-intensive production, not innovation. For example, decoupling of pollution from output may occur because of increased production, leading to economies of scale in fuel use, and productivity increases can lead to declining marginal abatement cost curves.

The case study of the Swedish NO_x charge, outlined in Box 3.2, provides a clear example where the use of indirect measures was helpful in the analysis, given that firm-level data on R&D expenditures was not available and the patenting effects could not be specifically linked to the introduction of the tax.[1] Despite this, the study's authors were able to effectively infer that innovation had occurred using firm-level analysis. First, the firms' marginal abatement cost curves shifted inward significantly following the introduction of the tax. This suggests that firms were able to meet given levels of emissions at less cost, through a combination of productivity improvements and innovation. While not being able to distinguish productivity gains from innovation gains, such a measure, in combination with other factors, suggests that innovation has been induced by the charge. Second, NO_x emissions became decoupled from power generation. Finally, even firms that did not install physical abatement technologies, such as end-of-pipe measures, still saw annual declines in emission intensities, suggesting that incremental process innovation was occurring within plants.

Box 3.2. **Case study: Sweden's NO$_x$ charge**

Sweden implemented a charge on nitrous oxide (NO$_x$) emissions in 1992 emanating from large combustion plants, typically firms generating power. NO$_x$ emissions, which include nitrogen dioxide (NO$_2$) and nitric oxide (NO), contribute to ozone smog, the formation of acid rain and particulate matter. They arise from high-temperature combustion. The Swedish charge was relatively high, compared to charges that other countries have implemented, but the revenues were recycled back to firms based on energy output.

The charge has been very successful in reducing NO$_x$ emissions from regulated firms, encouraging extensions of the charge to smaller facilities. Over the 1992-2007 period, total emissions of NO$_x$ from regulated plants remained relatively stable (even with the extension of the charge to smaller and relatively more polluting plants) while energy production for the same sample increased 77%, suggesting that the tax has been effective at decoupling production from NO$_x$ emissions. One of the first effects of the charge was that firms quickly adopted abatement equipment, with 62% of firms having mitigation equipment in 1993 compared to 7% in 1992. This equipment favoured cleaner production rather than end-of-pipe investments, which is to be expected with more flexible economic instruments. Relative intensities of NO$_x$ emissions for a number of firms actually increased over the period, generally resulting from switches to fuels that are more prone to NO$_x$ emissions but that help meet other environmental and policy goals.

The Swedish NO$_x$ charge did appear to have an effect on the level of patenting in NO$_x$ related areas. The 1988-93 period saw a significant jump in patenting levels, compared to periods before and after and places Sweden as one of the top inventors in this area, adjusting for population size. Although patenting in the post-1993 period is not as high, it still places Sweden as one of the top relative innovators in this area. However, being able to break out the interactions of tax *versus* pre-existing regulations, as well as considering the political economy angle that, because of the increase in patenting, a higher tax could be effectively applied, is difficult.

Yet, this does not suggest that innovation was not taking place. An important feature of the Swedish charge was the use of continuous monitoring devices, which helped firms recognise where and how NO$_x$ emissions formed and therefore how to optimally calibrate instruments and equipment to maximise the power-generation-to-emissions ratio. In looking at the chart to the right, which identifies marginal abatement costs curve for the energy sector

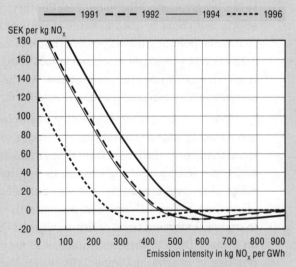

Marginal abatement cost curves of taxed emitters

StatLink ▩▩▩ http://dx.doi.org/10.1787/888932317977

over the initial years of the charge, it can clearly be seen that the cost to achieve a given level of abatement is falling. This is suggestive of innovative abatement methods as well as productivity gains in existing methods of abatement.

Moreover, there are annual declines in the emission intensities of firms, both for those that do adopt new physical mitigation technologies (3.2% decline) and for those that do not (2.9% decline). One would expect that the group of firms installing new physical mitigation technologies have ongoing declines: adoption by new members drives down the intensity in the short-run and ongoing efficiencies from better operating the equipment result in longer term declines. The decline for firms that do not adopt physical mitigation equipment suggests that new innovations in non-physical mitigation are being created and adopted, which are likely also to occur in firms that also adopt physical mitigation technologies. These feats are coupled with the decoupling of NO$_x$ emissions from power generation.

Therefore, while the patent data is somewhat ambiguous with respect to new technologies for NO$_x$ emission abatement, there is nevertheless innovation. These innovations require more indirect measurement methods but their importance should not be underplayed: they contribute significantly ongoing emission reductions and ongoing declines in abatement costs. For a more complete description of this case study, please see the summary in Annex A.

Source: OECD (2009b).

So the question remains: what indicators should be used when undertaking innovation analysis? Detailed R&D data provide a clear indication of firms' intentions to innovate, regardless of the outcome of that effort. Yet, R&D levels do not have good predicative value of patent levels, the success of that effort (Klienknecht *et al.*, 2002). Moreover, information on detailed R&D activities, especially by the private sector, is nearly impossible to obtain. Patent data can be a useful tool for inferring both inventive input and output where detailed R&D data is not available (Griliches, 1990). Indirect measures of innovation are also important for shedding light on the innovation story. Thus, no single available measure of innovation is perfect. While strides have been made to obtain better data sources, such as the EPO/OECD patent database, caution must still be exercised in drawing conclusions from innovation data and a wide variety of information should be sought.

3.2. Identifying the benefits and drawbacks of innovation

One of the challenges facing researchers and policy makers is how to encourage and measure innovation that is socially useful. Not all innovations have socially beneficial results. The innovations aimed at tax avoidance or which have no practical usage (*e.g.* developing a better telegraph machine in the 21st Century) provide no benefits to society and detract from efforts that could be used towards more useful outcomes. Some innovations, such as those that make polluting less expensive (think of new innovations that allow for cost-effective oil extraction of previously inaccessible locations) can even be considered bad (although useful) from an environmental perspective. At the same time, subjective valuations over the distinction between useful and non-useful innovations can present significant problems.

When looking at cross-country examples, one objective method to ensure that only economically useful innovations are used is to focus on patents that have been registered in more than one jurisdiction. Only those innovations that proved useful would justify the time and expense of patent registration in multiple countries. In addition, one can look to the effect of innovations on the costs to businesses. In the Swedish charge on NO_x emissions (described in Box 3.2), one can measure the effect of useful innovation on the declining marginal abatement costs of firms subject to the tax, as only useful innovation would have an impact. Despite these examples, it is very difficult to differentiate between useful and not useful innovation, especially when looking at inputs to innovation, such as R&D expenditures. Therefore, policy makers must realise that not all innovation is socially beneficial, but that means to identifying and only promoting useful innovations can be similarly problematic. Box 3.3 provides an interesting example.

Once an innovation has been developed, the environmental and economic impacts can be varied (and not always beneficial). Therefore, governments may wish to actively dissuade some innovations in the marketplace while promoting others, such as through the use of taxes. Figure 3.4 outlines potential government responses in the face of various combinations of economic externalities and environmental impacts from innovations.

The term economic externality is easiest to interpret in the upper half of the figure, where it is positive. This refers to the classic case for public support for inventions, because the economic benefits to society as a whole of a given invention are larger than what the potential inventors would manage to capture. One could, however, also envisage a situation where the benefits to society of a given invention being smaller that the benefits the inventor could obtain (the negative economic externality) – for example in situations where the prices in the economy are being distorted, so that the inventor earns "too much" on his invention.[2]

Box 3.3. **Is all innovation desirable? Innovation and the evasion of environmentally related taxation**

Many OECD countries differentiate diesel taxes by end use: full tax rates for on-road use and reduced or no taxes on off-road use (*e.g.* industry, agriculture, home heating). Since the fuel is nearly identical for either use, the possibility of tax evasion is high. In 2005, the price differential in many US states exceeded USD 0.13 per litre. Tax evasion is clearly not optimal: government revenues are reduced and evaders contribute to a deadweight loss. Marion and Muehlegger (2008) investigate the case of diesel taxation in the United States where, after October 1993, off-road diesel fuel was required to contain an inert dye to help authorities more effectively monitor compliance. In addition, dye was required to be added near the production source, reducing the monitoring effort of regulators.

This innovation in tax administration had a significant and immediate impact on fuel consumption, accounting for a wide range of other factors. Sales of diesel fuel (taxed) increased immediately by 25-30%, while fuel oil (a good substitute for diesel fuel and not taxed) had an immediate decrease. In line with expected economic theory, this effect is larger in states with higher tax rates.

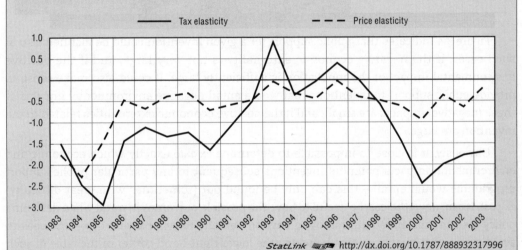

StatLink http://dx.doi.org/10.1787/888932317996

The authors performed additional analysis on the price and tax elasticities of diesel fuel. In the pre-dye period, these figures were statistically different suggesting that evasion was present. After the addition of dye, these values effectively converged. However, an interesting finding occurs when the elasticities were analysed on a yearly basis (see above chart). In the pre-1993 period, there is a persistent gap between the price elasticity and tax elasticity of diesel fuel. This suggests evasion, as only evaders would differentiate behaviour based on a tax change compared to any other type of price movement. With the introduction of the dye in 1993, the gap closes, and the tax elasticity becomes less than the price elasticity. Starting in 1998, however, the gap between the two elasticities re-emerges. This suggests that evaders have innovated and found new methods to avoid paying taxes, overcoming the obstacles of the dye. While innovation is important, it is clear that this type of innovation is not socially beneficial, resulting in a deadweight loss to the economy.

Figure 3.4. **Environmental impacts and economic externalities of innovations**

			Environmental impact			
			Positive		Negative	
			Large	Small	Small	Large
Economic impact	Positive	Large	"Ideal" case for public support of some sort – *e.g.* through public grants or preferential tax treatment.	Some support should be given.	Economic benefits outweigh the environmental drawbacks; still a case for public support.	More detailed assessment of costs and benefits required.
		Small	Some support should be given.	Some support should be given.	More detailed assessment of costs and benefits required.	The environmental impacts are too negative, support should not be given.
	Negative	Small	Environmental benefits outweigh economic drawbacks; still a case for public support.	More detailed assessment of costs and benefits required.	Support should not be given.	Support should not be given.
		Large	More detailed assessment of costs and benefits required.	The economic impacts are too negative, support should not be given.	Support should not be given.	No support should be given; application of such technologies should be curtailed, *e.g.* though taxes.

Figure 3.4 indicates that public support for a given invention could be justified also in such cases, if the negative economic externality is not very large, and if the positive environmental impact of the innovation is sufficiently large. It could also make sense to provide public support to inventions that would entail negative environmental impacts, if these (negative) impacts are small, and if the positive economic externalities related to the invention are large.

Obviously, it is close to impossible to determine *ex ante* exactly which economic and environmental impacts potential inventions subsequent to any particular public support programme would entail – this can only be found out (sometimes with great difficulty) *ex post*. It can, nevertheless, be useful to have the possible outcomes in mind when designing policy instruments aimed at promoting environmentally relevant inventions – and seek to avoid supporting inventions that belong in the lower right-hand corner of the table. If such inventions nevertheless are made, environmentally related taxes could be used to limit their wider diffusion.

3.3. Case studies of environmentally related taxation and the inducement to innovate

Clearly, innovation is important for effective environmental policy – but does taxation or do tradable permit systems (see Box 3.4 for a greater discussion on the similarities of these two instruments) actually play a role?

Before looking at taxes exactly, researchers have investigated the ability of general price changes to induce innovation within firms. In the environmental arena, oil prices, electricity rates, and other commodities have been used to study the impact that their prices have on demand, as well as the effect on innovation. Lichtenberg (1986 and 1987) finds that energy prices in the United States mainly in the 1970s did impact the relative level of R&D spending towards energy-related projects, drawing upon the significant price effects of the period. Popp (2001) finds that changes in energy consumption due to price changes can be disaggregated: two-thirds of the change in energy consumption results from price-induced

Box 3.4. **Similarities of environmentally related taxes and tradable permits**

When governments seek to address environmental challenges through market-based instruments, the debate is typically between taxes and tradable permits. The differences between taxes and tradable permits are, however, very small in theory, when it is assumed that there is a fair degree of certainty about the future. Specifically:

1. If an environmentally related tax set at rate per unit of emissions T leads to an emissions level Q, then alternatively regulating the same problem by issuing a quantity Q of tradable emissions permits will lead to a permit price per unit of emissions T (if the permit market is competitive).

2. The level and pattern of pollution abatement, as well as the incentives for innovation, will be the same under the two instruments. In both cases, the incentive firms face for abatement at the margin is T per unit of emissions, and firms would undertake abatement where the cost per unit is less than this incentive. In the diagram, the abatement undertaken reduces emissions to Q from the pre-regulation level U.

3. The abatement cost incurred by firms will be the same. The total abatement cost incurred by firms in reducing their emissions from U to Q is represented by the area labelled A under the marginal abatement cost schedule.

Properties 1-3 hold regardless of whether the permits are distributed free or sold (*e.g.* through an auction). In either case, the value of the last permit used is given by the abatement cost that would otherwise be incurred, and this is given by the marginal abatement cost at emission level Q, which is T per unit. The value of tradable emissions permits, therefore, is independent of the way in which the permits are distributed (so long as the permit market is competitive). Where permits are auctioned, there is a further point of similarity between an emissions tax and tradable emissions permits:

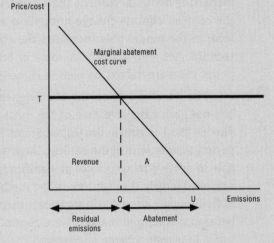

4. If the permits are sold in a competitive auction, then the auction revenue yield will be Q*T, which is the same as the tax revenue that would be collected from the environmentally related tax.

It is for these reasons that this publication addresses both environmentally related taxes and tradable permits, and case studies have been presented using both instruments. It should be noted, however, that real-world variations can cause differences between the two instruments. First, information is usually never perfect, requiring that policy makers rely on assumptions and have to factor in tolerances for risk about the errors of their assumptions. If the costs of increased abatement activities rise extremely quickly as abatement is undertaken (that is, the marginal abatement cost curve is steeper than the marginal damage curve), there is the possibility that a cap on emissions can provide high permit prices. In this instance, taxes may be a more appropriate instrument to balance the economic/environmental tradeoffs. Where it is believed that the marginal damage curve has a greater slope, the opposite may be true.

Second, compliance and administrative costs of the instruments have to be factored in. Third, the efficiency of permit markets are not always guaranteed, given concerns about market power, extent of participation, level of trading, and design constructs. Fourth, in a tax regime, new innovations would effectively lead to reduced total emissions – if the tax rate is not changed. With a cap-and-trade system, new innovations would not alter total emissions – as long as the cap is not modified – but permit prices would decrease. However, in principle, in both cases, the policy ought to be modified if new innovations reduce abatement costs – assuming it had been set at the optimal level before the innovation took place. In a tax regime, the tax rate ought to be reduced; in a cap-and-trade regime, the total number of permits ought to be reduced. Finally, there is an important difference in how a tax regime and a cap-and-trade regime interact with any other policy instruments that apply to the same environmental problem. Under a pollution tax, additional policy instruments could lead to further emission reductions; under a cap-and-trade regime, that is not the case. Since the cap is fixed, additional abatement will only lower the price of permits.

Source: OECD (2008).

factor substitution while the remaining one-third is because of price-induced innovation. Popp (2002) investigates energy prices on energy-efficiency technologies, finding that higher prices not only shifted firms away from energy-intensive processes but also induced innovation into new energy-saving methods. In addition, his work notes that there appear to be diminishing returns to R&D and that the supply of ideas (that is, the existing knowledge stock) is also critical. Furthermore, the price effect on innovation is rather quick: about one-half of the full innovation effect of energy price increases occurs within five years. Finally, Kumar and Managi (2009) and Crabb and Johnson (2010) find that long-term oil price rises do induce substantial technological progress.

Modelling specifically on climate change, OECD (2009a) finds that carbon pricing aimed at stabilising CO_2 concentration levels in the atmosphere would induce a more than three-fold increase in expenditures on energy R&D as a percentage of GDP (and four-fold for renewable energy R&D). As the stringency is increased, thereby leading to a higher carbon price, the level of R&D expenditures increases more than proportionally, given the increasing marginal costs of abatement. Despite these increases, the translated effects on the costs of climate change mitigation are small: forcing R&D to remain at the baseline level in the model only increases the costs slightly by 2052, assuming no breakthrough technologies. Yet, when backstop – or breakthrough – technologies are incorporated, the policy costs are halved, as seen in Figure 1.1.

The work to date on the effectiveness of economic instruments to induce innovation has not been extensive. One of the most widely analysed examples is the case of sulphur dioxide (SO_2) control in the United States in the 1990s. Burtraw (2000) finds that the tradable permit system (one of the earliest, large-scale schemes) in several north-eastern states was able to achieve its objectives at significantly less cost than *ex ante* analyses had suggested. Achieved largely through innovative methods, these cost reductions were achieved outside of traditional, patentable innovative means. Changes in production processes, organisational behaviour and input markets were central. For example, the flexibility brought about by the tradable permit scheme encouraged the expanded used of low-sulphur coal, facilitated by technical innovation and industrial reorganisation in the railroad sector following deregulation in the 1980s. New techniques in fuel blending were discovered. Impacted plants modified their organisational structures, shifting responsibility for the trading scheme from chemists to financial officers. These innovations were critical to the overall success of the programme, but many were clearly not patentable. Some analyses have even suggested that firms were better off after the introduction of the tradable permits system, though the large windfall gains from the grandfathering of permits likely contributed to this.

The potential for such results has led to discussion of the Porter hypothesis (Porter, 1991; Porter and van der Linde, 1995), which suggests that new environmental policies, including taxes, can act as a shock to induce firms to re-evaluate their operations. In doing so, innovations to address the new environmental policy can be found to better address pollution levels but that also increase the profitability of the firm, as firms have not previously explored all profitable opportunities. This win-win situation amounts effectively to a free lunch [or even a "paid lunch" as described by Jaffe and Palmer (1997)] for environmental policy: stronger protections for the environment and more profitable firms. Despite being popularised in recent years and the fact that some examples do exist, the overall empirical evidence for the Porter hypothesis is not strong (see Box 3.5).

Box 3.5. **The Porter hypothesis**

The seductiveness for policy makers of the Porter hypothesis – that the financial benefits to impacted firms of enhanced environmental regulation alone exceed their implementation cost, thereby increasing firm profitability – is difficult to ignore. However, compelling evidence is lacking, and there are concerns that the initial research was based on several cases out of hundreds of thousands of businesses (Palmer et al., 1995).

Recent theoretical work has suggested that some limited cases do exist where the Porter Hypothesis may be valid, given certain assumptions (see, for example, Popp, 2005; Greaker, 2003). However, both country-specific studies (Brännlund and Kundgren, 2009, for example) and reviews of the general empirical literature (such as by Ambec and Barla, 2006) generally find that environmental regulations have a negative impact on overall productivity of the firm, and that there is only mixed evidence on the linkage between financial and environmental performance.

There is stronger evidence for variants of the Porter hypothesis, as outlined by Jaffe and Palmer (1997). The "weak" form suggests that environmental regulation will stimulate environmental innovation, while the "narrow" form suggests that more flexible environmental policies provide greater incentives for innovation. This is set against the "strong" form that properly designed regulation may induce cost-saving innovation that more than compensates for the cost of compliance. Lanoie et al. (2010) attempt to analyse Porter hypothesis variants using an OECD survey of firms. First, they find strong evidence that environmental stringency is positively correlated with environmental R&D activities in firms. Moreover, environmental and business performance are positively correlated, confirming other findings that firms which seek out ways to be more efficient (such as with fuel use, for example) or have better environmental credentials to present to clients have better business performance as well. At the same time, the stringency of environmental policy is positively correlated with environmental performance and negatively correlated with business performance. What these findings seem to suggest is that firms are positively rewarded by undertaking activities that address both financial and environmental goals, such as energy efficiency. At the same time, the fact that businesses are adversely affected by the introduction of environmental policies by governments suggests that the strong version of the Porter hypothesis does not hold (although, it does suggest these policies are in fact targeting the emissions that would otherwise not be addressed by the private sector). The "weak" and "narrow" forms of the Porter hypothesis are explored elsewhere in this report, outside of the Porter hypothesis framework.

The suggestion that environmental policies can induce increased profitability among impacted firms poses some larger questions as well. In an open marketplace, the presence of increased levels of profitability (which theoretically should be low, as most benefits should be passed on to consumers) suggests that there may be constraints to the normal functioning of the market. In terms of the Porter hypothesis, the fact that profitability increased for selected firms may suggest that there are imperfections at play, such as market power. Policy makers should necessarily be concerned with the impact of environmental policy on the performance of firms; likewise, the potential for inducement to a persistent level of high profits by firms may suggest that there are negative considerations as well which policy makers must also take into account.

Although some significant examples do exist, such as the preceding example, there are few empirical studies to investigate the linkages between taxation, innovation and the environment. Empirically investigating innovation within firms and across sectors is difficult, especially investigating the potential linkages with flexible environmental policies. Thus, case studies that have been prepared specifically for this project will be examined to investigate this void.

As already mentioned, one of the most widely used environmentally related taxes within OECD economies is that levied on motor vehicle fuels, such as petrol and diesel. Almost always, motor fuel taxes are levied at the same time as regulations on vehicle makers regarding the emissions of vehicles are in force (and being tightened). These policies aimed at both producers and consumers are likely to have reinforcing incentives for innovation. A cross-country study, as outlined in Box 3.6, was undertaken to investigate how these different environmental policy instruments interacted and affected the level and type of patenting.

This cross-country analysis focused on petrol taxes, petrol prices and the regulatory stringency of tailpipe emissions and fuel efficiency in the United States, Germany and Japan. Regulatory stringency, especially for CO and NO_x emissions, appeared to have a positive effect on patenting in end-of-pipe technologies and better engine design, respectively. No effects on patenting were found to occur because of fuel efficiency limits. With respect to petrol prices and taxes, the results were more mixed. Petrol taxes appear to positively influence innovation in fuel efficiency measures that are not engine-related (such as aerodynamics and rolling resistance). However, the sign and significance of the coefficients are quite sensitive to the regression equation specification, and the signs on the tax and price coefficients are generally opposite (when the hypothesis would be that any price movement should generally have a similar effect). Additional work in this area could provide greater clarity to some of these issues.

The various case studies undertaken for this project highlight the difficulty in empirically testing the innovation impacts of environmentally related taxation. When large-scale taxes with high rates have been identified, such as those on motor vehicle fuels, there is significant noise from cross-country variation, underlying product price movements, and the interaction of many other environmental and economic policy instruments. These interactions can make it difficult to draw out clear conclusions about the specific impacts of environmentally related taxes on innovation. By contrast, environmentally related taxes levied on a smaller scale in single jurisdictions can seem to induce innovation without the triggering the expected indicators of innovations, such as patent counts.

In addition to the difficulty of teasing out the effects of environmentally related taxation in the data, there is also the issue of obtaining the data in the first place. Policy instruments require implementation periods and time for proper functioning of the tax or tradable permit scheme. Data collection is necessarily delayed, making *ex post* analysis of such measures follow significantly after. The case study of Korea's Emission Trading System, as outlined in Box 3.7, provides a clear example of some of these issues. Although the scheme was launched at the beginning of 2008, it will likely be many years before enough data exist to undertake large-scale analysis of the effectiveness of the scheme on environmental, economic and innovation grounds.

Box 3.6. **Case study: Cross-country fuel taxes and standards on patenting activity**

A wide range of economic and environmental policy tools have been used in OECD economies to address the fuel efficiency of, and air pollution emissions from, motor vehicles: regulatory limits on emissions, fuel efficiency standards, fuel composition, fossil fuel taxes, and even speed limits. Given the global nature of the vehicle market and the wide range of environmental instruments employed, this area provides a rich source to explore innovation impacts of motor fuel taxes compared to other instruments.

As part of this case study, standards on emissions of CO, HC, NO_x, and PM, standards on fuel efficiency, as well as petrol taxes and non-tax petrol prices have been investigated in a number of OECD economies. These standards were assessed against patent data divided by the three* main types of innovation that can be expected when these instruments are put into force:

● End-of-pipe innovations targeting specific air pollutants (such as catalytic converters).

● Engine-related production innovations (such as fuel injection and on-board diagnostic systems).

● Non-engine related production innovations (such as aerodynamic design and rolling resistance).

Of these three categories, patents over the 1965-2005 period significantly favoured engine-related innovations (72%) and end-of-pipe innovations (21%), with innovations in non-engine related innovations being quite small (7%). Innovations were significantly clustered in three countries (United States, Japan and Germany), comprising 89.2% of the related patents in the 19 OECD economies that have or had vehicle-producing facilities. The results of regressing the various standards and taxes on the three categories of patents across the three countries produces some interesting results in standard OLS regressions:

	End-of-pipe innovations		Engine-related production innovations		Non-engine related production innovations	
	(1)	(2)	(3)	(4)	(5)	(6)
CO standard (km/g)	9.30***	9.54***	−11.58	−9.24	−2.78**	−1.64
	(2.84)	(2.75)	(8.96)	(7.41)	(1.29)	(1.20)
HC standard (km/g)	−0.78	−0.95	−8.83***	−8.36***	−0.97***	−0.56*
	(0.61)	(0.71)	(1.91)	(1.91)	(0.28)	(0.31)
NO_x standard (km/g)	1.60	−2.93	57.05***	40.12***	11.57***	6.40***
	(4.07)	(5.27)	(12.83)	(14.12)	(1.85)	(2.30)
PM standard (km/g)	−0.38	−0.13	−6.25***	−5.54***	−1.60***	−1.31***
	(0.75)	(0.97)	(2.36)	(2.60)	(0.34)	(0.42)
Fuel efficiency standard (L/100 km)	−3.00***	−0.52	−4.49	0.59	0.29	0.18
	(1.13)	(1.28)	(3.55)	(0.86)	(0.51)	(0.56)
Petrol taxes	5.67	−209.04***	456.34**	−223.65	108.05***	88.27***
	(60.76)	(68.90)	(191.37)	(185.62)	(27.62)	(30.10)
Petrol prices	−67.01***	101.16***	−78.35	468.72***	−32.34***	−13.87
	(22.21)	(36.41)	(69.96)	(98.10)	(10.10)	(15.91)
Time fixed effects	No	Yes	No	Yes	No	Yes
Adjusted R^2	0.76	0.65	0.90	0.89	0.66	0.80

Note: All regressions had 108 observations and included controls for total patent counts and country fixed effects.

* Indicates $p < 0.05$.

** Indicates $p < 0.01$.

*** Indicates $p < 0.001$.

StatLink ᴍᴙᴾ http://dx.doi.org/10.1787/888932318034

Two disconcerting features of the above analysis stand out: i) the effect that time fixed effects have on the sign and significance of the variables, particularly petrol taxes and prices; and ii) the different signs of petrol taxes and prices. With respect to the first issue, the generally negative effect of petrol prices in regressions without time fixed effects suggests that rising petrol prices are unlikely to have a contemporaneous effect on inventions. Oil price spikes are usually unexpected and the first reaction by consumers is to reduce consumption of fuel by driving less and buying more fuel efficient cars from the

---

Box 3.6. Case study: Cross-country fuel taxes and standards on patenting activity *(cont.)*

existing stock, reducing emissions and therefore signaling to inventors less pressure for inventing new technologies that control emissions. This is different from petrol taxes which many times are debated publicly (or do not take effect right away) and are therefore less unexpected. With respect to the second issue, this cannot be easily explained away, especially for the latter, since one would have expected that the signs of petrol prices and taxes would have been the same (since they have the same effect on the consumer). These issues suggest that more work needs to be undertaken in this area and that caution must be used when drawing out conclusions from the above analysis.

Recognising these caveats, a number of interesting trends can be seen:

- First, fuel efficiency standards seem to have little discernable effect on patenting activity except for regression (1), even in non-engine related technologies that would be directly affected by such activities.

- Second, stricter regulatory standards on CO and NO_x emissions appear to have positive impacts on patenting in certain fields. The absence of stronger findings across pollutants may be due to the fact that there are significant tradeoffs in pollution abatement of this kind (for example, increasing abatement of NO_x in catalytic converters by changing the air-to-fuel ratio significantly increases the emissions of CO and HC).

- Finally, and most importantly to this project, is the significant variation in the petrol taxes and prices, due particularly to the inclusion (or not) of time fixed effects. It is comforting that petrol taxes in equations (5) and (6) have a positive and significant impact on patenting that is primarily related to fuel efficiency, although petrol prices are insignificant (it may be that the lagged effect of petrol prices may be more relevant given the unpredictability of oil price movements compared to the stability and predictability of excise tax levels). Clearly, petrol taxes appear to have an impact, but more research is needed to be able to draw out clear conclusions.

The results from this case study are generally consistent with the results from a similar analysis, which found that after-tax petrol prices were positively associated with engine design patents, while the effect of command-and-control environmental policy (represented by the mandate for on-board diagnostic systems in the United States) was insignificant. The reverse was true for end-of-pipe innovation: after-tax petrol prices had little effect while command-and-control policies were rather significant (Vries and Medhi, 2008).

* A fourth type of innovation was possible – input innovation relating to advancements in cleaner-burning fuels – but these patents only accounted for 0.1% of the patents of the four categories and therefore were not explored in the regression analysis.

Source: OECD (2009f).

Box 3.7. Case study: Korea's emission trading system

Korea has some of the worst urban air pollution among OECD economies. In response, the government introduced an emissions trading scheme in January 2008 that covers the majority of emissions of NO_x, SO_x, PM_{10} and VOCs. The first stage covered the largest emitters, with smaller emitters having been brought into the scheme in July 2009. The cap in the first year was set as the five-year average of emissions, with the cap falling gradually until 2014, when it reaches the specified limit. There is geographic differentiation, such that each city and provincial government within the scheme issues permits for that region, in recognition of the fact that local concentrations of some pollutants can vary significantly between regions.

Over the last few decades, and this decade in particular, Korea has made significant strides in air pollution technology, accounting for 23.1% of patents globally between 2000 and 2004, behind only the United States. Firms have also invested significantly in abatement equipment. Much of this data, however, is related to past policies and government actions. With the introduction of the emission trading scheme having only started recently, obtaining relevant data and being able to assess it against the introduction of the scheme is simply impossible at this stage. Additional work in this area should hopefully provide interesting material for future research.

Source: OECD (2009e).

Taken together, the case studies thus far identified as well as other case studies presented in the rest of this chapter and in Chapter 4 present a mixed bag on the effectiveness of environmentally related taxation. On the one hand, there are strong examples. The Swedish case study (see Box 3.2) highlighted the significant impact that taxation can have on stimulating multiple types of innovation and the many ways needed to looks for innovation impacts. In Switzerland (see Box 3.8), the tax on volatile organic compounds (VOCs) brought significant behaviour change and lead to many small-scale innovations that are difficult to capture in aggregate data.

On the other hand, there are case studies were the outcome is not so definitive. The United Kingdom's reduced tax rate (see Box 4.1) showed that firms facing less of a tax burden are less innovative, but not necessarily for the types of innovations stimulated by the tax. In a cross-country comparison of fuel taxes, prices and standards (see Box 3.6), the effects of taxes on motor vehicle innovation was inconclusive. In some cases, the confluence of multiple factors made breaking out the effect of taxes nearly impossible, such as with Israeli taxes on water (see Box 4.3) and the range of factors facing UK firms (see Box 4.6). In other cases, the tax had an effect on innovation but the design of the tax was such that the effect was actually negative for the development of innovation (although the effects on diffusion were positive), as with Japan's air pollution charge (see Box 4.2). What this suggests is that revealing the direct effectiveness of environmentally related taxation to induce innovation is not straightforward, being impacted by data issues as well as tax design. These issues will be discussed in the subsequent sections and chapters.

Many of the case studies undertaken for this report have focused on environmental damage caused by pollutants or emissions. This does not negate the fact that taxes on environmental damage caused by resource use or exploitation (*e.g.* water withdrawal, forestry, mineral extraction) could also be effective in mitigating environmental damage while spurring innovation into better resource efficiency and the development of potential alternatives. Israel's experience with water pricing provides an interesting example (see Box 4.3). Although some of these issues may be more complex given the potential interaction of royalties or taxes on super-profits (in the case of location-specific rents), taxes on the externalities should have a similar role to taxes on emissions.

3.4. Environmentally related taxation and different types of innovation

The innovation challenge related to the environment is vast. To meet some of the goals to which the global community has committed requires fundamental changes to the nature of supply and demand. These changes will not happen immediately but the presence of well-designed environmentally related taxation can begin to alter the trajectory of the current economic path to one that is more responsive to and innovative in addressing environmental issues. The innovations within the environmental realm that will result will be vast and can generally be categorised into three types, based on how they impact the innovator/adopter: product innovation, process innovation, and organisational innovation.

- Product innovation is the creation of new, or the enhancement of existing, end-products which, in the environmental area, constitute benefits to the environment. CFC-free aerosol deodorants or home water-saving devices typify this kind of innovation.
- Process innovations are innovations in the means of production that provide environmental benefits by reducing pollutants while continuing to produce the same final outputs. Making a power generation plant more fuel efficient is a typical example of process innovation.

- Organisational innovations are innovations that have an appreciable effect on pollution abatement but are themselves not a technology (such as the discovery of new means of structuring firms or analysing environmental performance). This is exemplified by the implementation of environmental accounting systems or the reorganisation of firms/ industries in response to environmental policy.

In respect of process innovation, there are two sub-categories that can be observed: end-of-pipe technologies and cleaner production technologies, as outlined in Figure 3.5. Both categories of innovation result in emissions reductions but they approach the problem from different angles. The type of environmental policy instrument in place can influence which process innovations occur.

Figure 3.5. **Types of environmentally related innovation**

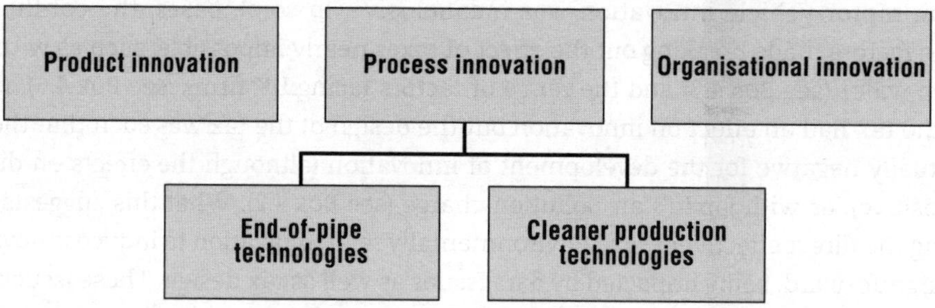

Source: Frondel et al. (2007).

- Cleaner production technologies seek to reduce the amount of pollutants created and emitted. By changing the means by which products are made, the goal is to reduce the creation of pollutants from the source. This can come from changing input sources to modifying the integrated production mechanism. For example, power plants switching from coal to natural gas can reduce the creation of emissions directly, calibrating vehicle engines to be more efficient through the use of on-board diagnostic systems can reduce the creation of emissions from driving, and eliminating chlorine from the pulp-and-paper industry can improve water quality.

- End-of-pipe technologies seek to reduce the amount of pollutant emitted, not necessarily also the amount created. They do not alter the production process to reduce the pollutants in the first place, but seek to address the pollutants once they have been generated. For example, "scrubbers" are used to capture and render less harmful emissions to air, typically from power generation station. NO_x and SO_x continue to be generated by the electricity production but an after-the-fact process seeks to reduce their actual emission. Moreover, carbon capture and sequestration seeks to capture the CO_2 emissions and store them underground to prevent their release to the atmosphere but has no effect on the creation of carbon dioxide.

Cleaner production technologies are generally considered more efficient, as they cover activities that can reduce input use and make the production process more efficient in addition to meeting environmental goals, while end-of-pipe technologies are generally only aimed at addressing environmental objectives. It is not possible that all environmental goals can be met by incremental production process changes. Power generation firms are likely only able to innovate so far in their production processes to reduce emissions of mercury,

SO_x, and other chemicals without fundamentally changing their production processes. As such, end-of-pipe technologies will continue to play a role in the overall mix of issues needed to address environmental challenges.

In the case studies identified in this report, there are examples of all forms of these innovations. In the Swiss case study on VOC emissions, paint makers developed new, low-VOC paints for the market in response to the emissions tax, reacting both to the financial impact of the tax on their products as well as changing demands in the marketplace for products that are more environmental and less damaging to human health. On the other hand, the nature of the tax and the agents which are liable to it can limit the potential for product innovation. In the case of the Swedish NO_x charge, the nature of the businesses liable to the tax – principally power generation firms that participated in a tax-recycling scheme – limited the effective possibilities for product innovations, as: i) there was a homogenous end product; and ii) the burden of the tax did not get passed along to consumers but was offset by the refunding mechanism.

In addition, organisational innovation is an important component of the actions that can lower the overall abatement costs of policy measures and are usually complements to process innovations. In the case study on the tax on VOCs in Switzerland, as described in Box 3.8, the paint making industry established a system to allow consumers to recycle their paints, providing an outlet to capture potential VOC emissions. In the interview case study of UK firms' R&D responses to a wide range of factors, as outlined in Box 4.6, it was found that firms' internal use of targets, both for energy quantity and greenhouse gas emissions, had a significantly positive impact on climate change-related R&D activities for process innovations,[3] suggesting that how firms organise themselves and operate positively influence their research priorities.[4] While one may suppose that the additional costs brought about by environmentally related taxation may induce firms to set targets to reduce these outlays, a direct linkage between environmentally related taxation and target setting requires further exploration in this context and is beyond the scope of this report.

Finally, the type of process innovations may be influenced by the role of environmentally related taxation when compared to other instruments. Using an OECD survey from 2003, Johnstone et al. (2008) analyse firms across seven major OECD countries. There is significant evidence that more flexible environmental policy instruments bring about abatement through cleaner production processes compared to end-of-pipe abatement, since such instruments give firms opportunities to exploit economies of scope in production and abatement activities. End-of-pipe abatement would only focus on abatement activities and are more related with the use of prescriptive approaches to environmental policy [Frondel et al. (2007) also use the same dataset and find similar results].

From the case studies, it can be seen that the vast majority of innovations identified can be classified as process innovations. The introduction of the Swedish NO_x charge, as described in Box 3.2, brought about significant process innovations, specifically related to cleaner production processes. Firms subject to the tax learned to optimise their operations, switched fuels, or installed better combustion technologies, typically being rules of thumb. Although a flexible policy instrument was used, many firms also adopted end-of-pipe technologies to reduce their emissions.

The Swiss VOC study suggests that firms created and adopted both cleaner production and end-of-pipe process innovations. Firms in all three studied industries took measures to reduce the usage of VOCs in their operations, through experimenting with VOC-free

Box 3.8. **Case study: Switzerland's taxes on VOCs**

Starting in 2000, the Swiss federal government instituted a tax on volatile organic compounds (VOCs) of CHF 2 per kilogram, rising to CHF 3 per kilogram in 2003. VOCs are solvents used in industries that require quickly evaporating substances, such as paint making and metal cutting. Besides human health effects, VOCs also contribute to ground-level ozone formation (summer smog). There is no agreed-upon definition of what constitute VOCs but substances generally included are: benzene, styrene, methylene chloride, perchloroethylene and tetrachloroethene. The intent is to tax emissions of VOCs within Switzerland to reduce these effects; as such, exports of VOCs or exports of goods containing them are exempted from the tax.

With the tax regime, emissions of VOCs have decreased significantly. In the 1998-2001 period, emissions on taxed products declined 12%; in the 2001-04 period, when the tax was fully implemented, emissions dropped a further 25%. This 33% decline is significant, but reductions in VOCs from non-taxed sources declined by 28% over the same period, largely due to reductions from automobile use.

This in-depth study focused on three industries – printing, paint making and metal cleaning/degreasing – all of which use significant amounts of VOCs. Through interviews, it was discovered that many firms were highly innovative, even though only a limited number of firms had formalised R&D programmes. In the paint making sector, product innovation occurred through the introduction of low-VOC (high solid) paints to the market. For the most part, the identified innovations occurred through the process of trial and error, such as looking to use less VOCs while maintaining the quality of printing jobs. The tax also spurred the creation of an industry-wide initiative by paint makers to offer recycling options for customers, an indication of organisational innovation.

There were significant variations in the reactions of firms to the new charge. Larger firms generally innovated and adopted new technologies rather quickly, while smaller firms, due to financial or informational constraints, were less likely to act. The role of officials in the cantons also varied, with some viewing their role as facilitative and administrative (and who helped with information and technology diffusion), compared to others who only viewed their role as tax administrator.

Source: OECD (2009c) and Banette (2009).

cleaners, new printers, etc. Moreover, firms also took measures to capture and recycle VOCs or use them in combination with other processes, such as co-generation.

3.5. Innovation degree: Incremental *versus* breakthrough technologies

In addition to the effects of environmental policy instruments on the different varieties of innovation, environmental policy instruments can also have differential effects on the degree of innovativeness. Firms are generally focused on those technologies and solutions that are closest to being market-ready, as those technologies have a more certain probability of success compared to ideas still at the blackboard stage. This is incremental innovation, which brings about better products and more efficient means of producing them through rather small technological advances. This type of innovation can play an important role in bringing about low-cost options to address environmental issues. Because it is only tweaking existing technologies, it generally cannot bring about transformational change. Environmentally related taxation and other market-based instruments provide incentives to accelerate market-ready innovative and to develop innovations that can be brought forward quickly.

For some environmental challenges, however, incremental innovation may only provide part of the solution. In terms of the climate change issue, it is believed that cuts of 80% of CO_2 emissions by 2050 may be necessary to stabilise GHG levels in the atmosphere. Given that there will be some free-riders that do not undertake as significant abatement activities, this effectively suggests the decarbonisation of industrial countries. Incremental innovation may simply not be enough.

Some technologies can provide a major leap forward in pollution abatement and may be critical to achieving environmental targets at more reasonable costs to GDP in the long-run. These technologies make an effective break with past technologies and offer a near completely different approach. Large-scale carbon capture and storage or carbon-free energy sources would be examples. These may not number as high as incremental innovation but a few breakthrough technologies can have a significant impact. Assuming the creation of breakthrough technologies in addition to incremental innovation into climate change models, for example, has shown that these significantly can have significant impacts on the estimated costs to GDP (OECD, 2009a; Bosetti *et al.*, 2009).[5]

Despite increased market pull due to environmentally related taxation, many of the impediments to innovation still exist – uncertainty, funding issues, difficulty in appropriability and others. These are typically more pronounced for longer-term (and therefore more likely fundamental and breakthrough) technologies (OECD, 2009g). For these types of innovations, the development costs may be greater, the time horizons longer, the uncertainty larger and the supply of investors small. Against these issues, environmentally related taxation will likely have little impact. The optimal environmentally related tax should only be focused on environmental externalities (and other associated externalities where the tax base is a proxy, such as accident externalities for fuel taxes) and not other market failures, such as those related to innovation (a tax that tries to account for more than environmental externalities would not be an optimal instrument and the rate may be so high as to face significant opposition and risk not being implemented at all due to political economy constraints). The result is that the incentives provided by a pollution tax may simply not be enough to encourage significant R&D aimed at technologies that may only be market-ready in several decades. Therefore, environmentally related taxation is likely to have a much larger impact on market-ready (and incremental) innovation over longer-term (and fundamental) research.

3.6. Constraints to innovation in response to environmentally related taxation

3.6.1. Firm-level constraints

Constraints at the firm level may prevent the full effect of environmentally related taxation in stimulating innovation. To start, firms may not be aware of all the opportunities available to them, as the induced innovation hypothesis suggests. The presence of search costs, incomplete information, organisational inertia, and other constraints suggest that firms are not constantly on the lookout for all potential opportunities to invest in innovation and therefore do not fully optimise with respect to their R&D budgets. Provided financial conditions are acceptable, firms could become somewhat complacent. Therefore, these indirect linkages between prices and innovation suggest that the assumptions around the complete optimisation of the firm may not hold in the real world (Jaffe *et al.*, 2002).

Firms' innovation budgets can also be allocated based on sub-optimal strategies (such as rule-of-thumb measures for how operations should run, when production practices are reviewed, and how often to engage outside analysis, for example). Sinclair-Desgagné (1999) suggests that firms introduce these mechanisms to help manage the vast amount of information facing them, from the market to government policy. Over time, they become unresponsive to the changing noise facing firms and can cause the creation of low-hanging fruit, where firms can undertake private actions that have both positive private and public net returns. With large shocks to the firm (such as the introduction of new environmental policies), firms re-evaluate their mechanisms take advantage of these low-hanging fruit and readjust their mechanisms to reduce the possibility of such scenarios in the future.

Other environmental instruments, in combination with shocks, such as taxes, can be useful in helping firms re-evaluate their innovation decisions. Arimura *et al.* (2007) find that environmentally related taxation itself does not bring about increased R&D expenditures in firms; environmentally related taxation induces firms to undertake environmental accounting, and the awareness of opportunities brought about by environmental accounting induces R&D expenditures. This two-stage effect is likely due to the information collecting mechanisms of the firms or the internal organisational structure. Relaxing the assumptions around the perfect optimisation of the firm likely accounts for a more realistic view of how firms, particularly small and medium-sized enterprises, approach innovation.

3.6.2. Environmentally related taxation and the resource constraint

Innovation generally has costs. Institutionalised innovation – researchers, labs, commercialisation trials – necessitates that firms devote funds to specific ideas. Even less formal innovation, such as allowing employees to tinker with existing processes, has an opportunity cost borne by the firm.

It is no surprise, therefore, that firms with fewer financial constraints (such as cash flow, ability to obtain financing) devote more resources to innovative activities (Savingnac, 2008). Where the funding is sourced is also important, with internal funding being more important for R&D expenditures, especially when compared against capital expenditures that utilise more external funding (Czarnitzki and Hottenrott, 2009). This can disadvantage small firms which are more affected by external constraints on R&D expenditures than larger firms. Finally, credit market constraints more readily affect cutting-edge innovation than routine innovation, the former being the driving force of technological progress (Binz and Czarnitzki, 2008).

For firms without the financial capacity to fund R&D activities internally, access to credit is critical. Many firms struggle to find adequate capital, given information asymmetries and uncertainty over outcomes. Yet, access to credit is an economy-wide issue that goes beyond R&D activities or environmentally related projects. It is for such reasons that governments typically try to help firms overcome some of these hurdles through preferential loan programmes or financial support of specific activities, such as subsidies to R&D.

Given these findings, one important question is the extent to which the imposition of environmentally related taxation affects firms' financial flexibility and therefore their decisions on innovation spending. A newly introduced environmentally related tax requires that firms must devote a greater part of their revenues to meeting their tax burdens. Provided that firms are not able to fully pass along their costs to consumers, firms may have less financial flexibility to undertake other activities, such as research and development, especially if innovation relies more on internal funding sources (Määttä, 2006).

In the case study on the UK's Climate Change Levy (CCL) (described in Box 4.1), all businesses are subject to taxes on energy; however, large, energy-intensive firms can negotiate agreements with the government for environmental targets in return for being subject to only 20% of the full CCL rate.[6] The analysis indicates that, despite facing a greater tax burden, firms subject to the full-rate CCL actually had a greater propensity to patent than firms subject to the reduced rate (this presupposes that successful innovation, such as patents, are correlated to R&D expenditures). This suggests that the additional costs associated with meeting the tax burden may not have impacted the firm's resources dedicated to its innovative efforts and that the inducement effect may generally outweigh any resource constraint to the firm.

3.6.3. Crowding-out, crowding-in and optimal R&D allocations

In addition to the level of R&D and other innovative activities that firms undertake, environmentally related taxation (or any other significant stimulus for that matter) should induce firms to evaluate the allocation of their R&D expenditures. These impacts work throughout the economy for both directly and indirectly impacted firms and can have significant impacts on overall levels and allocations of R&D and other investment decisions.

With respect to environmental policies, firms generally react in two ways, although the literature is quite sparse in this area. Firms (or the economy at large) can reallocate their fixed R&D budget to channel more resources towards innovations related to the new environmental policy. Known as "crowding-out", the reallocation places more resources towards the now more important pressing concern of environmental policies but at the expense of other areas of research and development activity due to a binding constraint around the availability of R&D activities. On the other hand, firms can maintain the level of resources in existing priorities and provide new, additional resources in response to the environmental stimulus, known as "crowding-in".

The desirability of crowding-in *versus* crowding-out is not always clear. If innovation is generally undersupplied in the market, crowding-in is desirable over the long run, as the level of innovation is below the societal optimum and crowding-in pushes the level of innovation closer to the societal optimum. On the other hand, if innovation is not generally undersupplied, crowding-in can lead to too much innovative activity in the economy at the expense of other productive activities. Of course, if crowding-out reduces R&D activities focused on environmentally harmful activities, this could be optimal as well.

Over the short run, crowding-in may also have limited effect on the actual level of innovation, as the ability of R&D activities to be scaled up in the face of new incentives is limited. The lags in implementing R&D activities and the limited supply of researchers can result in an inelastic supply in the short term. When governments increase R&D expenditures, for example, thereby increasing its price (the increased wage of researchers because of the labour-intensive nature of R&D), the level of private R&D is not very responsive, since much of the extra funding goes to bought-up salaries instead of increased innovative quantity (Goolsbee, 1998). Over the longer-term, pressures for increased R&D resources can encourage new researchers to enter the field, returning labour costs to a lower longer-term equilibrium.

In looking at whether crowding-in or crowding-out occurs, Goulder and Schnieder (1998) find that increased R&D towards climate change comes at the expense of both non-energy R&D and carbon-based energy R&D, with a result of a net decrease in overall R&D. On the other hand, Carraro et al. (2009) suggest a slightly nuanced approach. They look specifically at the effects of a global emissions trading system on R&D investments in the energy and non-energy sector. They find that the price of carbon permits brings about a shift in the relative R&D allocations towards more R&D spending targeted at energy. Due to the relatively small size of the energy R&D sector, however, the large increase in energy R&D does not fully offset the larger decrease in non-energy R&D. Yet, the increase in energy R&D does not come at the expense of non-energy R&D (that is, there is no crowding-out due to limited resources within firms for R&D activities). They suggest that the decline in non-energy R&D results from a non-energy output contraction due to the effect of carbon prices. Thus, while non-energy R&D declines simultaneously with energy R&D increases, there is no crowding-out effect, as firms are optimally allocating based on the new economic landscape following the introduction of climate policy, not because of financial constraints.

In addition, Popp and Newell (2009) look at potential crowding-out effects using general industry data and find that increases in energy R&D do not crowd out R&D in other sectors. Looking within sectors and using detailed industry data on R&D and financial performance, they find that crowding-out (as measured by patent output) does exist for firms with some interest in alternative energy (but not for the automotive sector). When a specific industry is looked at – refinery companies – green energy R&D crowded out innovations in refining and wells. These results are consistent with Gerlagh (2008) who finds that carbon-energy saving R&D crowds out carbon-producing energy, but has no impact on carbon neutral energy.

3.6.4. Firm size and market size

The scale of the market subject to environmentally related taxation is likely to impact the level of innovation. The costs of innovation efforts – some of which are likely fixed – can be significant. Firms must evaluate whether the costs of the investigative project into finding an innovative solution are outweighed by the expected benefits. For an impacted firm, these benefits are the cost-savings attributed to the implementation of the innovation within its own operation plus any expected revenues from licensing the innovation to other users. Firms not subject to the instrument could seek out innovations with the sole purpose of securing royalty revenue or launching a new product onto the market. These expected revenues are highly dependent on the potential size of the market.

In cases where the environmental challenges are global, such as climate change or ozone depletion, the innovation impacts are likely to be larger, given the significant size of the market. Global action to tackle these issues, therefore, creates a large market for innovations and should bring about lower-cost abatement than individual jurisdictions attempting to address these issues in isolation. This is tempered by the fact that triggered innovations may not be able to be applied easily across countries and may need to be adapted to local circumstances.

At the same time, individual firms face decisions about whether the resources dedicated to investigating potential options are cost-effective. In smaller firms, specialised knowledge about one part of the business may be lacking, compared to a larger firm that can have dedicated staff for specific areas. Moreover, innovations beyond things like better calibration of machines or new "tricks of the trade" can be more cost-prohibitive to smaller firms, given factors such as capital indivisibility. While some smaller firms may not actively seek out innovative solutions because of some of these factors, innovation adoption will happen naturally as existing equipment becomes outdated and the purchase of new equipment incorporates such innovations as standard features.

As seen in Box 3.8, Switzerland instituted a tax on volatile organic compounds (VOCs) in 2000. The Swiss market is relatively small, especially given the size of the overall European market in which it operates relatively seamlessly. In some ways, the size of the market constrains innovation. In the printing sector, the needs of Swiss firms were not large enough to affect the design of printing machines, which are typically made outside of the country. Despite this small market, there was considerable innovation following the implementation of the tax, mainly process innovations. Printing firms experimented with low-VOC or VOC-free inks and reducing VOC use in cleaning processes. In the paint making business, firms undertook a mix of innovations. Product innovations occurred (new low-VOC paints) as well as measures to recycle VOCs and new cleaning methods.

For smaller firms, the incentives for abatement were less. Firm-level interviews suggested that adoption of innovation was weaker in smaller firms, given financial constraints or lack of knowledge. Canton-level interviews suggested that many small firms had not innovated because the level of the tax is an absolutely and relatively small component of their overall operations. The knowledge and time required to experiment with new products was lacking with respect to smaller firms. SMEs were also more likely to hold to their traditional means of production and were quite reticent to change. The result is that the larger firms were the relative innovators while the smaller firms could be considered adopters and adapters of innovations.

Cantons also played a significant role with information and technology diffusion (Banette, 2009). Cantons all viewed their roles with respect to the implementation and administration of the VOC tax somewhat differently, ranging from merely being tax collectors and applying the law to being technical experts and actively promoting further abatement options. The effect was that cantonal officials had a significant role in helping firms subject to the tax, particularly SMEs, overcome some of the barriers to adoption of innovative processes for VOC abatement. The tax alone likely would not have been able to achieve the same level of results.

3.7. The adoption and transfer of environmentally related innovation

In addition to the creation of innovation, the adoption and transfer of innovation is important for achieving low-cost abatement. Diffusion brings about lower overall abatement costs, as the range of potential options for adopters of pollution abatement innovation is made larger. The extent of diffusion potential also increases the incentives for innovators to develop innovations, thereby inducing greater levels of innovation.

3.7.1. *The process of adoption*

Environmentally related taxation has a clear role to play in facilitating the adoption of environmentally related innovations across potential users. By taxing the emissions across all sources, all emitters now have increased incentives to adopt existing technologies to reduce tax payments. However, since in practice environmentally related taxation is most often instituted at a relatively low level, it may provide some additional incentives, but may not be enough to overcome some of the general barriers to technology adoption, especially at the household level.

The barriers that affect general innovation adoption also have pronounced effects on environmentally related innovations. In some cases, there must be a network of others users of the same innovation for it to reach its full potential or usefulness. This initial barrier can contribute to the issue of technology lock-in, whereby sometimes inferior technologies become the mainstream option only because they were first entrenched in the market. This could be illustrated by the need for a network of fuelling stations to exist before alternatively powered vehicles are adopted or by the inability of existing systems to handle (and meet base load demand with) small-scale power generators (such as wind mills) due to the intermittent nature of power generation of this type.

Second, consumers can have very high discount rates,[7] preferring sometimes to purchase lower-cost goods (with higher operating costs) than higher-price goods (with lowering operating costs). Jaffe and Stavins (1994) demonstrated that the upfront cost of home insulation was significantly more important than energy prices. Where the operating costs are based on environmentally harmful inputs, such as electricity, purchasing the lower-cost goods that have an overall more expensive life can have significant economic and environmental consequences. Moreover, Jaffe and Stavins (1995), using a model of home insulation investments based on realised data, model the effects of an energy tax and an installation subsidy (which are calibrated to have the same overall impact) on the expected take-up of investments to insulate homes. Their data suggests that, over a 10-year period, a 10% energy tax would raise the insulated value of a home (and therefore increase diffusion) by 2-6%, while a 10% installation subsidy would increase diffusion between 4-15%.

There are also problems that result from information asymmetries between two agents. The common example is the relationship between a house builder and a buyer. A house builder, knowing that utilising energy-efficient building techniques relating to insulation, sealing and windows, for example, can drastically reduce energy costs over the life of the house and more than account for the initial investment in these upgrades. Yet, he is likely unable to recover these costs from the buyer, as the buyer does not have the same knowledge of what was done and is not able to verify this independently, affecting the rate of technological diffusion. In a related manner, the issue of split incentives – that the agent bearing the costs of technology adoption is not the one reaping the benefits – similarly limits technology diffusion. A landlord is unlikely to make investments in energy-efficiency upgrades to her property if the energy bills are paid by the tenant; the tenant is unlikely to invest in similar upgrades unless the investments are portable once the tenancy ends or the tenancy is of sufficient duration (and known in advance).

Looking broadly, there are clear examples of the effect of innovation diffusion, including the Dutch food and beverage industry, which was subject to water effluent charges. Looking at the diffusion of biological water treatment technology, Kemp (1998) finds that the effluent charges were a significant positive factor in the diffusion of treatment technologies. Indeed, it is estimated that only around 4% of plants would have installed wastewater treatment equipment by the end of the period if the charge had remained at its (low) 1974 level, compared to the actual figure of over 40%. In addition, the French tax-subsidy scheme for NO_x and SO_2 emissions provides a platform to assess a plant's decision to install end-of pipe abatement equipment. Using panel data for 226 plants in three industries (iron and steel, coke and chemicals) for the period 1990-98, Millock and Nauges (2006) find that the total value of emissions taxes paid by the plant (for both pollutants) has a positive impact on its decision to invest in abatement equipment, even though the tax rates are very low. However, the magnitude of the effect varies considerably across the sectors and is only significant for the iron and steel sector.

The case studies also highlight the role of environmentally related taxation and innovation diffusion. As was seen in the Swedish case study, the imposition of the tax in 1992 had immediate impacts on the uptake of pollution abatement equipment for the impacted firms. In 1992, only 7% of regulated firms had NO_x mitigation technologies; one year later, 62% of firms had installed some form of mitigation technology, mainly in changes to their combustion methods. This is rapid diffusion of technology in response to a relatively high level of emission tax.

To overcome some of the specific effects mentioned in the preceding section, governments have sometimes adopted subsidies to promote diffusion of a particular technology. By reducing the cost of adoption, the aim is to speed up adoption in the early stages and then let the market drive demand thereafter. Subsidies on the initial price can help consumers with the "sticker shock" of purchasing more energy-efficient items, and possibly indicate additional information to the purchaser (Aalbers *et al.*, 2009). In the field of green energy policy, governments also use feed-in tariffs, which provide a per-kWh subsidy for applicable fuels. This instrument is intended to encourage adoption and overcome the learning-by-doing barriers faced by new technologies by promoting scale effects. However, such policies can have unintended consequences for innovation. In the context of climate change, subsidies for the adoption of existing, alternative technologies can create lock-in effects, reducing the incentive for R&D efforts into newer and non-subsidised technologies, leading to a societal welfare loss (Kverndokk *et al.*, 2004). The tax system can be used in other ways to also promote technology diffusions, which will be discussed in Chapter 4.

3.7.2. *The process of technology transfer*

Inasmuch as technology diffusion is important within countries, retaining innovative outputs in one country will not serve the global community where environmental problems cross boundaries. Transfer of technologies and patents can lower global abatement costs. They can also encourage countries to engage in stronger environmental protection by making the initial cost of policy action less. Lovely and Popp (2008), for instance, find that over time countries implement environmental policies at lower levels of per capita income.

Environmental problems are often unique, requiring specialised knowledge and solutions; therefore, transferring technical solutions may be difficult. Even when jurisdictions face similar problems, recipients of technology need to have the scientific base to accept, understand, and modify the innovation to work well within the new jurisdiction (Johnson and Lybecker, 2009). The transferors must also be comfortable with sending proprietary information abroad and the patent system thus plays a key component. Strong intellectual property protections encourage greater transfer of knowledge, particularly among developed countries. Since new innovations also typically push against existing legal boundaries, greater predictability of legal/intellectual property regimes to new innovations would encourage transfers.

Inventors typically respond to domestic incentives for undertaking invention but, where a country is behind others, it typically uses foreign innovations (*e.g.* patents) as a starting basis (Popp, 2006). Even where the country is not behind, other countries can be important sources of innovation. For example, the United States was the first to introduce strict vehicle emissions standards but the majority of related patents came from outside the country (Lanjouw and Mody, 1996). Many times, however, there are some differences between countries and therefore there is a need for adaptive R&D. This adaptive R&D recognises that foreign innovations are not a perfect fit given the different domestic conditions, suggesting that diffusion across countries will be slower than within countries (Pizer and Popp, 2008).

In addition, the technical capacity of senders and the absorptive capacity of recipients of technology transfer are critical to determining flows of technology. These features suggest the ability and willingness of countries to engage in transfer. Looking at the transfer of wind power technologies and the role of the Kyoto Protocol's Clean Development Mechanism (CDM) and other factors, Haščič and Johnstone (2009) find that recipient countries' technological capacity is two to three times as important as the CDM. Moreover, the role of the source country's supply is three to eight times as important as the CDM, indicating that these forces must be considered when looking at the overall effectiveness of policy tools.

The choice of environmental instrument which countries adopt, whether it be taxation or something else, can impact on the transfer of technology across jurisdictions. Johnstone and Haščič (2010) demonstrate that flexibility in environmental policy increases the range of innovations created domestically and transferred as well as the extent of innovations that are imported. With less flexible policy arrangements, countries limit the range of innovations that can be profitably used by their industries. Two countries with differing prescriptive regulatory approaches have little interest in sharing innovation, as their industries are focusing on meeting regulatory outcomes that may have little overlap. Since there is a significant potential for mismatch, technology transfer does not occur.

On the other hand, for two countries with flexible approaches, their industries are trying to reach the same outcome: emission reductions. Since the range of possibilities to achieve this is great, the innovations could be potentially useful to firms in either country and thus there is significant scope for diffusion. Even if one country has a flexible approach whereas others do not, the country with a flexible approach is able to potentially benefit from all the innovations taken place in other countries. The reverse does not hold. It is for these reasons that the direction and level of innovation diffusion is impacted by the choice of instrument of governments.

Despite these concerns about adaptive R&D, solutions to global environmental challenges focus increasingly on the role of emerging economies, given their rapidly increasing populations and per capita wealth, where the capacity for innovation may be significantly less than developed countries. Many of these countries will be late adopters, following on the work previously done in developed countries. In these cases, such late adopters are able to move quickly, building upon the groundwork undertaken by early adopters, with Hilton (2001) showcasing this in an analysis of early and late adopters in the phase-out of leaded petrol. What this suggests is that the ability to facilitate both the development and the diffusion of innovation is critical to bringing about environmental gains not only in more technologically advanced economies but also those that are still emerging.

3.8. Conclusions

The theoretical basis for environmentally related taxation to induce innovation is strong. Taxes, especially those levied directly on the pollutant, provide incentives for the creation of innovation because there are incentives for its adoption in order to minimise tax payments. Moreover, these innovations span the range of potential innovations: product innovation, process innovation (both end-of-pipe and cleaner production) and organisational innovation. Non-tax-based instruments are generally not as potent. Taxes, in addition to the creation of innovation, are also important in facilitating the transfer of innovations across countries.

Assessing the effectiveness of environmentally related taxation to induce innovation in practice begins with measurement. This is especially important for taxes: compared against the more limited scope of complying with a given regulation, the very wide range of innovations that can be stimulated by a tax makes it difficult to search for and assess the results. With patent data, setting the criteria of where to look for potential innovations can be very broad. Metrics of innovation beyond patents are needed, therefore, such as inputs to innovation (R&D expenditures) or indirect measures of innovation output (such as the effects on marginal abatement cost curves).

Taking measurement issues into consideration, the empirical evidence for the innovation impacts on environmentally related taxation is strong but not completely conclusive. The data constraints noted above do pose challenges, especially when doing broad-based analysis using patents. The cross-country study on fuel taxes and the reduced rates of the UK's energy tax highlighted some of these issues. Nevertheless, more narrowly focused studies that used alternative measures of innovation have provided some robust results. The Swedish and Swiss cases, for example, clearly showed the innovation impacts of those taxes.

The case studies highlighted some additional findings as well. The fact that environmentally related taxation imposes a cost on firms that can reduce their profitability does not appear to translate into reduced innovation outputs. The innovation potential does seem to be increased with market size, especially with respect to patenting. Finally, in looking at the types of innovations induced, not just the quantity, it is clear that taxation brings about a full range of innovations, including new products and enhanced production techniques. Yet, environmentally related taxes (like most other environmental policy alternatives) may not have a strong influence on innovations of a more fundamental nature compared to those that are more market-ready.

Notes

1. It is interesting to note that the patents in Sweden related to NO_x emissions were nearly evenly split between cleaner production and end-of-pipe innovations, whereas patents across a wide range of countries for NO_x abatement generally favoured end-of-pipe measures, which could be attributed to the more flexible nature of Sweden's approach.

2. For example, if agricultural prices are kept artificially high through some (non-environmental) subsidies and/or border protection measures, an invention that leads to larger agricultural production could in this sense be said to entail negative economic externalities.

3. The same was not found for product innovations. As innovative products typically reduce the emission of pollutants from the end user, and not the producing firm, the effect of targets should have a more indirect effect.

4. Conversely, it could hold that firms that have already innovated and are in the process of implementing the innovation set self-imposed targets to help guide the implementation process.

5. It should be noted, however, that the incorporation of breakthrough technologies into climate change models typically occurs through the assumption that these technologies (also called backstop technologies) will occur based on the projected climate for innovation. This assumption is used to show the effect that these technologies *may have*, but does not necessarily suggest that they will in fact occur.

6. The CCA discount is scheduled to be reduced to 65% from the current 80% as of 1 April 2011.

7. These high discount rates may simply reflect the fact that consumers very much prefer consumption in the present period compared to future periods, not that there are necessarily market distortions or failures.

References

Aalbers, Rob *et al.* (2009), "Technology Adoption Subsidies: An Experiment with Managers", *Energy Economics*, No. 31, pp. 431-442.

Ambec, Stefan and Philippe Barla (2006), "Can Environmental Regulations be Good for Business? An Assessment of the Porter Hypothesis", *Energy Studies Review*, Vol. 14(2).

Arimura, Toshi H., Akira Hibiki and Nick Johnstone (2007), "An Empirical Study of Environmental R&D: What Encourages Facilities to be Environmentally Innovative?", in Johnstone, Nick (ed.), *Environmental Policy and Corporate Behaviour*, Edward Elgar and OECD, Cheltenham, UK.

Banatte, Sam (2009), "Taxe incitative sur les composes organiques volatils (COV) : rôle des cantons dans les effets sur l'innovation", Mémoire de master, Université de Neuchâtel and Institut de hautes études en administration publiques, Switzerland.

Binz, Hanna L. and Dirk Czarnitzki (2008), "Financial Constraints: Routine Versus Cutting Edge R&D Investment", *Working Paper*, No. 08-005, Centre for European Economic Research.

Bosetti, Valentina *et al.* (2009), "The Role of R&D and Technology Diffusion in Climate Change Mitigation: New Perspectives Using the WITCH Model", *Working Paper*, No. #274, Fondazione Eni Enrico Mattei.

Brännlund, Runar and Tommy Lundgren (2009), "Environmental Policy without Costs? A Review of the Porter Hypothesis", *International Review of Environmental and Resource Economics*, Vol. 3(2).

Burtraw, Dallas (2000), "Innovation under the Tradable Sulfur Dioxide Emission Permits Program in the US Electricity Sector", Discussion Paper, No. 00-38, Resources for the Future, Washington DC.

Carraro, Carlo, Emanuele Massetti and Lea Nicita (2009), "How Does Climate Policy Affect Technical Change? An Analysis of the Direction and Pace of Technical Progress in a Climate-Economy Model", Nota di Lavoro 08.2009, Fondazione Eni Enrico Mattei.

Crabb, Joseph M. and Daniel K.N. Johnson (2010), "Fueling the Innovative Process: Oil Prices and Induced Innovation in Automotive Energy-Efficient Technology", *Energy Journal*, Vol. 31(1), pp. 199-216.

Czarnitzki, Dirk and Hanna Hottenrott (2009), "R&D Investment and Financing Constraints of Small and Medium-sized Firms", *Small Business Economics*, http://dx.doi.org/10.1007/s11187-009-9189-3.

Frondel, Manuel, Jens Horbach and Klaus Rennings (2007), "End-of-Pipe or Cleaner Production? An Empirical Comparison of Environmental Innovation Decisions across OECD Countries", *Business Strategy and the Environment*, No. 16, pp. 571-584.

Gerlagh, Reyer (2008), "A Climate-Change Policy Induced Shift from Innovations in Carbon-Energy Production to Carbon-Energy Savings", *Energy Economics*, No. 30, pp. 425-448.

Goolsbee, Austan (1998), "Does Government R&D Policy Mainly Benefit Scientists and Engineers?", *American Economic Review*, Vol. 88(2), pp. 298-302.

Goulder, Lawrence H. and Stephen H. Schneider (1998), "Induced Technological Change and the Attractiveness of CO_2 Abatement Policies", *Resource and Energy Economics*, No. 21, pp. 211-253.

Greaker, M. (2003), "Strategic Environmental Policy: Eco-dumping or a Green Strategy?", *Journal of Environmental Economics and Management*, No. 45, pp. 692-707.

Griliches, Zvi (1990), "Patent Statistics as Economic Indicators: A Survey", *Journal of Economic Literature*, Vol. 28(4), pp. 1661-1707.

Haščič, Ivan and Nick Johnstone (2009), "The Clean Development Mechanism and International Technology Transfer: Empirical Evidence on Wind Power Using Patent Data", presented at the Workshop on "Globalisation and Environment", Kiel Institute for the World Economy, September 2009, available at http://papers.ssrn.com/sol3/papers.cfm?abstract_id=1493241.

Hilton, F. Hank (2001), "Later Abatement, Faster Abatement: Evidence and Explanations From the Global Phaseout of Leaded Gasoline", *The Journal of Economic Development*, No. 10, pp. 246-265.

Jaffe, Adam B., Richard G. Newell and Robert N. Stavins (2002), "Environmental Policy and Technological Change", *Environmental and Resource Economics*, Vol. 22, pp. 41-69.

Jaffe, Adam B. and Karen Palmer (1997), "Environmental Regulation and Innovation: A Panel Data Study", *The Review of Economics and Statistics*, Vol. 79(4), pp. 610-619.

Jaffe, Adam B. and Robert N. Stavins (1994), "The Energy Paradox and the Diffusion of Conservation Technology", *Resource and Energy Economics*, No. 16, pp. 91-122.

Jaffe, Adam B. and Robert N. Stavins (1995), "Dynamic Incentives of Environmental Regulations: The Effects of Alternate Policy Instruments on Technology Diffusion", *Journal of Environmental Economics and Management*, No. 29, pp. 43-63.

Johnson, Daniel K.N. and Kristina M. Lybecker (2009), "Challenges to Technology Transfer: A Literature Review of the Constraints on Environmental Technology Dissemination", *Working Paper*, No. 2009-07, Colorado College.

Johnstone, Nick and Ivan Haščič (2010), "Environmental Policy Design and the Fragmentation of International Markets for Innovation", in V. Ghosal (ed.), *Reforming Rules and Regulations*, MIT Press.

Johnstone, Nick, J. Labonne and C. Thevenot (2008), "Environmental Policy and Economies of Scope in Facility-level Environmental Practices", *Environmental Economics and Policy Studies*, Vol. 9(3), pp. 145-166.

Kemp, Réne (1998), "The Diffusion of Biological Wastewater Treatment Plants in the Dutch Food and Beverage Industry", *Environmental and Resource Economics*, Vol. 12, pp. 113-136.

Kleinknecht, Alfred, Kees van Montfort and Erik Brouwer (2002), "The Non-Trivial Choice between Innovation Indicators", *Economics of Innovation and New Technology*, Vol. 11(2), pp. 109-121.

Kumar, Surender and Shunsuke Managi (2009), "Energy Price-Induced and Exogenous Technological Change: Assessing the Economic and Environmental Outcomes", *Resource and Energy Economics*, No. 31(4), pp. 334-353.

Kverndokk, Snorre, Knut Einar Rosendahl and Thomas F. Rutherford (2004), "Climate Policies and Induced Technological Change: Which to Choose, the Carrot or the Stick?", *Environmental and Resource Economics*, Vol. 27, pp. 21-41.

Lanjouw, Jean Olson and Ashoka Mody (1996), "Innovation and the international diffusion of environmentally responsive technology", *Research Policy*, No. 25, pp. 549-571.

Lanoie, Paul *et al.* (2010), "Environmental Policy, Innovation and Performance: New Insights on the Porter Hypothesis", forthcoming in *Journal of Economic Strategy and Management*.

Litchtenberg, F.R. (1986), "Energy Price and Induced Innovation", *Research Policy*, No. 15, pp. 67-75.

Litchtenberg, F.R. (1987), "Changing Market Opportunities and the Structure of R&D Investment", *Energy Economics*, No. 9, pp. 154-158.

Lovely, Mary and David Popp (2008), "Trade, Technology and the Environment: Why do Poorer Countries Regulate Sooner?", *NBER Working Paper*, No. #14286, available at *www.nber.org/papers/w14286*.

Määttä, Kalle (2006), *Environmental Taxes: An Introductory Analysis*, Edward Elgar: Cheltenham, UK and Northampton, US.

Marion, Justin and Erich Muehlegger (2008), "Measuring Illegal Activity and the Effects of Regulatory Innovation: Tax Evasion and the Dyeing of Untaxed Diesel", *Journal of Political Economy*, Vol. 116(4), pp. 633-666.

Millock, K. and C. Nauges (2006), "Ex post Evaluation of an Earmarked Tax on Air Pollution", *Land Economics*, Vol. 82(1), pp. 68-84.

OECD (2004), "Triadic Patent Families Methodology", *STI Working Paper*, No. 2004/2, OECD, Paris.

OECD (2008), *Environmentally Related Taxation and Tradable Permit Systems in Practice*, OECD, Paris, available at *www.olis.oecd.org/olis/2007doc.nsf/linkto/com-env-epoc-ctpa-cfa(2007)31-final*.

OECD (2009a), *The Economics of Climate Change Mitigation: Policies and Options for Global Action Beyond 2012*, OECD, Paris, *http://dx.doi.org/10.1787/9789264073616-en*.

OECD (2009b), *Innovation Effects of the Swedish NO_x Charge*, OECD, Paris, available at *www.olis.oecd.org/olis/2009doc.nsf/linkto/com-env-epoc-ctpa-cfa(2009)8-final*.

OECD (2009c), *Effects of the VOC Incentive Tax on Innovation in Switzerland: Case Studies in the Printing, Printmaking and Metal Cutting Industries*, OECD, Paris, available at *www.olis.oecd.org/olis/2008doc.nsf/linkto/com-env-epoc-ctpa-cfa(2008)35-final*.

OECD (2009d), *Indicators of Innovation and Transfer in Environmentally Sound Technologies: Methodological Issues*, OECD, Paris.

OECD (2009e), *A Case Study of the Innovation Impacts of the Korean Emission Trading System for NO_x and SO_x Emissions*, OECD, Paris.

OECD (2009f), *Fuel Taxes, Motor Vehicle Emission Standards and Patents Related to the Fuel Efficiency and Emissions of Motor Vehicles*, OECD, Paris, available at *www.olis.oecd.org/olis/2008doc.nsf/linkto/com-env-epoc-ctpa-cfa(2008)32-final*.

OECD (2009g), "Policies for the Development and Transfer of Eco-Innovation: Lessons from the Literature", Paper for the OECD Global Forum on Environment on Eco-Innovation, 4-5 November 2009, OECD, Paris, available at *www.oecd.org/dataoecd/21/36/43811507.pdf*.

OECD (2009h), *Innovation in Firms: A Microeconomic Perspective*, OECD, Paris, *http://dx.doi.org/10.1787/9789264056213-en*.

OECD (2009i), *OECD Patent Statistics Manual*, OECD, Paris, *http://dx.doi.org/10.1787/9789264056442-en*.

OECD (2010a), "Gross domestic expenditure on R-D by sector of performance and source of funds", *OECD Science, Technology and R&D Statistics* (database), *http://dx.doi.org/10.1787/data-00189-en*.

OECD (2010b), "Government budget appropriations or outlays for RD", *OECD Science, Technology and R&D Statistics* (database), *http://dx.doi.org/10.1787/data-00194-en*.

Palmer, K., W.E. Oates and P.R. Portney (1995), "Tightening Environmental Standards: The Benefit-Cost or the No-Cost Paradigm?", *Journal of Economic Perspectives*, Vol. 9(4), pp. 119–132.

Pizer, William A. and David Popp (2008), "Endogenizing Technological Change: Matching Empirical Evidence to Modeling Needs", *Energy Economics*, No. 30, pp. 2754-2770.

Popp, David (2001), "The Effect of New Technology on Energy Consumption", *Resource and Energy Economics*, No. 23, pp. 215-239.

Popp, David (2002), "Induced Innovation and Energy Prices", *American Economic Review*, Vol. 92, pp. 160-180.

Popp, David (2005), "Uncertain R&D and the Porter Hypothesis", *Contributions to Economic Policy and Analysis*, 4(1), article 6.

Popp, David (2006), "International Innovation and Diffusion of Air Pollution Control Technologies: The Effects of NO_x and SO_2 Regulation in the US, Japan, and Germany", *Journal of Environmental Economics and Management*, No. 51, pp. 46-71.

Popp, David and Richard G. Newell (2009), "Where Does Energy R&D Come From? Examining Crowding Out from Environmentally-Friendly R&D", *NBER Working Paper*, No. #15423, October 2009, available at *www.nber.org/papers/w15423*.

Porter, M.E. (1991), "America's Green Strategy", *Scientific American*, April, p. 168.

Porter, M.E. and C. van der Linde (1995), "Toward a New Conception of the Environment-Competitiveness Relationship", *Journal of Economic Perspectives*, Vol. 9(4), pp. 97-118.

Rassenfosse, Gaeten de and Bruno van Pottelsberghe de la Potterie (2009), "A Policy Insight into the R&D-Patent Relationship", *Research Policy*, No. 38, pp. 779-792.

Savignac, Frédérique (2008), "Impact of Financial Constraints on Innovation: What can be Learned from a Direct Measure?", *Journal of Economics of Innovation and New Technology*, Vol. 17(6), pp. 553-569.

Sinclair-Desgagné, Bernard (1999), "Remarks on Environmental Regulation, Firm Behaviour and Innovation", Scientific Series Paper, Centre interuniversitaire de recherche et analyse des organisations (CIRANO), Montreal, Canada, available at *www.cirano.qc.ca/pdf/publication/99s-20.pdf*.

Vries, Frans de and Neelakshi Medkhi (2008), "Environmental Regulation and International Innovation in Automotive Emissions Control Technologies", *Environmental Policy, Technological Innovation and Patents*, OECD, Paris.

Chapter 4

Tax Design Considerations and other Tax-based Instruments[*]

> This chapter considers how the design of environmentally related taxes – the level of the tax, the extent of the tax base and the predictability of the rate – influences the ability to induce innovation. It also explores the effect of measures to address political economy considerations. Attention is then turned to other potential tax-based measures, such as accelerated depreciation allowances and R&D tax credits, to address the environmental and innovation challenges. The chapter concludes with a discussion of potential instrument combinations to achieve an optimal outcome.

[*] The statistical data for Israel are supplied by and under the responsibility of the relevant Israeli authorities. The use of such data by the OECD is without prejudice to the status of the Golan Heights, East Jerusalem and Israeli settlements in the West Bank under the terms of international law.

How environmentally related taxation is designed can have a significant impact on its environmental effectiveness. This same range of factors – from the level of the tax to its implementation and administration – can play an important role regarding the innovative impacts of the instrument.

4.1. Identifying the appropriate level of the tax

4.1.1. The initial level of the tax

A well-defined environmentally related tax should be set at the Pigouvian level (that is, where the tax equates the marginal damage from pollution with the marginal cost of pollution abatement). Where the tax is on a proxy to the environmental damage, such as a motor vehicle, other externalities need to be considered when setting the rate. The rate is influenced by a number of factors: society's wealth, society's valuation of the environment, the extent of the damage, the advent of new technologies and processes that address the environmental challenge, the actual efficacy of policies in addressing the environmental problem and the potential reversibility and/or tipping point of the environmental challenge. With tradable permits, much the same information is necessary, but it is used to assess the optimal quantity of pollutants that should be permitted. Many environmental challenges persist over very large time horizons, centuries with respect to climate change, for example, and therefore policies must be attuned to these dynamics.

But the simple Pigouvian level of the tax is determined exogenously to its broader effects on the economy. In a general equilibrium sense, a tax on pollution is effectively a factor tax and therefore interacts with pre-existing factor taxes. These interactions can have some significant effects and can result in the optimal level of the tax and the Pigouvian level of the tax being different. Goulder (1995), for example, finds that pre-existing distortions should lead to a lower level of an optimal environmentally related tax. Consideration for other externalities, political economy issues and the general revenue raising needs of governments are also important factors in determining the final rate. A fuller discussion is presented in Chapter 5.

From an innovation perspective, there are additional considerations to account for in considering the optimal level of the tax. Parry (2005) suggests that the type of innovation to be created should influence the level. If the technology in the economy is all within the public domain (and therefore there is no cost to access the technology), the level of the emission tax should hover around the Pigouvian level. Where the technology is private (and a monopolist charges royalties to access the information), the license fee would be too high to encourage optimal diffusion of the technology, suggesting that a reduction in the tax rate would reduce the royalty fee and improve diffusion.

One of the largest issues facing environmental economics is the issue of uncertainty, which is typically larger for environmental issues than other issues (Pindyck, 2007), given the significant informational constraints and issues present. The difficulty of obtaining, or complete lack of, such information makes it extremely difficult for policy makers to quantify these effects and translate them into appropriate tax rates or quantity targets.

One would naturally expect that a higher rate of environmentally related taxation would induce greater levels of innovation. In the case study on the UK's Climate Change Levy (CCL) [and its companion Climate Change Agreements (CCA)], described more fully in Box 4.1, some firms were subject to a full rate of the CCL, while other firms were subject to an 80% reduction in return for agreements to meet specific targets, typically regarding energy efficiency. Accounting for firms' characteristics that might encourage CCA participation, it was found that firms subject to the reduced rates within CCA were

---

Box 4.1. Case study: Concessions in the UK's Climate Change Levy

The United Kingdom introduced the Climate Change Levy (CCL) in 2001, which placed a tax on electricity (GBP 0.43 per kWh), coal (GBP 0.15 per kWh), natural gas (GBP 0.15 per kWh) and liquefied petroleum gas (GBP 0.07 per kWh) used by businesses. Large and energy-intensive firms entering into a Climate Change Agreement (CCA) would be subject only to 20% of the CCL in return for meeting agreed-upon targets for energy consumption in order to mitigate potential competitiveness impacts from countries without such taxes [see Pearce (2006) for further discussion of the political economy considerations of the CCL].

Analysis was undertaken to explore the differential economic, environmental, and innovation impacts of firms subject to CCAs *versus* firms subject to the full CCL. To address biases regarding the types of firms that enter into CCAs, an instrumental variable approach was employed.

With respect to environmental outcomes, CCA firms increased their emission intensities by more than 20% compared to firms subject to the full CCL, both in relation to output and to costs. CCA firms also significantly increased their use of electricity compared to full-rate CCL firms, consistent with the higher tax rate on electricity. The overall effect on carbon emissions was similar. This is understandable given the nature of the CCL. Since the CCL is a tax on energy – and therefore the implicit carbon price of the tax varies significantly by fuel – there may be incentives for firms to switch into fuels which are taxed at a lower rate but which produce significantly higher levels of CO_2 emissions (or just less incentive to switch to cleaner fuels). On firms' economic performance, there were no observable differences between CCA firms and full-rate CCL firms with respect to employment, output, or total factor productivity.

With respect to innovation, the analysis suggests that CCA firms are up to 16 percentage points less likely to patent overall than full-rate CCL firms given the low incentive provided by a discounted tax rate. A concern, however, stems from the fact that when the same analysis was done solely on climate-change-related patents in place of patents overall, the differences between the two sets of firms do not seem to be as apparent. One would have presupposed that the innovation incentive would have been stronger for climate change-related innovation than innovations in general. This may be caused by the significant difficulty of researchers in identifying specific patents related directly to climate change-related innovation, especially innovations resulting from taxes. A broad discussion of the difficulty of linking environmentally related patents and taxation is provided in Box 3.1.

Therefore, this analysis suggests that reduced rates of the Climate Change Levy have had negative environmental impacts and firms subject to the full-rate CCL have not weathered more adverse economic consequences. The innovative effects of the tax suggest that patenting may be greater for firms facing the full tax rate but that classification of the data for climate-change related patents makes strong conclusions difficult.

Source: OECD (2009f).

---

significantly less likely to patent – up to 16 percentage points – than those firms subject to the full rate of the CCL. This difference in propensity to innovate occurs for overall innovation, as measured by total patent counts. Potential patent classification issues could account for the fact that this result did not hold when only looking at the effect of the CCL and CCA on climate change-related patents.

4.1.2. *Impacts of predictability and intertemporal rates on the propensity to innovate*

In addition to the issues that policy makers face when setting the initial level of the tax (or the quantity of permits), ongoing changes to the parameters used to set the initial rate raise questions about whether and how the rate should change in response. As new information comes to light, such as regards the impact of the environmental damage or society's willingness to undertake more/less abatement, policy makers face potential dilemma as to the trade-off between ensuring that environmentally related tax rates reflect the best possible information with the value of predictability for environmental and innovative effectiveness.

When contemplating whether to undertake actions to reduce their environmental impact in the face of environmentally related taxation or other policies, polluting agents obviously face uncertainty about the future. Purchasing new technologies can create lock-in for the firm, as a new technology just over the horizon could provide significantly more benefits. The firm may also believe that the policy environment might change, such as rates of environmentally related taxation or the market price of tradable permits. These factors affect the expected return on investment and can therefore affect investment decisions and levels of innovative activities.

Such issues present significant uncertainty and will impact how an affected firm reacts. The firm will likely scan the future and decide whether to act now (in any number of ways) or wait until a future time period when there is more information (and thus the firm is able to make a better decision). Dixit and Pindyck (1994) explain that the flexibility to wait and decide upon a course of action in the future is a source of value to firms today. This "real options literature" suggests that firms place significant value on their ability to change course. This can be by delaying action now and taking a decision in the future when more information may be present or changing course in the future by selecting now a path with low sunk costs. This action may lead to higher costs in the future but the option to wait on a decision may be worth more in the present. When uncertainty surrounds large investments (whether it be a capital investment or investments in R&D), this flexibility is particularly useful for firms. For example, a firm looking to construct a power generation plant today must weigh all the potential factors in the future: input prices, construction costs, carbon taxes, new technologies, demand, etc. Elevated levels of uncertainty lead to less action now, as the value of waiting for better information (or less uncertainty) has increased.

Uncertainty can come in two forms. One is market-based risk, such as the input prices of production or the expected price that a firm will be able to fetch for its final product. Some of these risks can be more easily offset in financial markets, such as through the use of forward contracting or financial instruments. Where the policy instrument is a tradable permit (and therefore a *de facto* input to production), hedging of these instruments can provide some additional predictability. In these circumstances, it should be noted that the ability to create more predictability over future prices through hedging of tradable permits has different effects on adopters and creators of innovation. Adopters of innovation can undertake strategies to provide certainty over their future prices and therefore their future costs and savings.

Innovators (that are not also adopters), however, are not directly bound by the prices of tradable permits and are not able to control their adopters' prices either. An unchanging tax rate, on the other hand, provides the same stability to both innovators and adopters.

The other form is policy-based risk. Governments can abruptly introduce, change or repeal policies that have a significant impact on the operating conditions under which firms operate. Political dynamics or new information on the damage of pollution can cause significant changes in policies that may have been implemented with long-term stability in mind. Using a cross-country perspective, OECD (2009b) finds that the stability of environmental policy (including taxes, regulations and other instruments) is positively associated with environmental patents in the areas of air, water and waste. This effect is distinct from the effect of the stringency of environmental policy, which is also found to be important.

Reedman et al. (2006) use the real option methodology to assess firms' technology adoption behaviours in the face of a carbon tax. When the level and implementation date of the carbon tax are known, firms in the Australian electricity market should invest more in low-carbon technologies, whereas uncertainty of these parameters suggests that decisions on these investments should be delayed until more information on costs are known. Baker and Shiitu (2006) find that optimal R&D expenditures for energy technology in the face of uncertainty vary. For the most part, R&D expenditures for both conventional and alternative energies decrease with increases in uncertainty of a carbon tax. However, if firms are sufficiently flexible and the probability that the tax will be high enough to make alternative power generation profitable, R&D may increase concomitantly with the increased risk.

A clear example of government policy unpredictability is the production tax credit offered to wind power in the United States. Over a decade-long period, from 1999 to 2009, the tax credit was renewed six times, either having expired or coming months away from expiring each time. This significant unpredictability over the presence of the subsidy resulted in significant variation in wind power additions to the American energy grid. However, the variation in the investment level was not due to the underlying financial viability of wind power (and therefore the absence of the credit) but rather due to the uncertainty about the rate and how it impacted bargaining between energy companies and wind power firms over rates (Barradale, 2008).

In the case study on factors affecting climate change innovation in the United Kingdom, as described in Box 4.6, the effect of the EU ETS, the European Union's trading system for greenhouse gas emissions, was investigated on the innovative behaviour of interviewed firms, among a range of other variables affecting firms' operating environment. While the presence of firm-level greenhouse gas targets, customer and investor pressure and the general climate change orientation of the firm were positively linked with greater climate change R&D propensity (both product and process), a correlation does not appear to exist with participation in the EU ETS. The fact that permit prices have been trading at rather low prices may have reduced the incentive to undertake inventive activity. It is also conceivable that the volatility of permit prices and the uncertainty surrounding the parameters of subsequent phases of the EU ETS, such as the third phase which is to start in 2013, have caused firms to opt to wait until a future period to undertake innovative activity (this does not necessarily mean that they have waited to undertake abatement activities).

Japan's experience, as outlined in Box 4.2, provides a much stronger example of the effect of uncertainty of environmentally related taxation on innovation. Starting in the 1970s, SO_x emissions were taxed based on an exogenously determined level of compensation that was to be paid to victims of air pollution. As emissions declined and compensation increased, tax rates skyrocketed before the system was eventually reformed. Because rates increased significantly in the early years and there was recognition that such a system was politically unsustainable, firms undertook very little innovative activity, as seen in the count of related patents. Firms still continued to adopt new technologies to reduce their tax payments (and meet other regulatory requirement concurrently in place) but development of innovation was curtailed.

It is important to note that predictability does not necessarily imply that the tax remains constant over a long period; it means that the rate stays in a comfortable range around its expected (and credible) path. That is, the tax rate can be considered predictable, even if it is planned to gradually rise or fall, provided that this is foreseen by governments and industry.

4.1.3. *Innovation impacts on intertemporal tax rates and emission levels*

If policy makers have done their job well, the optimally set environmentally related tax should induce innovation. By allowing firms to achieve given levels of abatement at lower cost, innovation therefore implies that the marginal cost of abatement curve makes an inward shift. For policies that are intended to adapt to ongoing developments, innovation coupled with no change to the marginal damage from the pollutant, suggests that the optimal tax rate should therefore be reduced (for tradable permit systems, the quantity of permits should be trimmed) in the face of an inward-shifted MAC curve, as seen in Figure 4.1.

Figure 4.1. **Innovation impacts with taxation and tradable permits**

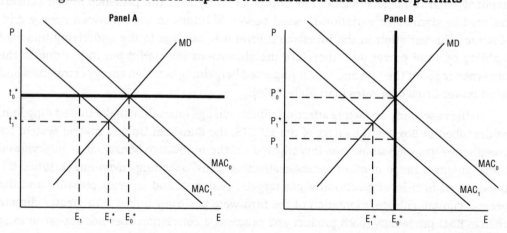

In the case of an environmentally related tax (Panel A), the initial tax is optimally set at $t_0{}^*$, so as to equate the marginal abatement cost curve (MAC_0) with the marginal damage curve (MD) in the original period to obtain an optimal level of emissions ($E_0{}^*$). With the advent of an innovation, the available options for abatement to firms expand, resulting in an inward shift of the marginal abatement cost curve to MAC_1. With the tax rate fixed, emission levels drop significantly to E_1. In a world of ever-vigilant environmental policy, the tax would be lowered to

Box 4.2. **Case study: The uncertainty of Japan's charge on SO$_x$ emissions**

Japan has a long history, starting in the 1960s, of seeking to control emissions of sulphur oxides (SO$_x$) which are generally created through the combustion of oil and coal for power generation among others, and cause respiratory problems. Regulations relating to emission rates, fuel usage, and smokestack height, for example, were all put into place and contributed to significant declines in emission levels and in ambient concentration levels.

At the same time, victims of air pollution-related diseases were seeking compensation from governments and industry. As a result, a charge on SO$_x$ emissions was enacted in 1973 and put into practice in 1974, with the proceeds being used to compensate air pollution victims. The rate was not based on the marginal damage of an extra unit of pollution in the present but based on the amount of revenues needed to compensate victims injured from historical emissions of SO$_x$ as well as other kinds of pollutants. As the number of victims and their compensation grew significantly and emissions rates continued to drop, the rates of taxation per unit of emission skyrocketed, as seen in Panel A below. In many of the first few years, rates were increasing significantly every year. In 1987, reforms were brought in to attempt to limit the tax rates, as firms' charges could have constituted nearly seven times the price of fuel, based on using high-sulphur (three per cent) oil in Osaka.

Panel A: SO$_x$ tax rates

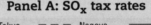

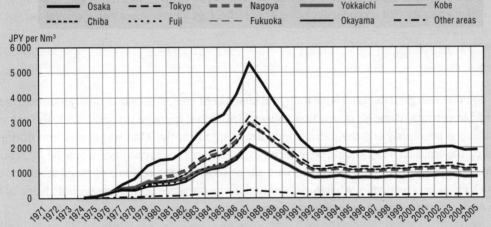

Panel B: Total patent activity related to SO$_x$ abatement

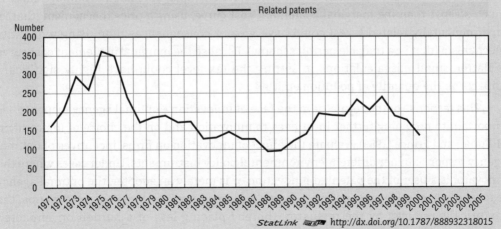

StatLink ⟨⟨⟨ http://dx.doi.org/10.1787/888932318015

Box 4.2. **Case study: The uncertainty of Japan's charge on SO_x emissions** (cont.)

Over this time period, there was significant adoption of abatement technologies, particularly flue-gas desulphurisation (a type of end-of-pipe technology that reduces the sulphur content of combustion), among regulated firms who sought to reduce their tax liability. At the same time, however, Panel B demonstrates that patent activity related to SO_x emissions was actually declining as tax rates were increasing. This suggests that the tax did not provide an environment where undertaking innovative activities was profitable. There are a couple of potential reasons for this:

● First, with tax rates rising quickly and reaching incredibly high levels, it became apparent that the current system was fundamentally flawed. There was significant political pressure to reform the system. This lack of credibility over the entire system may have significantly deterred investments in long-term R&D efforts.

● Second, the technologies which were developed in the 1970s due to stringent legal regulations and pollution control agreements between government and industry in dense industrial areas were nearly sufficient to bring about the subsequent emissions reduction in other areas in the 1980s. The compensation levy contributed more to the diffusion of SO_x abatement technologies developed earlier than to the development of them.

Therefore, the Japanese experience underscores the importance of reasonable predictability of the tax rate in the long run, supported by certainty of the policy environment, in order to create a climate that is conductive not only to technology adoption, but also the development of innovation.

Source: OECD (2009h).

re-equalise marginal demand with marginal abatement cost. The case is nearly identical with tradable permits (Panel B): innovation causes MAC curve to move inwards. With the cap on emissions previously set, the permit price drops significantly. With responsive and optimal policy, the emissions cap should be reduced to E_1^*, where the permit price would be P_1^*. Thus, with an unresponsive policy environment, innovation in the presence of taxes leads to too much emissions reduction, whereas innovation in the presence of tradable permits leads us to no emission reductions, but large price declines.

Differences may arise when the slopes of the marginal abatement cost and marginal damage curves differ (Weitzman, 1974). Where, for example, the marginal damage curve is much steeper than the marginal abatement cost curve, using a price mechanism (that is, environmentally related taxes) could have greater consequences than using a quantity mechanism (that is, a tradable permit scheme). Because the marginal abatement cost curve is flatter, small miscalculations in setting the tax level could have highly significant impacts on the quantity of pollution emitted.

This response of the regulator in the face of innovation – optimal agency response – is important to providing further incentives to innovation and on the choice of instrument. Milliman and Prince (1989) evaluate the effect of innovation on firms' incentives of optimal agency response under different environmental policy instruments. Under cases where the innovator is a user of the patented innovation or is merely an adopter, the optimal agency response is best under an emissions tax or auctioned permit. Under these scenarios, the tax/price is reduced by the regulator, thereby placing less of a burden on impacted industries. For a third-party innovator who is not directly subject to the environmental policy, emissions taxes are the least optimal. The optimal regulator would reduce taxes under this scenario – providing reduced incentives for further abatement without providing any relief to the innovator, since they are not subject to the environmental policy. On the other hand, command-and-control approaches would provide the greatest incentives to this innovator, since the optimal agency response in the face of innovation is to strengthen the policy, providing additional benefits for the innovator.

It is important to note that the presence of technical change may not always lead to an inward shift of the marginal abatement cost curve. The type of innovation may have differing impacts on the movement of the marginal abatement cost curve (Amir et al., 2008 and Bauman et al., 2008). End-of-pipe innovations will always lead to a downward shift of the marginal abatement cost curve, since there is no advantage to them but to reduce pollution. On the other hand, production process innovations may encourage pollution expansion because these innovations also impact the underlying cost function of the firm, and thus the new innovation may encourage expansion of production. This may, in fact, lead to an outward shift of the marginal abatement cost curve. The implication being that, where tax rates are sticky (and therefore unlikely to be changed in the presence of innovation), command-and-control or quantity options may be more effective in this respect.

Another interesting feature of technological change on the marginal abatement cost curve is the effects of intermediate innovations vis-à-vis longer-term innovations (Baker et al., 2008). These intermediate innovations (intermediate innovations in the context of climate change would be less-emission intensive carbon sources but not carbon-free sources) can initially lead to downward shift of the MAC curve for low levels of abatement but, as abatement reaches high levels, the marginal cost curve shifts outward. The authors present a simple example to illustrate. Suppose that there are three power sources in the economy: coal (high emissions), natural gas (fewer emissions), and nuclear (zero emissions). With no environmentally related taxation, coal has the lowest total production cost per unit, followed by natural gas and then nuclear; the imposition of a carbon tax reverses the order: nuclear, followed by natural gas and coal are now the least expensive. Therefore, an economy where electricity is sourced only from coal would start to shift into nuclear. No natural gas plants would be built. Now, with an innovation in the natural gas plant that generates a lower after-tax production cost than nuclear, the MAC curve would shift inward as electricity generation moves from coal to natural gas. This occurs for low levels of abatement in the short term. However, where significant abatement needs to occur, such as with the near decarbonisation of economies for climate change, even a full switch into natural gas would not achieve enough abatement. Therefore, natural gas production would need to give way to the nuclear option at high levels of abatement. The marginal cost of switching from natural gas to nuclear is now greater than switching from coal to nuclear, leading to an outward shift of the MAC curve at high levels of abatement.

The effects of innovation on the marginal abatement cost curve occur in an economy where other factors are changing as well. The scale effects of economic growth are pushing the MAC curve outwards (such that achieving a set amount of emissions costs more in a growing economy than a stagnant economy). This economic growth may also be having income effects on the marginal damage curve, such that it is shifting leftwards as people would be willing to pay more to achieve a certain environmental improvement the richer they become. Thus, the effect of innovation on the marginal abatement cost curve has to be considered against the overall effect on the other factors in determining the optimal taxation rates.

It is interesting to note that models of optimal carbon tax prices can differ significantly. Most foresee a rising carbon price in relation to rising temperatures and increasing marginal abatement costs as the low-hanging fruits are picked, leaving expensive options until a future period. On the other hand, innovation impacts could cause a declining carbon price, one that even reaches zero in the distant future (Acemoglu *et al.*, 2009). By implementing an optimal strategy of a carbon tax and green R&D subsidies, the kick-start provided to R&D activities into green technologies creates a process where firms are more and more likely to invest in green R&D (because of past investments and the fact that they are getting better at it). Due to this snowballing effect, the government policy stimuli can be rolled back, such that the optimal tax is zero several decades out. Such analysis is within a highly stylised model and cannot be used for specific policy advice but nevertheless underscores the potential power of innovation in this arena.

On the other hand, environmentally related taxation may have an indirect positive impact on pollution levels given the feedback effect. The presence of environmentally related taxation on a specific pollutant encourages innovation to reduce the emissions of that pollutant (such as efficiency measures). Such innovations reduce the demand for the underlying product. The effect of the reduced demand of the underlying good is a decrease in its price. The lowered price would have a scale effect, encouraging greater consumption and thereby impacting the overall level of emissions.[1] A rebalancing of the tax rate may be optimal. Such feedback effects are greater where a tax on a pollutant is levied on emissions that are highly correlated to an underlying good (such as carbon dioxide and fossil fuels) or where a tax on a proxy to pollution is levied (such as on fuel). Where the level and extent of the tax are large, the feedback effects are expected to be greater.

Political economy dynamics may make adjusting the instrument in the face of innovation difficult, even though the optimal response of governments to the inward shift of the MAC brings about the same price/emissions combination. On the one hand, reducing tax rates may be seen as a "reward" to polluters and the political will to bring about these changes could prove difficult with citizens. One mitigating feature of taxes is that many environmentally related taxes have been set at fixed levels in the initial years. Inflation and economic growth eat away at their effective bite over the years, leading to a *de facto* continual price decrease over years. In the Swedish case of a charge on NO_x emissions, described in Box 3.2, the tax was implemented at SEK 40 per kilogram in 1992. The tax rate was not modified until 2008, resulting in a real decline of the tax rate of around 20%. Such a design feature can weaken the arguments for reducing headline rates of environmentally related taxation in the presence of innovation.

On the other hand, reducing the total number of permits in the face of innovation can have a range of political economy angles, depending on the way in which it is carried out. First, simply revoking some pollution permits or effectively devaluing them[2] could be considered an expropriation of property rights, as is the case in some jurisdictions. Second, if the time periods between rounds of auctioning are short, governments can wait and simply offer fewer permits in the subsequent round. Finally, if the time periods between rounds are longer, governments could enter into the market with the goal of buying permits in order to retire them. The second option would likely pose the fewest political economy issues from either industry or citizens, although the effect of short time periods undermines the benefits from predictability.

4.1.4. Lead-in time for environmentally related taxation

Announcing a new tax (or an augmentation to an existing one) with an immediate effective date provides immediate incentives for abatement but experimenting with new techniques, installing new equipment, making new products, or switching inputs all require time to fully implement. Therefore, an environmentally related tax that is announced and implemented in a relatively close time period provides little, if any, opportunity for firms to abate during the time immediately following its introduction. The effect is that polluters are subject to the tax on their emissions in the current period (which are based on historical behaviour) as well as those in the short term (given the inability for capital assets to be quickly replaced or for processes to be changed).

Credible announcements that environmentally related taxation will be implemented in the near future (one to two years, for example) instead of in the very short term can still provide firms with the abatement incentives of the tax without collecting revenue based effectively on pre-tax production arrangements. It can also provide the incentive for increased investments in innovation activities without the revenue effect. Such a lead-in may also to help ease the implementation of environmentally related taxes that have strong constituencies arguing against its introduction.

The Swiss VOC emission case study, as described in Box 3.8, utilised such an approach. The law entered into force in January 1998, with the tax coming into force two years later. This implementation period was prompted by suggestions from industry as well as from the need for relevant government authorities to build competencies and infrastructure for effective tax administration. In response to the credible future imposition of taxes, some abatement started in 1998. One interviewed firm even adopted expensive incineration equipment with high operating costs in the mid-1990s because of the expectation of taxes on VOC being introduced.

4.1.5. Competitiveness concerns and political economy dynamics

As outlined in OECD (2006), there are political economy considerations that impact the design of environmentally related taxation. Exemptions, rate reductions or other measures feature into a wide range of taxes that have been implemented in OECD countries. Concessions are typically made to address distributional concerns related to environmentally related taxation. As home heating and transportation typically consume a larger percentage of the budget of low-income households, there are concerns that the burden of these taxes falls disproportionately on those households least able to afford it.

Moreover, concessions are also made to emission-intensive, and therefore more potentially trade-exposed, sectors in order to address potential competitiveness issues against jurisdictions not levying such taxes. Where a country has levied an environmentally related tax in advance of its peers who are also facing the same environmental challenge, some of the benefits will spill over to them, since not all the information or innovation can be perfectly captured. Thus, the abatement cost for followers may be less than for the initiating country, suggesting that a lower tax would help reduce the spillovers and therefore the competiveness differences between the initiating and following countries. On the other hand, where the initiator is a developed country and the followers are developing countries, this may be a more desirable result. Rosendahl (2004), for instance, suggests that since environmental technologies are first developed in industrialised countries, optimal environmental taxes should be higher in developed countries than developing countries, thereby creating incentives for learning in developed countries that can then benefit late adopters in developing countries.

These competitiveness concerns also manifest into environmental concerns, as firms sometimes can relocate and continue polluting at business-as-usual level. In climate change, this "carbon leakage" is a concern expressed often by industry. However, as the size of the market grows (either through expanding the reach of policies or co-ordinating policies among countries), the potential for such leakage diminishes quickly (OECD, 2009c).

In many jurisdictions, the use of basic resources, such as water and home heating fuel, are fraught with competitiveness and distributional concerns. The use of progressively tiered pricing can provide basic levels of the good at a low price but increased prices on larger usage continue to provide significant abatement incentives on the margin. Israel, for example, applies block tariffs for all users of water – residential, industrial and agricultural.[3] Prices for households, for example, were ILS 3.93 per m^3 for the first eight cubic metres per month, ILS 5.50 per m^3 for the next seven and ILS 7.65 per m^3 thereafter in 2008 (OECD, 2009i). Much the same structure occurs for agriculture, with the added incentives of lower prices for using saline or recycled (treated sewage) water. The experience of Israel's water prices, described in Box 4.3, shows that the environmental effectiveness of water pricing can vary significantly across sectors, as price elasticities are much lower for households than for other users. The innovation impacts from such water pricing are difficult to disentangle, however, given the coexistence of public investments, regulations, information campaigns and the like.

In Sweden, the policy surrounding the introduction of the NO_x charge included a provision to refund the charge, less a small fraction for administration, based on the firms' energy output. This refund mechanism offsets some of the impacts of the charge, with cleaner-than-average firms receiving a net payment and dirtier-than-average firms making a net payment. Given the decoupling of the tax base from the refund base, the incentive to abate largely remains; however, such a mechanism may have a small negative impact on the inducement of innovation for firms that are also creating pollution, as described in Box 4.4.

While such a refunding mechanism may have some small negative innovation impacts,[4] this has to be weighed against the political economy angle that such a high tax rate (when compared to other jurisdictions that implemented such charges, such as France's charge at about one one-hundreth of that of Sweden) may never have been able to be implemented without such refunding. The Swedish refunding provision may also have led to a tax rate that is higher than even a level suggested by associated environmental externalities. Sweden's neighbour, Norway, has introduced a similar tax on NO_x emissions but at nearly half the level and without a refunding provision.

In practice, the fear of some energy-intensive or trade-sensitive businesses fleeing to seek out lower-tax jurisdictions may not be as warranted as some predict. Box 4.1 presents a case study on the United Kingdom's Climate Change Levy which provides rate reductions for some emission-intensive firms and sectors. In comparing firms subject to the full rate against those subject to the reduced rate, it was found that firms facing the full rate were no more economically disadvantaged than those facing the reduced rate, despite the concerns that those paying the full rate would be less competitive – both against their domestic competitors paying the reduced rate and against international competitors not subject to the CCL at all.[5] None of the measures – levels of employment, real gross output and total factor productivity – drops when firms pay the full rate of the CCL instead of a

---

Box 4.3. **Case study: Water pricing in Israel**

Situated in an arid location and faced with a growing population, income growth, and more frequent droughts, the conservation of water is a significant concern in Israel. As such, government policy has been focused on creating strong incentives for its more efficient use (in addition to expanding supply through desalination). As a result of a number of factors, including strong government policy, water demand in Israel averages around 300 m^3 per year, compared to an international norm closer to 1 700 m^3 per year (Mason, 2009).

One of the most visible signs of this determination has been through the use of increasing water prices for domestic, commercial and agricultural uses, which can have the same effect as taxes. The comparison of agricultural to residential prices highlights the different effects between the sectors. In agriculture, farmers face an escalating three-block tariff rate, based on their quota. Between 1995 and 2005, for example, average real prices of agricultural water increased by over 68% to USD 0.33 per m^3. Despite significant increases over the longer period as well, the value of agricultural production per m^3 of water has more than trebled since 1958. The pricing structure has encouraged more efficient use of water, such as through drip irrigation, as well as substitution (using recycled and treated sewage water sources, which are around USD 0.20 per m^3). On the other hand, per capita use in households continues an upward trajectory, despite prices having been increased significantly as well. Declines are generally only seen when complemented with water saving campaigns (which promote water saving through information campaigns, communication projects, and enforcement of existing regulations).

However, it is difficult to disentangle the effects of prices and other policies on the Israeli achievement with water conservation. This is especially true when differentiating between innovation and productivity impacts. In many cases, price changes were put in place simultaneously with significant public expenditures to help with technology adoption, regulations on service provision, and marketing campaigns. Nevertheless, the effect of these policies has triggered a considerable amount of innovation – in water-saving technologies, supply enhancing technologies, etc. – and the significant increases in water prices that have been recently seen provide a fertile area for additional future study.

Source: OECD (2009i).

---

reduced rate. It appears, however, that the lower incentive faced by firms because of the lower tax burden led to less innovative performance, as measured by patent counts, than firms subject to the full rate.

These political economy considerations have generally focused on the individual firm. However, the design of certain environmentally related taxes can lead to different innovation impacts based on individual and collective innovation propensities. With pollution taxes (with no additional features), there are incentives for individuals and collections of individuals (such as industry associations) to invest in innovation that benefit all participants. In agriculture, industry associations typically collect check-off fees to fund R&D activities because the expected benefits of these innovations accrue to all farmers but are beyond the means of farmers to fund individually. Similarly with pollution taxation, groups of firms affected by the tax (or even the entire group) have incentives to pool resources and undertake collective R&D activities, where economies of scale may make individual actions uneconomic.

Box 4.4. **Tax refunding and the impact on innovation**

Many times a firm subject to an environmental tax has an incentive to innovate in order to both use the innovation within their own firm and to license it to others. With refunding, the incentive may be somewhat blunted as the adoption of innovations not only reduces the tax liable for pollution but also the expected refund, especially with growing market concentration. Thus, expected return to innovation is reduced.

With an environmentally related tax, innovating firm j tries to minimise their costs (C_j) which is influenced by their emissions (e_j), their output (q_j) and the abatement technology (k_j), considering their R&D costs (D_j), their expected royalty revenue from a new innovation (R_j) licensed to m non-innovating firms, and their tax payment (t^*e_j). With a refund mechanism, the refund also affects the firm's cost structure, and is based on the emissions of all n firms within the industry. Thus, the innovating firm's cost function can be thought of as:

$$C_j = c_j(e_j(k_j), q_j, k_j) + D_j(k_j) - R_j(k_j) + te_j - t\frac{q_j}{Q}\sum_{i=1}^{n}e_j(k_j) \tag{1}$$

The firm's royalty price is simply: $R_j(k_j) = m(k_j)P_m(k_j)$. If the cost equation is differentiated with respect to technology k_j and set to zero to find its minimum, the rearranged equation takes the form:

$$\frac{\partial D_j}{\partial k_j} = -\frac{\partial c_j}{\partial k_j} + \frac{\partial R_j}{\partial k_j} + t\frac{q_j}{Q}\sum_{\substack{i=1 \\ i \neq j}}^{m}\frac{\partial e_i}{\partial k_j}, \text{ where } Q = \sum_{i=1}^{n}q_i \tag{2}$$

From equation (2), the factors affecting the optimal level of R&D expenditures can be divided into three terms. The first term is the cost effect of the abatement technology on the innovating firm itself – the greater the costs, the lower the optimal level of R&D expenditures. The second term is the effect of the innovation on the royalty revenues received from the non-innovating firms – the greater the expected royalty revenues, the greater the optimal level of R&D expenditures. Finally, the third term is the effect of the innovation on the firm's refund (N.B.: with no refunding, the equation would simply be the first two terms). As $\partial e_i/\partial k_j < 0$, the effect of the refunding mechanism is to reduce the optimal level of R&D investment compared to a scenario where no refunding takes place. Since a small share of output is constant for changes in abatement technology k_j, as $q_j/Q \to 0$, the third term does not disappear. Thus, the refunding effect does have some negative impact on the overall level of innovation inducement, and does increase with growing market concentration.

The effect for an external firm, not producing output q, but innovating and selling abatement technology k_j to the firms subject to the tax, is slightly different. It will set the price for the innovation as the reservation price of the last licensing firm. However, non-adopting firms will be affected by the technology adoption of other firms because of the effect on the refunding mechanism. Therefore, the price that the innovator will set for the abatement technology is the difference between the costs of adoption with the costs of non-adoption. In this case, the effect of the refunding charge does not have a significantly negative role, assuming that no firm has significant market power.

Thus, the theory sheds light on how refunding mechanisms can affect the innovation incentives of environmentally related taxation. Where the innovator is one of the firms subject to the tax, there will always be some dampened incentive to innovative, even when there are many, non-co-operating firms. With an external firm, a perfectly competitive market suggests that refunding has no discernable impact on the incentives to abate. In the Swedish example, no firm had more than 12% of the total output, suggesting that the marketplace was relatively competitive, with no single player having a dominant position with respect to production levels.

Source: OECD (2009e).

The inclusion of additional design features can have profound impacts on collective innovation incentives. Under Japan's SO_x charge, the amount of revenue to be raised is exogenously determined, with the emissions level deciding the level of the tax. The individual firm has incentives to reduce emissions in order to reduce its individual tax liability. On the contrary, there is no incentive for the collective of firms facing the tax to invest in abatement innovation. Results flowing from such investment and taken up by all firms would reduce the level of collective emissions, which would only result in an automatic augmentation of the rate. The individual Japanese firm under such a scenario would be not better off.[6] In the same way, under the Swedish system of refunding, any collective innovation that results in additional abatement by all firms reduces the amount of tax collected but also reduces the amount refunded, thereby leaving the individual firm no better off. Under both of these different design systems (and to some extent cap-and-trade systems), innovation is only profitable at the individual level because emission reductions due to technological progress occur relative to the emission levels of other individual firms.

4.2. The extent of the tax base

For a polluting firm subject to a tax considering only its own production, incentives will exist to undertake innovative activities to reduce its own tax burden. Its decision will be made on a number of factors centred around the return on investment when implemented within the firm, such as the level of the tax and availability of resources.

Firms are also conscious of the market outside of their firm. For polluting firms, an innovation developed in-house can be sold to others in a similar position. For third-party firms, the return on investment from their innovation can come solely from the license of the intellectual property (IP) to other firms, either through the licensing of the IP directly or in machinery where the technology is embedded. In such circumstances, the range of actions covered by the tax can have considerable impacts on the incentives for innovation. The extent of the tax base is dependent on two basic factors: i) the size of the jurisdiction levying the tax (or jurisdictions, where the same pollutant is taxed in a similar fashion); and ii) the proportion of pollutants taxed within a jurisdiction (for example, a tax on CO_2 emissions *versus* a tax on CO_2 emissions only from particular sources).

The scale of opportunities, as determined by the extent of the tax base, is critical: the greater the tax base, the greater the opportunities to profit from innovations, and therefore the greater the level of innovation resulting from the imposition of the environmentally related tax. Where the extent of the tax is relatively small, the incentives for innovative behaviour can be limited due to fewer opportunities to recoup any expenses. Policy makers in smaller jurisdictions must additionally balance how the limited extent of the tax will influence the propensity of firms to innovate at a given tax level. It is for these reasons that cross-country co-operation in pollution abatement can provide the greatest returns. Not only can a wide range of countries act to abate and address global environmental challenges with fewer free-rider problems, the potential for greater innovation is present.

The Swedish tax on NO_x emissions, as described in Box 3.2, was progressively extended to smaller firms in the years following its implementation. This continued extension of the market provided greater opportunities for innovative behaviour for emitting firms and external firms. The analysis on the patenting of related innovations in Sweden could not find a definitive relationship between the tax and the intensity of patenting (nor on the extension of the tax base on patenting activity). The Swedish market for innovative NO_x abatement technology, even with more potential firms as clients, may still be below a certain threshold necessary for more intensive innovative activity.

4.3. Administering the tax

The administration of environmentally related taxation is, generally, a relatively straightforward procedure – the rate is applied to the quantity of pollutants emitted (or a proxy thereof). This is in contrast to the process undertaken through regulatory procedures, where administrators can have discretion when administering environmental programmes. This discretion can provide uncertainty for businesses over what the final form and effect of the administered measure will be. Sometimes it is in differential interpretations of the regulations across officials, different levels of enforcement across polluters, or a greater scope for judgement.

While this phenomenon is more applicable to command-and-control instruments, environmentally related taxes can also be subject to this problem. Complex tax rules can create such discretion, although environmentally related taxes are generally much less complex than corporate or personal income tax systems. One exception is when environmentally related taxation is mixed with negotiated agreements. For example, many countries offer discounts on energy or carbon taxes to energy-intensive industries in return for agreements to undertake specific measures or meeting agreed-upon targets. Such processes introduce: i) uncertainty for the private sector due to the negotiated processes regarding the stringency of the measures and the means to measure compliance; and ii) opportunities for tax reductions without much environmental benefit in return.

The administration of environmentally related taxation may also play a unique role for the firm in that it provides a source of information that would not have been previously collected. For firms, complying with the requirements of the corporate tax system is unlikely to provide additional information about the health and profitability of the firm, since this information is already collected for other purposes and is central to managing a well-functioning enterprise.

Environmentally related taxation is likely different, especially with respect to taxes levied directly on pollutants. Without the environmental policy, there is little financial reason to track and account for specific pollutants (firms and households likely do track taxes on proxies to pollution, as these tax bases can have significant relevance). Therefore, the imposition of pollution taxation provides information to polluters that before was typically unknown.[7] For some pollutants, this administration can come at a significant cost to the firm (and to the administering agency). The use of proxies to the pollutant is therefore generally used in such circumstances. A fuller discussion of the advantages and disadvantages of using proxies is found in Section 4.4.

This information on emissions not only helps tax administration but also helps the firm to better understand how pollutants are created, especially where the formation of pollution is not linear with input use. The information collected is not so valuable that firms will collect it even in the absence of environmental policy. Nevertheless, the information gleaned can help mitigate some of the effects of the tax and help firms find inexpensive ways, such as behaviour change or instrument calibration, to reduce pollution.

In the Swiss case study of the VOCs, firms were required to keep detailed accounts, outlining flows of VOCs and goods containing them, in order to precisely indentify the emissions of VOCs for which they were liable. More importantly, the Swedish case study identified the continuous monitoring devices that firms had to install in order to track NO_x emissions. The initial cost of these devices were EUR 30 000 to EUR 36 000, with annual inspection and maintenance costs running approximately EUR 12 000 per year. Through these devices, firms were able to try out new combinations of fuel, temperature and other settings to determine the effects on NO_x emissions, given that the formation of NO_x varies significantly between sources and is determined by a wide range of variables. This information was critical to firms in understanding how specific actions and new processes could lead to inexpensive emissions reductions.

An interesting side note is that innovations can also be used to aid tax administration. The development and deployment of new technologies can help to overcome some of the barriers to monitoring emissions and administering taxes on disparate and hard-to-monitor sources.

4.4. Tax-based policy instruments

Thus far, this publication has focused heavily of the role of taxes on pollution and proxies to pollution. Yet, the tax system can also be used in other ways to achieve environmental and innovation aims. This range of tax measures can be placed into three categories, as outlined in Figure 4.2.

Figure 4.2. **Categories of tax-based measures**

Discouraging the environmental bad	Inducing the environmental good	Inducing innovation
Placing a cost on environmentally harmful activities.	*Providing positive incentives to undertake actions that will help achieve environmental objectives.*	*Providing positive incentives to undertake actions to increase innovation (broadly or targeted towards the environment).*
• Taxes levied on pollution. • Taxes levied on proxies to pollution.	• Accelerated depreciation for abatement capital. • Reduced VAT rates on less environmentally harmful goods and activities.	• Measures to reduce the cost of innovation (such as R&D tax credits, accelerated depreciation for innovation capital, and enhanced allowances for R&D labour costs or reduced taxes on R&D labour). • Measures to increase the returns to innovations, such as reduced corporate tax rates certain types of income.

The first category deals with much of what this report has already discussed. These measures place a price on the environmentally harmful activities and therefore encourage pollution abatement and innovation – the stick of environmental policy. In the second category, governments can also try to encourage environmentally beneficial actions by reducing their cost – the carrot of environmental policy. Measures in this category include targeted reductions in the rate of value added tax for certain appliances and accelerated depreciation for investments in environmentally related capital. In the third category, governments can use the tax system to try to encourage supplemental innovation, with measures such as R&D tax credits and lower corporate tax rates on the returns to innovation. Instead of focusing on providing punitive aspects to pollution, the final two categories are targeted to provide benefits for innovating and adopting clean investments and therefore have many characteristics similar to subsidies.

It should be noted that countries' corporate tax systems have many pre-existing features and distortions compared to a benchmark system. Some of these features bring a tax system closer to optimality, while others exacerbate distortions. The tax measures below are intended to provide guidance for policy makers who are contemplating such measures within existing tax frameworks and therefore this section does not delve into fundamental issues of corporate taxation.

4.4.1. Taxes on pollution

Taxes levied directly on pollution (or tradable permit schemes, see Box 3.4) are generally considered the most effective instruments available to environmental policy makers, as outlined in Chapter 3. They provide an incentive to undertake abatement action and to innovate for the entire range of activities that contribute to reductions in pollution. As seen from earlier sections of this chapter, these effects are highly dependent on design factors.

Despite the benefits that such taxes have, their administration can pose some difficulties. Measuring exact emissions, especially from mobile sources such as transportation, can pose a significant challenge to administrators. Monitoring systems and new tax collection infrastructure may be needed. Compared to taxes on proxies to pollution, where the tax infrastructure may already exist and measurement technologies may not be needed (such as with petrol where VAT and excise taxes are already levied), taxes on pollution can therefore be more burdensome for governments. Affected polluters could also face additional costs of compliance because of the need to purchase monitoring equipment and hire additional staff to manage tax compliance.

4.4.2. Taxes on proxies to pollution

Taxes on proxies to pollution are typically levied on goods or actions that generally lead in a subsequent step to pollution; as such, they include a wide spectrum of actions. Motor vehicle fuel taxes are a common example, as they tax the fuel, not the actual pollutants that are emitted when it combusts. While combustion of motor vehicle fuel is well known to cause a wide variety of undesirable pollutants, a fuel tax only encourages abatement and innovation in the use of fuel – not necessarily of the undesirable emissions that result from its combustion. By restricting the areas in which the tax provides incentives to abate, one does also on the incentives to innovate. Thus, advancements in cleaner engine design or end-of-pipe mechanisms, such as catalytic converters, are not made more attractive with an undifferentiated fuel tax. Moreover, the initial correlation between the pollutant and the taxed proxy has the potential to break down over the longer-term as the innovation and abatement effort is placed into reducing the use of the proxy.

Taxes on proxies to pollution can embody incentives that can lead to greater short-run pollution and may retard innovation incentives. Environmentally related taxation on the first purchase of motor vehicles, for instance, is based on the fact that driving leads to pollution. The increased cost of cars will induce people to have fewer cars (and, based on the structure of the tax, opt for less pollution-intensive vehicles). Such a tax is indifferent, however, to the actual amount of driving that occurs, the type of driving, or (in most cases) the type of fuel that is used. Thus, the car tax provides a one-time incentive to purchase a less polluting car but, after the purchase, provides no incentive for further abatement.[8] Moreover, following the introduction of the new tax, new cars (which are typically much cleaner) are now even more expensive compared to used cars (which typically pollute significantly more). Thus, consumers have greater incentive to purchase used cars or maintain their existing cars longer, potentially increasing pollution in the short run compared to a baseline scenario [Johnstone *et al.* (2001) provide an interesting study of Costa Rica].[9] Since newer vehicle vintages typically embody new technologies, such as more fuel efficient design or additional end-of-pipe pollution devices, one-off motor vehicle taxes can also act as a *de facto* tax on environmentally related technologies.

Despite some of the potential drawbacks of taxes on proxies to pollution in terms of efficiency and ability to induce innovation, the simplicity of administering them makes the implementation more attractive. Taxes on proxies to pollution are typically levied where monitoring is easier and collection points are fewer (or where taxes are already levied). In the case of motor vehicles, the installation and verification of monitoring exact carbon, nitrogen and other emissions for each tailpipe combined with creating tax collection system would likely be prohibitively expensive. Fuel taxation, on the other hand, can be levied relatively easily at fewer points where other taxation (*i.e.* VAT) is already charged to consumers.

In addition, taxes on proxies to pollution can overcome some hurdles to environmental policy, especially where the timelines are quite long. In the case of driving, emission taxes (or even taxes on fuel) to be paid on future driving may not adequately be considered by consumers when purchasing vehicles. One-off motor vehicle taxes (that differentiate by environmental performance) can account for these distortions by highlighting consumption choices.

Much of the discussion thus far has characterised environmentally related taxation (and tradable permits) as a non-prescriptive economic instrument. In practice, taxes can be constructed to have prescriptive features. Most often this occurs when taxes are levied on proxies instead of the pollution itself. Like car registration fees in several other OECD countries, the French bonus-malus system, for example, instituted in 2007 and 2008, taxes the registration of new cars based on their rated grams of CO_2 emissions per kilometre driven.[10] For ranges below a certain threshold, levels of subsidy are provided to purchasers of cleaner cars. Cars with higher emissions per kilometre face increasing bands of taxes, topping out at EUR 2 600. Such taxes may provide clearer signals to consumers between classes of cars, but do not necessarily provide incentives for incremental improvements in emissions that do not result in a change in the tax band in which a vehicle is placed.[11]

4.4.3. Accelerated depreciation allowances for abatement capital

The overall tax system – not just environmentally related taxation – may influence the type and level of innovation. The corporate income tax system can provide differential incentives to the types of investments that firms make, possibly changing the diffusion of existing innovations. In addition, these differences can alter the demand for new innovations in various areas and therefore encourage the creation of some types of innovation over others.

The corporate tax system is organised such that eligible business expenses can be deducted from revenues in order that only net income is taxed. Expenses on inputs to production that are fully utilised in the current period are immediately deducted from income. Capital assets are more complex as the usefulness of these assets are not depleted in only one period – they provide benefits to the firm over a longer horizon. In order to align the benefits of the asset with the costs of the asset over a multi-year period, the asset should be depreciated over the expected useful life of the asset with the expected usage in each year providing that year's deduction against revenues.

If the tax code provides for depreciation allowances greater than the economic or even general rate of depreciation,[12] an indirect subsidy for capital expenditures exists. This occurs because firms are provided a tax deduction in an earlier time period than the actual depreciation takes place.[13] As taxation is deferred to a future period, the use of accelerated depreciation acts similar to an interest-free loan. Therefore, the benefits accorded to capital subject to accelerated depreciation lower the cost of its acquisition, thereby increasing demand and/or allowing firms to reallocate some funds for other productive (and profitable) ventures. House and Shapiro (2008) show that temporary enhancements to the depreciation schedule can induce significant investments in technologies covered by the accelerated depreciation, specifically those with longer tax recovery periods. For a more detailed analysis, see Box 4.5.

As accelerated depreciation provisions enacted to specifically encourage environmentally beneficial actions provide benefits to "good" actions, there are clear similarities to the issues faced with subsidies in general. These include selecting applicable assets classes (and thereby favouring some types of abatement investments over others), subsidising activities that would have taken place regardless, and effectively requiring public revenues to be raised from other sources to compensate for the negative fiscal effect. In addition, countries' use of accelerated depreciation for targeted areas is many times part of a short-term policy response.

Comparing against a standard or existing corporate tax system, the provision of accelerated depreciation in targeted sectors can impact on the type of investment, not only the level. Yale (2008) suggests that accelerated depreciation provisions can additionally favour capital expenditures (which are typically for end-of-pipe technologies) compared to non-capital expenditures (which are typically for cleaner production technologies), distorting investment decisions.[14]

For example, a coal-burning power-generation firm is deciding how to abate sulphur dioxide emissions. Two options exist: purchase and install a scrubber that has a 30-year lifespan or switch from high-sulphur (and cheaper) coal to low-sulphur (and more expensive) coal. Assume that the benefits of each option are equivalent under the prevailing tax system: the firm is no better off paying for the scrubber and depreciating it over its useful life according to economic rates of depreciation compared against switching to a more expensive fuel and incorporating this additional cost into its annual tax deduction for expenses. With the introduction of accelerated depreciation for "green" technologies, the tax benefits to installing a scrubber are now greater because this has a large upfront capital cost compared to ongoing operating costs. This encourages greater investment in, and diffusion of, end-of-pipe technologies over cleaner production abatement technologies. By encouraging diffusion, such measures help promote some of the learning-by-doing and learning-by-using effects. Yet, encouraging certain technologies or technologies aimed at certain fields can create technology lock-in.

This tax-based measure is of interest as a number of countries have implemented accelerated depreciation as a feature of their environmental policies. In the United States, Sansing and Strauss (1998) note that, for the SO_x programme of tradable permits, the tax depreciation period for pollution abatement expenditure was 60 months, much less than the useful life of such equipment. In Canada, eligible capital that generates clean energy or conserves energy is eligible for accelerated depreciation at 50% per year on a declining basis (Canadian Department of Finance, 2010). The Netherlands' Vervroegde Afschrijving van Milieu-investeringen (VAMIL) scheme provides favourable depreciation rates for

Box 4.5. **Effects of accelerated depreciation allowances for environmentally friendly investments**

In some cases, countries have used accelerated depreciation allowances as a way of giving additional incentives to adopt new environmental technologies that are embodied in capital, i.e. where the adoption requires new investment in particular assets.

To illustrate the effects, it can be useful to consider the financial outcome for a firm increasing investment by one unit in the current period, followed by a similar reduction in the subsequent period. Absent any taxation, the additional corporate revenue would be $p\dfrac{\partial F}{\partial K}$ where F refers to the production function F(K,L). If the additional investment is financed by borrowing, then the additional costs would be the sum of interest and economic depreciation of the asset: $r + d$. The first-order condition for profit optimisation is $p\dfrac{\partial F}{\partial K} = r + d$.

Introducing taxation, the additional net corporate revenue would be $(1-\tau)p\dfrac{\partial F}{\partial K}$. The net financing costs would be $(1-\tau)r$ as, in a typical OECD-country tax system, interest expenses can be deducted from the corporate income tax base. The economic depreciation of the investment asset would, as before, be d. Finally, assume as a benchmark case that the tax allowance for depreciation would match exactly the economic depreciation of the asset, meaning that net financing costs are reduced by τd. Putting these elements together, the first-order condition now becomes $(1-\tau)p\dfrac{\partial F}{\partial K} = (1-\tau)r + d - \tau d$. However, this reduces to $p\dfrac{\partial F}{\partial K} = r + d$ showing that, in the benchmark case with interest deductibility and a tax depreciation schedule identical to economic depreciation, there are no distortions to investment decisions from corporate income taxation: at the margin, businesses face the same investment incentives as they would in the absence of corporate income taxation.

The effects of a targeted accelerated depreciation allowance schedule for environmentally friendly investments can now be assessed. For these investments, the last term in the first-order condition becomes larger as more than the economic depreciation can be deducted. Even though this higher level of early depreciation would be matched by less depreciation at later periods, the investing firm would still have an advantage in net-present-value terms from the reduction in tax liabilities being advanced in time. Consequently, accelerated depreciation increases the net return and thereby the incentive to invest in the class of assets benefiting from the accelerated depreciation allowances. In the context of full interest deductibility, accelerated depreciation for particular investments effectively implies an indirect subsidy compared with other investments subject to the same corporate income tax system but with tax allowance for depreciation matching economic depreciation.

selected technologies that have been approved by the government (European Commission, 2009). In the United States, a 50% accelerated depreciation allowance is available for qualified reuse and recycling property and qualified cellulosic biofuel plant property (United States Internal Revenue Service, 2010). In addition, Mexico provides a 95% depreciation allowance in the first year for capital investments in solar, wind and geothermal energy (KPMG, 2007).

Spain (as described in Box 4.7) instituted a tax credit for qualifying environmentally related capital investments, thereby lowering the effective cost of such investments, similar to accelerated depreciation allowances. The analysis of this initiative showed no linkages between its introduction and patenting in related areas, some of which may have been affected by the fact that investments that were also required by law qualified for this deduction.

The case study involving interviews of UK firms, as outlined in Box 4.6, examined the relationship between an accelerated depreciation provision (or "Enhanced Capital Allowance scheme" that provides 100% expensing against taxable profits) and the economic and innovative performance of firms. The use of the accelerated depreciation provision was positively correlated with total factor productivity in firms, suggesting firms using the provision updated their capital stock and realised productivity gains. The provision did not,

---

Box 4.6. **Case study: Factors influencing UK firms' patenting activity**

In a companion study to that described in Box 4.1, managers of UK firms were interviewed on a wide range of factors using an innovative survey methodology. The goal was to determine how environmental policy instruments, intrafirm organisational behaviour and other marketplace pressures influenced firms' environmental, economic and innovative performance. The interview data was linked to outside data on firm performance, size, energy usage, etc.

First, the factors influencing energy intensity within firms were analysed. The presence (and relative stringency) of energy quantity targets within the firm was strongly associated with lower energy intensity and higher firm total factor productivity. Moreover, the more stringent firms were with their investment criteria (that is, the higher the hurdle rate set for investing), the more energy intensive the firm. Finally, participation in the UK's Enhanced Capital Allowance scheme, which provides an immediate expensing of the full cost of investments in energy- and water-saving projects, was linked to lower energy intensity and higher firm productivity. Variables related to participation in the EU ETS and Climate Change Agreement (CCA) participation were not significant.

Second, using much the same approach, the interview responses were analysed against the firm's innovation propensity, as measured by firms' responses to their R&D intensity in general, as well as towards climate change-related product and process innovation.* It was found that the presence of targets and the level of pressures from investors and customers were positively correlated with greater R&D propensity. EU ETS participation and accelerated depreciation were found to have little effect, potentially due to the low prices and significant unpredictability surrounding the permitting system. The stringency of CCA participation was also not found to be significant, though likely due to unobserved heterogeneity (for more robust and reliable results surrounding this policy, refer to the companion study, as described in Box 4.2).

Therefore, it is interesting to note the differences between the two analyses. The Enhanced Capital Allowance had significant impact on lowering energy intensity of firms but no discernable impact on the innovative nature of firms. This underscores the fact that the capital depreciation allowances induce adoption and diffusion of innovations but, utilised alone, provide few incentives for invention.

* It should be noted that this contrasts from the approach taken in the companion case study. In the companion report, a more quantitative approach was taken by analysing the number of climate change-related patents attributable to certain firm factors, with some issues raised over the correctness of the patent data. This approach, on the other hand, focuses on managers' responses regarding these questions, which may help provide a more holistic view of a firm's innovative potential and sidestep some of the data problems found in the patent search.

Source: OECD (2009g).

---

however, have any statistically significant impact on firms' propensity for innovation (broad-based or climate change-related). This suggests that accelerated depreciation may provide additional incentives to adopt capital-intensive, pre-existing technologies but does not necessarily induce actions into the development of new technologies and processes.

4.4.4. Environmentally related reductions in value added taxes

Similar to accelerated depreciation provisions, the tax system can encourage other types of environmentally beneficial actions (through capital investment) through reduced rates on consumption taxes. A number of countries have implemented reduced VAT rates to encourage consumption of less environmentally harmful products, usually energy-efficient appliances.[15] By reducing the after-tax price to the consumer, such policies both make the product more competitive compared to other products while also highlighting and promoting the energy-efficient models.

The general view relating to VAT tax policy is that a standard rate with few, if any, reductions is the optimal design to promote efficiency and reduce distortions in an economy (OECD, 2009j).[16] Rate reductions reduce the revenues of government (necessitating the levelling of other, and likely more distortionary, taxes) while increasing the administrative complexity of the system, both for business and tax administrators. In many cases, rate reductions are implemented to address perceived distributional issues by providing reduced rates on food, energy, and other staples. However, these reductions benefit both richer and poorer citizens and other forms of distributional policies are likely to be much more effective.

Rate reductions can be effective at encouraging the adoption of existing innovations among customers. In case studies of the European appliances market, the European Commission (2008) suggests that moving from countries' standard VAT rates to reduced VAT rates would lead to some significant changes in market share. "B" category refrigerators and freezers would see market shares decline by twenty points in favour of the more environmentally friendly A and A+ machines. Consumption pattern changes in washing machines and dishwashers are predicted to be slightly less. With the concurrent existence of the EU ETS – a tradable permit system with a hard cap – any emissions reductions due to this initiative will have no impact on the overall level of emissions over the current trading period.

From the innovation perspective, this instrument is likely of limited value and previous studies (Copenhagen Economics, 2008; European Commission, 2008) have not been able to clearly identify innovation creation impacts. The reduced rate can encourage firms to develop new models to take advantage of the increased demand likely to be induced by the lower after-tax price. Where the tax reduction is not passed through, the increased profit margin can be an additional incentive. However, almost exclusively, the reduced VAT is on categories of products that are already on the market, providing incentives to potentially scale-up existing production or transfer the innovation to new models, but not strong incentives to develop new innovations. Moreover, once products have met the criteria for reduced-rate VAT, there is no policy incentive to make the products more efficient, providing no additional incentive for innovation creation.

While reduced VAT rate may provide little incentives for direct measures of innovation, such as new patents, the adoption of innovations and the associated increases in manufacturing may nevertheless in turn produce learning-by-doing effects, both simply from the scaling-up of production and the extending of energy-efficient innovations to new models. Yet, in some countries, the penetration of energy-efficient appliances already approaches 80% with only a standard rate, leaving little room for additional changes in consumption patterns and potentially creating a free-rider problem (Copenhagen Economics, 2008). For this reason in countries where such reduced rates exist, constant surveillance of the reduced rate criteria is also important, as technological advances can render impotent the reduced rate advantage.

Yet, the overall environmental impacts are not necessarily positive from these measures, as there are drawbacks that limit the effectiveness of this instrument. Such reductions subsidise consumers that would have purchased environmentally friendly goods regardless. By reducing the after-tax price of the product, the policy may also bring about additional consumption. Consumers may now spend the same money but purchase a larger appliance or may now be tempted to enter the market and purchase an additional appliance (such as buying a new fridge for the kitchen and moving the old one to the basement). Increasing energy efficiency could even make people less conscious of the energy draw of appliances. Although per unit energy consumption is better, this effect could actually increase absolute energy consumption.

Finally, the structure of value added taxes means that such features can only effectively be targeted to consumers and not to firms. Value added taxes are levied each time a good or service is transferred. Firms are also able to claim credits for all taxes that they pay to others. The tax incidence of VAT falls squarely on final consumption, when the full value added is taxed (and taxed only once because of the refunding of taxes paid at other stages). Therefore, reduced rates of VAT on inputs to production (such as more energy efficient dishwashers for a restaurant) will provide no additional incentive, since businesses do not pay VAT.[17]

4.4.5. *Tax measures to reduce the costs of innovation*

The tax system can also be used to provide stronger incentives for innovation which can hopefully lead to newer, lower-cost solutions to environmental challenges by seeking to reduce the costs of undertaking innovation. This can occur via three channels:

- First, governments can provide accelerated depreciation allowances for innovation capital, such as testing facilities and prototype capital. These measures seek to provide a benefit to firms to purchase depreciable capital by allowing depreciation at a more rapid rate. Generally, the issues surrounding this instrument are similar to those discussed in Section 4.4.3.

- Second, governments can focus on reducing the labour costs of innovation activities. This can occur through reducing employers' tax burden on labour, such as through enhanced allowances for R&D labour costs or reductions in payroll taxes or employers' social security contributions for innovation-related staff. These two general approaches can lower the after-tax costs of undertaking innovative activities (regardless of the outcome of that innovation) but can have differential effects on the factors of production used in the innovative activities (*i.e.* capital *versus* labour).

- Third, R&D tax credits can lower the after-tax costs of innovation from both capital and labour expenses by providing a tax credit for all eligible R&D-related expenses. While tax credits, tax allowances and rate reductions are different,[18] they all contribute to reducing the costs of undertaking innovation. Coupled with the fact that R&D tax credits are used significantly in OECD and are more general, the rest of this section will focus exclusively on this instrument.

R&D tax credits are growing in popularity among OECD governments, with 21 OECD countries having R&D tax credits in 2008, an increase from 18 in 2004. In some countries, these measures can provide significant additional financial incentives to undertake R&D activities, as evidenced by the implicit tax subsidy rates in Figure 4.3, since R&D tax credits are usually the main tax policy instrument to target innovation. Small and medium-sized enterprises (SMEs) face even higher tax subsidies in countries such as Canada and the Netherlands.

Figure 4.3. **Tax subsidy for R&D in OECD countries**

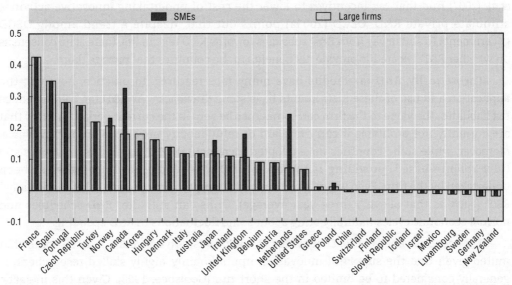

Notes: For year 2008. The tax subsidy is defined as one minus the B-index, the present value of before-tax income necessary to cover the initial cost of R&D investment and to pay corporate income tax, so that it is profitable to perform research activities. It should be noted that the B-index is but one measure of incentive towards R&D in a country and should be considered alongside other measures, such as the income tax rate, depreciation allowances, other tax credits and general R&D policy. For more information on the B-index, see Warda (2009).
1. The statistical data for Israel are supplied by and under the responsibility of the relevant Israeli authorities. The use of such data by the OECD is without prejudice to the status of the Golan Heights, East Jerusalem and Israeli settlements in the West Bank under the terms of international law.
Source: OECD (2009k).

StatLink ⟨≋ᴵˢᴾ⟩ http://dx.doi.org/10.1787/888932317483

For governments considering using R&D tax credits as an instrument of environmental policy, one of the most important issues to consider is that R&D tax credits only provide additional incentives to invent. These measures do not provide additional incentives to adopt or use the technology, as there is no economic incentive to abate non-priced emissions with an R&D tax credit policy alone. This has significant impacts on the environmental benefits of such new technology (low if diffusion is low) but also on the expected return to innovators (low if diffusion is low as well). An example to consider is carbon capture and storage. If R&D tax credits are the main policy instrument of governments, there would be few incentives to innovate in this area. Even if the innovation reduced the cost to near zero, a carbon-emitting firm would have little incentive to invest in or adopt this technology, since there is no rationale to do so without other environmental policy instruments in place. However, technologies that stimulate reduced carbon emissions through energy efficiency would be additionally stimulated by this measure because of pre-existing market prices for energy.

The empirical literature is sparse on the efficacy of R&D tax measures targeted at environmentally related outcomes in inducing additional R&D expenditures. General surveys of the literature find that R&D intensity has a long-term price elasticity around unity, although short-term responses are considerably less and variations can be large across countries (Bloom et al., 2002; Hall and van Reenen, 2000). Since the social rates of return are generally much higher than the private rates of return, this can have large impacts of the economy. Looking at R&D broadly, Guellec and van Pottelsberghe de la Potterie (2003) find that R&D tax incentives and direct funding bring about additional investments in R&D by the business community. Wu et al. (2007) find that tax incentives to lower the cost of undertaking innovative activities stimulate private R&D. OECD (2009a) points out that, despite a trend among OECD governments to shift away from R&D subsidies towards R&D tax credits, there is no consensus on whether R&D tax credits provide net benefits to the countries that provide them.

Theoretically, R&D tax credits have many nice features. In practice, however, the application and use of these instruments have some drawbacks. As R&D tax credits reduce tax liabilities, the value to the firm can sometimes be zero in the current period for firms that are currently not profitable if carry-forward or refundability provisions are not available. Moreover, while some countries have adopted incremental R&D tax credits, many are still fully volume-based, providing tax relief to R&D that would have nevertheless been undertaken by the firm. Even where governments attempt to only provide relief to additional R&D (such as that above a three-year average), firms can still alter R&D schedules and normal economic cycles naturally bring about some fluctuations in R&D spending.

The responsiveness of R&D activities to the imposition of credits may be somewhat muted given that the supply of innovative inputs (mainly highly skilled researchers) is generally considered to be limited in the short-run (Goolsbee, 1998). Given this inelastic supply, government incentives to increase R&D activity may serve more to increase the cost of R&D activities through wage increases than actually increase the quantity of R&D being undertaken.

R&D activities can be considered quite mobile across jurisdictions. The introduction of R&D tax credits may increase innovation activity in one jurisdiction but at the expense of R&D in another, less tax-favourable jurisdiction. Wilson (2007), for example, finds that R&D tax credits increase R&D activity strongly within a given US state but that the activity comes completely at the expense of R&D activities in other states. For jurisdictions concerned solely about domestic economic growth, whether the R&D is additional or relocated is of not great concern.[19] For environmentally related innovation, which has significant public benefits, however, it is the overall, global level that is of concern.

Akin to some of the issues with accelerated depreciation allowances and reduced rates of VAT, issues of instrument targeting are present. Governments must decide what acceptable measures to be included are and administration of the tax credit can be difficult for both governments and taxpayers. Firms also have an incentive to re-classify normal businesses expenses as R&D expenses. The issue of administration may be more pronounced for R&D tax credits targeted at specific outcomes. For such targeted R&D activities, such as those in the environmental realm, tax credits may spur the level of R&D in the environmental arena, but may come at the expense of R&D in other areas through crowding-out.

The government of Spain instituted two provisions within the corporate income tax system: a research and development and technological innovation (R&D&I) tax credit and a tax credit for eligible investments in environmental capital, as described in Box 4.7. Assessing the effectiveness of R&D tax credits is a difficult task, typically given the lack of

Box 4.7. **Case study: Environmentally related tax credits in Spain**

The corporate income tax system in Spain contains two provisions of note: i) a tax credit for eligible current and capital research and development and technological innovation (R&D&I) expenses (with a higher rate for investments above the previous two-year average); and ii) a tax credit for eligible environmental investments (EI) aimed at reducing air and water pollution, as well as industrial waste. The EI tax credit can be used for a wide range of investments, including those to meet regulatory requirements. An analysis was undertaken to look at the environmental impacts of the R&D&I tax credit and the innovation impacts of the EI tax credit.

First, in evaluating the environmental impacts of the R&D&I tax credit, it was found that the proportion of firms claiming an EI tax credit in the year (or two) after claiming an R&D&I tax credit was systematically greater than the proportion of firms claiming an R&D&I tax credit in the year after claiming an EI tax credit. This suggests that the R&D&I tax credit may have had a positive effect on environmental investments (and potentially environmental innovation, as firms seek to implement the fruits of their R&D efforts), compared to the reverse scenario, as seen in the table below.

EI tax credit after R&D&I tax credit

Use of R&D&I tax credit		Companies with EI tax credit			Million of euros claimed through EI tax credit		
2000	2003	2001 or 2002	2004 or 2005	% change	2001 or 2002	2004 or 2005	% change
Yes	No	192	136	−29.2	4.8	3.8	−20.3
No	Yes	338	395	16.9	18.6	26.7	43.7

StatLink ⟶ http://dx.doi.org/10.1787/888932318053

Second, in evaluating the innovation impacts of the EI tax credit, very little supportive evidence was found. There does not appear to be an increase in relevant patenting with the introduction of the tax credit. In addition, measures supporting cleaner production processes are generally more conducive to innovation than end-of-pipe technologies. Of measures supported by the EI tax credit, however, 68% were end-of-pipe technologies, with the remainder being cleaner production technologies. This proportion is significantly higher than Spanish firms' environmental investments overall, suggesting that the tax credit had a role in influencing the type of technology, but not necessarily its innovativeness.

Therefore, this study suggests that the R&D&I tax credit may encourage more environmental investments. Yet, there is little evidence that the EI tax credit encouraged innovative behaviour. By providing tax credits for actions that are needed to meet existing regulatory standards, not only are governments subsidising pollution abatement, but the scope for innovation is also limited compared to a tax credit focused on abatement activities beyond that required and therefore more driven by cost savings. The findings from this case study are somewhat limited, however, as the additional environmental innovation impacts of the R&D&I tax credit could not be assessed.

Source: OECD (2008).

a counterfactual. Looking at the environmental impacts of the R&D&I tax credit in this instance, the proportion of firms applying for an environmental investments tax credit after having applied for an R&D&I tax credit was significantly higher than the contrary, suggesting that the R&D&I tax credit may have led to environmental innovations that were subsequently implemented.

4.4.6. *Tax measures to increase the returns to innovation*

In addition to measures to help reduce the costs of undertaking innovation, governments can also use the tax system to help increase the gross after-tax returns to innovation (OECD, 2010). Like the measures outlined above, the provision of additional features that reduce government revenues may necessitate the raising of other taxes or tax rates that could increase distortions elsewhere in the economy.

One such feature to stimulate innovation through the corporate income tax system is to exempt from taxation (or provide reduced rates to) the returns to innovation, such as the royalty income streams from intellectual property (IP). Ireland does this by exempting patent income from the corporate taxation base. Along the same lines, countries can also reduce the tax burden on the transfer of IP by reducing the capital gains tax on IP sales, such as in France and Greece. These two approaches focus on IP that is created by one firm and then transferred to others.

Taking a broader approach, governments can increase the after-tax returns from innovation by providing reduced rates to all returns to innovation, including IP developed and used within the same firm. The Netherlands' Innovation Box (which has replaced the Patent Box programme in 2010) seeks to do just this. Returns to IP (for which a patent or a special R&D qualification has been granted) are subject to a corporate tax rate of 5%, instead of the statutory rate of 25.5% (at which some deductions can still be made). The returns can include royalty payments, capital gains on sale and internal revenues from the use of the innovation. Of course, a generally applied, lower corporate tax rate to all firms for all types of activities also increases the after-tax returns to innovation, along with all other components of the firm.

How these measures affect the innovation and adoption incentives will vary. The tax measures mentioned in this section are typically applied for all types of innovation. These measures are typically not specified by type of outcome and doing so may add significant complexity to existing administrative systems. Given also the few examples where such practices are in place and the fact that the effects of these measures on innovation and diffusion can be quite diverse (both because of the measures and how firms react), these measures are not further discussed in Section 4.5.

As alluded to at the beginning of this section, there may be structural features of countries' corporate tax systems that provide adverse incentives for some types of innovative activities (OECD, 2010). For example, in many countries, the costs of self-developed intellectual property can generally be immediately expensed and may benefit from additional tax benefits. The same IP purchased outside the firm is generally capitalised and depreciated over time, creating a differential between self-development and purchasing. Moreover, the mobile nature of IP may mean that tax planning activities can shield such income from domestic taxation. These fundamental issues of corporate tax systems are beyond the scope of this paper.

4.5. The choice of tax instrument

The choice of environmentally related tax instrument can have a large impact on the resulting innovation (and environmental) impacts, as seen from the preceding discussion. Different tax instruments provide differing levels of incentive for both innovation invention and adoption. The following structure will attempt to compare the innovation impacts of the various instruments. If one takes a firm that emits pollutants (and sells products that can pollute), its total emissions can be thought of as being composed of the following components:

Figure 4.4. **Determinants of emissions and scope for innovation**

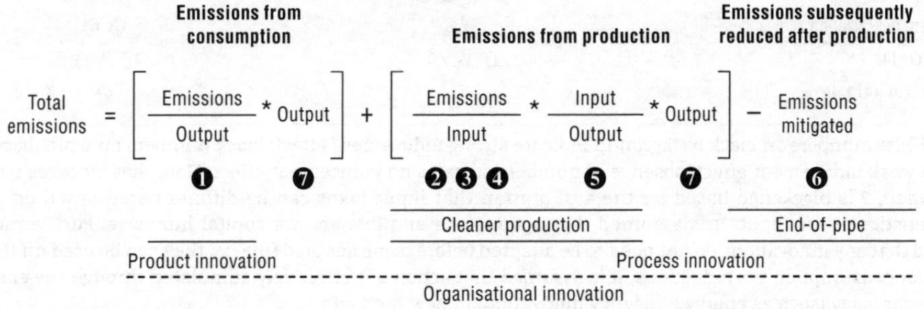

There are three factors that determine its direct and indirect emissions: how much its outputs pollute when used, how much the firm itself pollutes when making the outputs, and how much the firm does to negate its emissions from production after the pollution has already been created. Figure 4.4 also outlines, below the equation, the various types of innovations that can be used to reduce emissions for each component. The numbers represent specific actions that can be taken to reduce emissions and are outlined here:

❶ Create new products for consumers that generate fewer emissions when used. For example, firms could offer to consumers more energy-efficient appliances that reduce carbon emissions, or paints with a high solid content that release fewer VOCs into the atmosphere.

❷ Use less emission-intensive inputs (of the same type). For example, a power generation firm could switch from high-sulphur to low-sulphur coal.

❸ Use less emission-intensive inputs (of a different type). The same power generation firm could generate power from natural gas instead of coal, which will likely require more structural modifications to the existing capital stock.

❹ Reduce pollution intensity per unit of input (without modifying inputs). For example, the same plant could also optimise their equipment to reduce NO_x emissions per unit of fuel (which remains the same) but not impact the overall fuel usage per kWh. An example related to cars would be on-board diagnostic systems.

❺ Reduce input use per unit of output. For example, a power generation firm could make their overall plant more efficient for fuel use without affecting the amount of NO_x emissions per kWh by insulating to minimise heat loss. This occurs through reduced use of fuel per usable kWh, not reduced emissions per unit of fuel.

❻ Finally, undertake end-of-pipe/remedial measures. For example, an aluminium producer could reduce CO_2 emissions by using carbon capture and storage to prevent emissions from entering the atmosphere even though they have already been created.

❶-❻ Organisational innovation cannot be linked exclusively to one area in the equation above, as they typically affect the general orientation of the firm. As such, they tend to act as complements to other types of innovations within the firm.

❼ Of course, the firm (and the consumer) could simply produce (and consume) less.

Each of these alternatives is a way in which emission levels can be reduced in the economy. The choice of environmental policy instrument has a direct bearing on which actions are stimulated. Table 4.1 outlines each of the five main tax measures and the

Table 4.1. **Inducements for innovation by tax instrument**

	Invention propensity	Adoption propensity
Taxes on pollution	❶❷❸❹❺❻	❶❷❸❹❺❻
Taxes on proxies to pollution	❶❷④❺	❶❷④❺
Accelerated depreciation allowances	③⑤	❸❺
R&D tax credits	①③⑤	①③⑤
Reductions in VAT rate	①	❶

Notes: White numbers on black background indicate strong inducement effect; black numbers on white background indicate weak inducement effect; absence of number indicates no inducement effect. Note that for taxes on proxies to pollution, 2 is blackened based on the assumption that input taxes can be differentiated based on physical characteristics of the input. It is assumed that actions one and two are not capital intensive. Furthermore, it is assumed that any innovations do not need to be adapted before being adopted (that is, they can be used off the shelf). Finally, for consumption tax reductions, it is assumed that action 5 is effectively stimulated through the embedding of this technology (such as energy efficiency innovation) in new products.

strength of the innovation creation and adoption incentive that they have on each emission reduction possibility. The table assumes that each instrument is implemented in isolation by governments.

From the preceding table, it is clear that some instruments encourage a wider range of actions (and therefore provide greater incentives for innovation) than others. Taxes on pollution provide incentives for all six of the potential abatement measures, as levying the tax directly on the pollutant does not exclude any potential abatement measure and provides the greatest range of incentives for invention and technological change. As the incidence of the tax moves further from the actual pollutant, the range of potential measures for abatement decreases. Taxes on proxies to pollution provide much the same incentives, except where the abatement actions become disconnected from input use. Thus, taxes on proxies to pollution have no impact on actions four and six, in line with findings that taxes on pollution encourage relatively greater end-of-pipe abatement than taxes on proxies to pollution.

Accelerated depreciation allowances encourage greater investment in physical capital. Such an instrument does not affect mitigation measures that are generally not capital intensive, such as actions one, two and four. Even for capital-intensive measures, an accelerated depreciation allowance as the sole policy instrument provides no incentive for abatement unless it is through the greater rationalisation of other inputs (such as fuel) which have a positive price in the market. For this reason, action six is not stimulated by this instrument.

Similarly, generally available or environmentally targeted R&D tax credits alone cannot provide incentives for mitigation, unless these help reduce the cost of existing processes or create new products (without a price on carbon, R&D that significantly reduces the cost of carbon capture and storage, for example, would still have no economic rationale to be adopted). As such, only actions one, three and five are stimulated for invention and adoption. Assuming that the invention can be used off the shelf (that is, no adaptive R&D expenditures are required between firms), the R&D tax credit provides no additional incentive for adopting the innovation once it has been created, unless the R&D addresses the use of something that has a pre-existing market price.

Finally, reductions in value added taxes for "green" purchases provide direct incentives for consumers to adopt new innovations, as they lead to a direct and identifiable price reduction *versus* non-reduced goods and services. The incentives for firms to invest in innovation activities are less strong, as the firm receives no direct benefit from the consumption tax reduction (although it will benefit from increased demand and can increase its prices somewhat) and these measures are frequently temporary. Moreover, such rate reductions are usually based on a standard, such as the EnergyStar programme, and once those standards have been met, there is no incentive to make further investments.

4.6. Creating a policy package: Combinations of environmental and innovation instruments

Before governments decide how to act in this area, it is important to consider pre-existing features of a jurisdiction's tax and innovation systems. Such features can already be exerting a powerful influence on the environmental and innovation objectives that are desired. From the environmental perspective, for example, ongoing tax preferences to fossil fuel production or the under-pricing of resource royalties can have powerful influences that can undermine concurrent environmental efforts. From the innovation perspective, on the other hand, a tax system that has restrictive rules on tax losses or the shifting of tax credits across time periods can potentially discourage very risky (and therefore potentially very rewarding) innovation in the environmental arena. It is critical, therefore, that these features are well understood before governments attempt further actions to address additional issues.

Once having taken account of various pre-existing constructs, governments must then decide how to move forward. A number of studies have demonstrated that market-based instruments are vastly superior to command-and-control approaches (Downing and White, 1986; Milliman and Prince, 1989). The theory suggests that the more flexible and less prescriptive the instrument, the more room there is for firms to find the lowest-cost abatement.[20]

More recent work has sought to empirically rank the various instruments available to governments in terms of economic efficiency and innovation inducement. Using the specific example of the US electricity sector and climate change mitigation, Fischer and Newell (2008) present a ranking of instruments that are best able to achieve both criteria: 1) emissions tax/charge; 2) emissions performance standard; 3) fossil fuel tax; 4) share requirement for renewable energy; 5) subsidy for renewable energy; and 6) subsidy for research and development.

In practice, many countries' environmental policies consist of a wide number of different tools; the OECD's *Instrument Mixes for Environmental Policy* (2007) highlights some examples. Whether multiple instruments are needed and how these tools interact can play a critical role in evaluating the policy's overall innovative and environmental performance. However, in only a few cases would one instrument by itself achieve an optimal level of emissions reduction most efficiently. The complexity of markets and the multiple policy issues suggest that a mix of instruments, co-ordinated to work together to address each one's gaps, may be the best means of achieving overall economic compliance and would achieve the social optimum at significantly lower cost. For example, an information campaign about energy efficiency in appliances could reinforce a carbon tax. Regulations could help address pollution issues where the damage is not geographically uniform. However, where mixes do occur, they should provide the greatest flexibility to achieve the intended outcome while minimising the overlap of similar instruments (such as a carbon tax and a tradable permit system over the same activities)[21] (OECD, 2007).

Drawing all of this information together, the question remains: what should governments do to ensure that environmental challenges are addressed at lowest cost?

One option is for governments to solely use taxes on environmentally harmful activities. Environmentally related taxation corrects the negative externality and addresses the oversupply of pollution. While there are some implementation, political economy and other issues that may suggest that taxes cannot fully achieve the intended outcome alone, they do have strong impacts on environmental effectiveness at low cost. Because they price pollution, they also stimulate innovation into new means to reduce environmentally harmful activities if they are well designed. These forces only raise the return on the creation and adoption of innovation to a level that is consistent with the incentives on other market goods; they do not overcome, or even directly address, the specific issues facing the innovation externalities. The innovation market failure remains.

A second option is the utilisation of other tax-based instruments alone, such as accelerated depreciation allowances or reduced rates of VAT. These measures provide incentives for the adoption of existing innovations where the incentives may not get passed along for the development of new innovations, since these are typically short-term policy measures. However, these tools typically pick only certain types of abatement activities to stimulate, leaving significant areas of environmental damage unaffected. They face many of the same issues for policy makers as the design of subsidy as well: encouraging consumption, violating the polluter-pays principle and requiring more government intervention into which types of abatement are stimulated. Because of this, governments should be cautious in using tax concessions as a tool of environmental policy.

A third option is for governments to solely use the tax system to promote innovation as a means to address environmental challenges. In doing so, the intent is to encourage the development of technologies that can then be widely adopted. R&D tax credits reduce the after-tax cost of undertaking these activities and therefore can expand the range of innovations and technologies available, as well as reducing their cost. The drawback with using an innovation-only policy is that for many environmental innovations, R&D tax policies provide little to no incentive to adopt the resulting innovations, since there is no cost to undertaking the environmentally harmful activity in the first place. Moreover, environmental policy focused solely on R&D instruments will forego early and low-cost abatement activities in favour of technologies on the horizon. This delay adds significantly to the cost of achieving a set environmental outcome as much more expensive abatement is needed in the future (Duval, 2008).

In addition, the degree of public *versus* private benefit of the innovation must also be considered. The greater are the private benefits (where the private rates of return are closer to the social rates of return) of the potential innovation, the less vast are the positive externalities. Innovators are better able to reap the returns, and the spillovers to society are less. This suggests that there is a stronger role for governments in addressing innovation priorities that have high social, but low private, returns. Basic innovation is consistently flawed by long time horizons, significant uncertainty and less tangible end results. These features create greater unwillingness of private industry to engage, even though they can be highly useful and are the main driver behind breakthrough technologies (see Section 3.3 on the benefits of breakthrough technologies on the costs of environmental actions). Popp (2006a) finds that patents derived from basic government R&D are cited more frequently in patent filings than private patents, and that the offspring of these government patents are subsequently cited 30% more, indicating the importance of basic R&D in filling an important knowledge gap as well as facilitating knowledge transfer. R&D tax credits may not produce enough of the basic R&D needed for optimal environmental policy.

To underline these points, Popp (2006b) finds that relying solely on taxes or R&D efforts to address the climate change issue will not be sufficient. Subsidies to R&D were more effective in inducing R&D expenditures than the introduction of a carbon tax set at the socially optimal rate. However, for the reasons mentioned above, the R&D subsidies alone do not bring about significant environmental benefits from the business-as-usual scenario; emissions (and the resulting atmospheric temperature) continue to rise significantly. Combining an R&D subsidy with an optimal carbon tax provides for somewhat larger R&D expenditures than R&D subsidies alone and significantly more than an optimal carbon tax alone, as outlined in Table 4.2.

Table 4.2. **Welfare effects of taxes and R&D subsidies**

	% increase in R&D spending in energy efficiency	% increase in R&D spending in backstop technology	% of maximum welfare gain
Optimal tax and R&D subsidies	13.7	24.7	100
Optimal tax only	1.2	7.6	95
Optimal R&D subsidies only	10.3	13.5	11

Notes: For first two columns, increase in R&D spending is projected for 2025 compared against a business-as-usual scenario. Here, backstop technology refers to carbon-free energy sources.
Source: Popp (2006b).

StatLink ⟨⟩ http://dx.doi.org/10.1787/888932318091

Yet, those same instruments that bring about large increases in R&D expenditures do not necessarily translate into similarly large welfare gains for the economy. In the model, the imposition of an R&D-only policy would have little relative effect on an economy's welfare compared to a carbon tax, which itself is nearly identical to the combination of an optimal carbon tax and R&D subsidy policy in combination, consistent with Popp (2006c). In addition, modelling undertaken by the OECD suggests that an R&D-only policy towards climate change would not – at any plausible level – be able to stabilise the concentration of CO_2 in the atmosphere (OECD, 2009c).

Given that environmental and innovation policies utilised alone are unlikely to provide an optimal outcome for society, some combination of instruments will be needed. At the simplest level, the presence of two separate, but related, market failures surrounding innovation in the environmental arena suggests instruments that target each. To target the environmental externality, taxes levied directly on the pollutant can be quite effective. Other environmental instruments may be needed where taxes are ineffective, but care is required to ensure that overlaps and differing incentives do not undermine the overall policy effectiveness.

The issue now turns to the optimal policies for innovation with the presence of taxes. Many countries have general innovation policies that are used to support higher levels of innovative activities and overcome the unique issues facing innovation development. In addition to legal protections, R&D tax credits, R&D subsidies and broader investments in education and infrastructure are used. This publication is not able to delve into whether countries' overall innovation policies are adequate and correctly address all the various issues retarding the optimal level of innovation. However, if environmental innovation is similar to other innovation and countries have adequate measures in place, then the combination of taxes and general innovation policies should properly address the two externalities. No additional government intervention may be needed.

However, the important question remains whether environmental innovation is similar to other types of innovation. If not, some differential governmental action may be required. Jaffe *et al.* (1995) suggest that the environment poses unique challenges to innovation and that additional (and directed) action beyond general innovation policies are required. If the social costs of environmental damage are low in the current period but expected to rise significantly, this could suggest that starting R&D now (ahead of what the market would provide even with a standard innovation toolset and environmentally related taxes) is important. Given its dynamic nature, innovating now and changing the innovation path can lead to lower costs in the future.

In addition, innovation brought about by taxes and R&D tax credits is typically incremental in nature. For some environmental problems, such as particulate matter in urban areas, the effects are manageable (though obviously not desirable) and tipping points are not apparent (those points at which the problem becomes irreversible). Incremental innovation provides an adequate approach to addressing these issues. In other cases, such as climate change, there are significant and identifiable targets that have to be met to avoid large-scale environmental problems. Over the course of the next 40 to 50 years, this entails a significant decarbonisation of economies. To avoid significantly high costs, breakthrough technologies (such as large-scale carbon-free energy sources) are necessary. Incremental innovation (such as better energy efficiency) will simply not be enough to achieve the environmental goals at the lowest cost to world growth. The incentives provided by taxes and R&D credits encourage such incremental innovation but they may not be enough to overcome all the barriers to an optimal innovation output: financing constraints, uncertainty, significant appropriability of basic research, very long timelines and others.

Since some of the features of environmental innovation may be different from other types of innovation, the reliance on environmentally related taxes and general R&D tax credits (and other general features of innovation policy) may not be enough. Providing additional R&D tax credits targeted towards specific types of environmental outcomes would be unlikely to address the fact that such measures may not stimulate basic innovation into breakthrough technologies. For this reason, targeted efforts towards innovation in key areas, such as through R&D grants, may entail greater administrative costs but can focus R&D efforts to needy areas, providing direct support to worthy projects. These projects should be at the basic level of research and encourage actions that other actors in the economy are unlikely to undertake. By targeting fundamental but use-inspired innovation, the differential needs of environmental innovation can be addressed and which can lead to further innovations at a more useable level for emissions reductions. Such actions will most likely be outside the tax system and, therefore, beyond the scope of the publication.

Undertaking more directed approaches to basic, environmentally related innovation is not entirely the solution. One issue is that innovation is not so clearly siloed that governments can simply direct resources at a different level of the silo to target a different stage of innovation in the same field. Innovation is network-based, dynamic and draws upon innovations in a range of other fields. Especially at the basic level, it may be difficult to assess what is environmentally related. Moreover, other "instruments" are also vitally important to a vibrant culture of innovation: a robust patent system, solid legal framework, efficient tax administration, and a society that promotes and is receptive to innovations, among many others.

4.7. Conclusions

Merely putting in place environmentally related taxation does not guarantee success; what the tax looks like and how it is implemented crucially affect its outcome. One of the least surprising outcomes is that the higher is the level of the tax, the greater are the incentives for innovation. In the United Kingdom, lower rates of energy taxes were correlated with lower rates of innovation. Increasing the extent of the tax also increases the incentives for innovation as the potential market for innovation (selling to other firms/consumers) expands significantly. This is particularly true for third-party innovators. Even the announcement of a tax can have impacts on innovation.

The predictability of the rate and the overall credibility of the policy play an important role. Without these features, investments with long-term payoffs (such as new capital or into R&D) are much less likely to be made. Japan's charge on SO_x emissions provides an enlightening example of the effect of the scheme's uncertainty on firms' innovation efforts over the long run. Political economy considerations can also have strong impacts on innovation, especially those targeting the potential effects on pollution-intensive firms. Revenue recycling is a tool used in a number of circumstances to maintain the marginal abatement/innovation incentive while minimising the actual effect on firms' profitability and competitiveness *vis-à-vis* less polluting industries. Such mechanisms can have small negative effects on innovation for the firm but large negative incentives for innovation at the collective level. Japan's charge, although not consisting of revenue recycling, incorporated a similar feature.

Environmentally related taxes levied directly on the pollutant are not the only types of environmental policy tools within the tax system. A common feature is to shift the tax base to a proxy to the pollution, such as motor fuel or motor vehicles, as monitoring and administration are much easier. Such taxes on proxies to pollution reduce the scope of potential innovations that are stimulated, leading to a less efficient outcome.

Three other instruments try to encourage the good instead of trying to discourage the bad. Accelerated depreciation allowances and reduced rates of VAT for environmentally related goods provide incentives for the adoption of existing technologies but their translated effect onto innovation creation is much more limited. They also share many of the drawbacks of subsidies: lack of targeting and subsiding actions that would have taken place regardless. R&D tax credits, on the other hand, stimulate innovation creation but do little to stimulate innovation adoption. Without a positive price on the environment, the innovations stimulated by the R&D tax credit would have no positive financial impact on polluters (except where they also help reduce costs of actions that already have a positive price). These other tax instruments may play a role in the green innovation story but have additional drawbacks to taxes directly levelled on pollution. Governments should be cautious in their use and fully evaluate their expected environmental benefits against the loss in tax revenues.

Creating a policy package that adequately addresses all environmental issues is beyond the scope of this publication but analysing the current tax and innovation constructs for potential features that could undermine new measures is critical. This is especially true for features of the tax system. Looking more specifically at the focused areas of taxation and innovation, some general lessons can be drawn when implementing new policies. Environmentally related taxation in most cases can well address the negative externality of too much pollution. Similarly, broad-based innovation policies that attend to the positive externalities of innovation (such as R&D tax credits) will generally also do well for the environment. The innovations needed for achieve environmental goals have an additional component that may differ from other types of innovation. The need for breakthrough technologies that come from advances in fundamental research may be more pronounced in the environmental realm; therefore, additional targeting of innovation policy (such as through R&D grants) to basic environmental innovation priorities may be warranted.

Notes

1. It should be noted that the rebound effect resulting from regulatory measures are likely much greater than from taxes. With regulatory measures on vehicle fuel efficiency, for example, reduced fuel costs can lead to additional driving. With taxation, any rebound effect due to tax-induced fuel efficiency is tempered by the fact that additional driving faces taxation.

2. For example, legislating that one pollution permit is no longer equivalent to one tonne of carbon dioxide emitted, but 0.9 tonne.

3. The statistical data for Israel are supplied by and under the responsibility of the relevant Israeli authorities. The use of such data by the OECD is without prejudice to the status of the Golan Heights, East Jerusalem and Israeli settlements in the West Bank under the terms of international law.

4. It should be noted that in the empirical assessment of the Swedish charge, the potential impact of the refunding mechanism on innovation was could not be tested given the lack of a counterfactual.

5. While firms subject to the reduced rate had to commit to actions and/or targets in environmental areas, such as energy efficiency, it is generally believed that these constraints were not particularly onerous on the impacted firms.

6. The individual firm would likely be worse off, since its tax liability would be the same but it has also contributed to the collective costs of R&D expenditures.

7. In the European Union, however, all major firms are obliged to measure many sorts of emissions, including NO_x, on a continuous basis.

8. It should be noted that one-off motor vehicle taxes are usually combined with recurring taxes on vehicles, motor vehicle fuel taxes, and other charges, such as tools and road pricing schemes.

9. Higher one-off taxes on new vehicles will also likely increase the resale value of used cars, thereby limiting this impact.

10. For a full description of motor vehicle taxes related to CO_2 emissions in OECD, see OECD (2009d).

11. It should be noted that the bands within the French bonus-malus scheme are intended to be gradually shifted downwards, thereby providing increasing incentives for fuel efficiency innovations.

12. The economic rate of depreciation refers the change in the present value of an asset over a set time period. The change in value can occur either because of physical depreciation or a change in the market value of the asset.

13. Immediate expensing is simply an extreme form of accelerated depreciation.

14. It should be noted the under most existing tax systems, expenses such as for marketing and for employee training are immediately deductible even though they create profits in the future as well.

15. Many of these also exist alongside rate reductions where there are clear negative environmental outcomes (*e.g.* domestic energy, meat).

16. Some theory suggests that optimal VAT taxation should place higher rates on more price inelastic goods (so-called Ramsey taxation) but the administrative issues render this highly unwieldy. For a fuller discussion, see Heady (1993).

17. Exemptions in the VAT structure provide exceptions to this logic. Zero rates of VAT, however, are just an extreme form of reduced VAT rates.

18. Tax credits differ from tax allowances or accelerated depreciation allowances in that tax credits are generally independent of the nominal corporate tax rate. Tax credits provide a deduction from taxes payable while tax or depreciation allowances provide a deduction from net income for tax purposes. Tax credits can also be made to be refundable.

19. The zero-sum effects of tax competition among jurisdictions are likely of concern, however.

20. Others, such as Bauman *et al.* (2008), suggest that market-based instruments may not always be better than command-and-control instruments in spurring innovation. In cases where a production process innovation leads to an increase in the marginal benefit of emitting another unit of pollution, emissions could in fact rise. This rise in emissions would have different effects: with a command-and-control tariff, there would be no effect, whereas an emissions tax would tax each additional unit of pollution. Therefore, it is ambiguous in cases such as this whether an emissions tax is always better than command-and-control instruments at spurring innovation.

21. With an emissions cap, taxes levied on exactly the same emissions will not induce greater abatement, as long as the cap is fixed. The effect of the tax will simply be to cause a drop in the price of the tradable permits.

References

Acemoglu, Daron *et al.* (2009), "The Environment and Directed Technical Change", *NBER Working Paper*, No. #15451, October 2009, available at *www.nber.org/papers/w15451*.

Amir, Rabah, Marc Germain and Vincent van Steenberghe (2008), "On the Impact of Innovation on the Marginal Abatement Cost Curve", *Journal of Public Economic Theory*, Vol. 10(6), pp. 985-1010.

Baker, Erin, Leon Clarke and Ekundayo Shittu (2008), "Technical Change and the Marginal Cost of Abatement", *Energy Economics*, No. 30, pp. 2799-2816.

Baker, Erin and Ekundayo Shittu (2006), "Profit-maximizing R&D in Response to a Random Carbon Tax", *Resource and Energy Economics*, No. 28, pp. 160-180.

Barradale, Merrill Jones (2008), "Impact of Policy Uncertainty on Renewable Energy Investment: Wind Power and PTC", United States Association for Energy Economists WP 08-003, available at *http://ssrn.com/abstract=1085063*.

Bauman, Yoram, Myunghum Lee and Karl Seeley (2008), "Does Technological Innovation Really Reduce Marginal Abatement Costs? Some Theory, Algebraic Evidence, and Policy Implications", *Environmental and Resource Economics*, Vol. 40, pp. 507-527.

Bloom, N., R. Griffith and J. van Reenen (2002), "Do R&D Credits Work? Evidence from a Panel of Countries, 1979-97", *Journal of Public Economics*, No. 85, pp. 1-31.

Canada, Department of Finance (2010), *Budget 2010: Leading the Way on Jobs and Growth*, Department of Finance, Ottawa, available at *www.budget.gc.ca/2010/pdf/budget-planbudgetaire-eng.pdf*.

Copenhagen Economics (2008), *Reduced VAT for Environmentally Friendly Products*, for DG TAXUD, European Commission, available at *http://databankmilieu.nl/pdf/pr-31492k17b.pdf*.

Dixit, A. and R. Pindyck (1994), *Investment under Uncertainty*, Princeton University Press: Princeton.

Downing, P.G. and L.J. White (1986), "Innovation in Pollution Control", *Journal of Environmental Economics and Management*, No. 13, pp. 18–29.

Duval, Romain (2008), "A Taxonomy of Instruments to Reduce Greenhouse Gas Emissions and Their Interactions", *OECD Economics Department Working Paper*, No. #636, OECD, Paris.

European Commission (2008), "The Use of Differential VAT Rates to Promote Changes in Consumption and Innovation", 25 June 2008, available at *http://ec.europa.eu/environment/enveco/taxation/pdf/vat_final.pdf*.

European Commission (2009), "Environmental Compliance Assistance Programs for SMEs – VAMIL and MIA", available at *http://ec.europa.eu/environment/sme/cases/cases06_en.htm*.

Fischer, Carolyn and Richard G. Newell (2008), "Environmental and Technology Policies for Climate Mitigation", *Journal of Environmental Economics and Management*, No. 55, pp. 142-162.

Goolsbee, Austan (1998), "Does Government R&D Policy Mainly Benefit Scientists and Engineers?", *American Economic Review*, Vol. 88(2), pp. 298-302.

Goulder, Lawrence H. (1995), "Effects of Carbon Taxes in an Economy with Prior Tax Distortions: An Intertemporal General Equilibrium Analysis", *Journal of Environmental Economics and Management*, No. 29, pp. 271-297.

Guellec, Dominique and Bruno van Pottelsberghe de la Potterie (2003), "The Impact of Public R&D Expenditure on Business R&D", *Economics of Innovation and New Technology*, Vol. 12(3), pp. 225-243.

Hall, B. and J. van Reenen (2000), "How Effective are Fiscal Incentives for R&D? A Review of the Evidence", *Research Policy*, No. 29, pp. 449-469.

Heady, Chris (1993), "Optimal Taxation as a Guide to Tax Policy: A Survey", *Fiscal Studies*, Vol. 14(1), pp. 15-41.

House, Christopher L. and Matthew D. Shapiro (2008), "Temporary Investment Tax Incentives: Theory with Evidence from Bonus Depreciation", *American Economic Review*, Vol. 98(3), pp. 737-768.

Johnstone, Nick *et al.* (2001), "The Environmental Consequences of Tax Differentiation by Vehicle Age in Costa Rica", *Journal of Environmental Planning and Management*, Vol. 44(6), pp. 803-814.

KPMG (2007), *Taxes and Incentives for Renewable Energy*, available at *www.kpmg.com/Global/en/ IssuesAndInsights/ArticlesPublications/Documents/Taxes-Incentives-Renewable-Energy.pdf*.

Mason, Yael (2009), Israeli Ministry of Environmental Protection, presentation to the OECD Global Forum on Environment on Eco-Innovation, 4-5 November 2009, Paris, France.

Milliman, Scott R. and Raymond Prince (1989), "Firm Incentives to Promote Technological Change in Pollution Control", *Journal of Environmental Economics and Management*, No. 17, pp. 247-265.

OECD (2006), *Political Economy of Environmentally Related Taxes*, OECD, Paris, *http://dx.doi.org/10.1787/ 9789264025530-en*.

OECD (2007), *Instrument Mixes for Environmental Policy*, OECD, Paris, *http://dx.doi.org/10.1787/ 9789264018419-en*.

OECD (2008), "Taxation, Innovation, and the Environment – The Spanish Case", OECD, Paris, available at *www.olis.oecd.org/olis/2008doc.nsf/linkto/com-env-epoc-ctpa-cfa(2008)38-final*.

OECD (2009a), "OECD Work on Innovation – A Stocktaking of Existing Work", *STI Working Paper*, No. 2009/2, OECD, Paris.

OECD (2009b), "Environmental Policy Framework Conditions, Innovation and Technology Transfer", *Working Paper*, ENV/EPOC/WPNEP(2009)2, OECD, Paris.

OECD (2009c), *The Economics of Climate Change Mitigation: Policies and Options for Global Action Beyond 2012*, OECD, Paris, *http://dx.doi.org/10.1787/9789264073616-en*.

OECD (2009d), *The Scope for CO_2-based Differentiation in Motor Vehicle Taxes*, OECD, Paris.

OECD (2009e), *Innovation Effects of the Swedish NO_x Charge*, OECD, Paris, available at *www.olis.oecd.org/ olis/2009doc.nsf/linkto/com-env-epoc-ctpa-cfa(2009)8-final*.

OECD (2009f), *Econometric Analysis of the Impacts of the UK Climate Change Levy and Climate Change Agreements on Firms' Fuel Use and Innovation Activity*, OECD, Paris, available at *www.olis.oecd.org/olis/ 2008doc.nsf/linkto/com-env-epoc-ctpa-cfa(2008)33-final*.

OECD (2009g), *Survey of Firms' Responses to Public Incentives for Energy Innovation, including the UK Climate Change Levy and Climate Change Agreements*, OECD, Paris, available at *www.olis.oecd.org/olis/2008doc.nsf/ linkto/com-env-epoc-ctpa-cfa(2008)34-final*.

OECD (2009h), *The Impacts of the SO_x Charge and Related Policy Instruments on Technological Innovation in Japan*, OECD, Paris, available at *www.olis.oecd.org/olis/2009doc.nsf/linkto/com-env-epoc-ctpa-cfa(2009)38-final*.

OECD (2009i), *The Influence of Regulation and Economic Policy in the Water Sector on the Level of Technology Innovation in the Sector and its Contribution to the Environment: The Case of the State of Israel*, OECD, Paris, available at *www.olis.oecd.org/olis/2008doc.nsf/linkto/com-env-epoc-ctpa-cfa-rd(2008)36-final*.

OECD (2009j), "Base Broadening and Targeted Tax Provisions", *Working Paper for Working Party*, No. 2 on Tax Policy Analysis and Tax Statistics, CTPA/CFA/WP2(2009)23, OECD, Paris.

OECD (2009k), *OECD Science, Technology and Industry Scoreboard 2009*, OECD, Paris, *http://dx.doi.org/10.1787/ sti_scoreboard-2009-en*.

OECD (2010), *The OECD Innovation Strategy: Getting a Head Start on Tomorrow*, OECD, Paris, *http://dx.doi.org/ 10.1787/9789264083479-en*.

Parry, Ian W.H. (2005), "Optimal Pollution Taxes and Endogenous Technological Change", *Resource and Energy Economics*, No. 17, pp. 69-85.

Pearce, David (2006), "The Political Economy of an Energy Tax: The United Kingdom's Climate Change Levy", *Energy Economics*, No. 28(2), pp. 149-158.

Pindyck, Robert S. (2007), "Uncertainty in Environmental Economics", *Review of Environmental Economics and Policy*, Vol. 1(1), pp. 45-65.

Popp, David (2006a), "They Don't Invent Them like They Used to: An Examination of Energy Patent Citations over Time", *Economics of Innovation and New Technology*, Vol. 15(8), pp 753-776.

Popp, David (2006b), "R&D Subsidies and Climate Policy: Is There a 'Free Lunch?'", *Climatic Change*, No. 77, pp. 311-341.

Popp, David (2006c), "Comparison of Climate Policies in the ENTICE-BR Model", *The Energy Journal*, Special Issue, No. #1, pp. 11-22.

Reedman, Luke, Paul Graham and Peter Coombes (2006), "Using a Real-Options Approach to Model Technology Adoption under Carbon Price Uncertainty: An Application to the Australian Electricity Generation Sector", *The Economic Record*, Vol. 82(S1), pp. S64-S73.

Rosendahl, Knut Einar (2004), "Cost-effective Environmental Policy: Implications of induced technological change", *Journal of Environmental Economics and Management*, No. 48, pp. 1099-1121.

Sansing, Richard C. and Todd Strauss (1998), "How Tax Policy Can Thwart Regulatory Reform: The Case of Sulfur Dioxide Emissions Allowances", *Journal of the American Tax Association*, Vol. 20(1), pp. 49-59.

United States, Internal Revenue Service (2010), "How to Depreciate Property", Publication 946, Cat. No. 13081F for tax year 2009, available at *www.irs.gov/pub/irs-pdf/p946.pdf*.

Warda, J. (2009), "An Update of R&D Tax Treatment in OECD Countries and Selected Emerging Economies, 2008-2009", mimeo.

Weitzman, Martin L. (1974), "Prices *vs.* Quantities", *The Review of Economic Studies*, Vol. 41(4), pp. 477-491.

Wilson, Daniel J. (2009), "Beggar thy Neighbor? The In-State, Out-of-State, and Aggregate Effects of R&D Tax Credits", *The Review of Economics and Statistics*, Vol. 91(2), pp. 431-436.

Wu, Yonghong, David Popp and S. Bretschneider (2007), "The Effects of Innovation Policies on Business R&D: A Cross-National Empirical Study", *Economics of Innovation and New Technology*, Vol. 16(4), pp. 237-253.

Yale, Ethan (2008), "Taxing Cap-and-Trade Environmental Regulation", *Journal of Legal Studies*, No. 37, pp. 535-550.

Chapter 5

A Guide to Environmentally Related Taxation for Policy Makers

This chapter provides a broad overview to policy makers about the considerations surrounding environmentally related taxation. Taxes are assessed against other potential policy instruments before turning to fundamental tax design considerations. The chapter also delves into the use of revenues derived from such taxes and the political economy considerations present during implementation. It finishes with a discussion about why taxes should be central to countries' environmental policy approaches but that taxes alone may not be enough to fully and cost-effectively address the environmental challenges.

Environmental challenges are placing increased pressures on governments to find mitigation measures that address the environmental damage being done in a method that does not cause significant harm to current and future economic growth. Governments have a range of tools at their disposal: regulations, policies to encourage environmental innovation, subsidies to pollution abatement, and environmentally related taxation. Implementing the right policies at the right time is crucial.

Of late, significant attention has been focused on taxes (and similarly on tradable permits).[1] Taxes have many positive features, such as economic efficiency, environmental performance, raising revenues for governments and being transparent. To gain all these effects, however, design considerations are important and the political economy aspects have to be considered. Environmentally related taxation can be, and has been, used in a wide range of circumstances: landfilling or incineration of waste, local and global air pollutants, discharges to water and many others.

If this chapter can be boiled down to four key pieces of advice in using environmentally related taxation, the guidance would be:

- while not the only instrument, taxes should be strongly considered as a significant component of environmental policy;

- tax bases should be as broad as possible, providing few (if any) exceptions;

- governments should not be afraid to impose tax rates that will in fact achieve the environmental objective, especially when the tax falls directly on the pollutant; and

- competiveness and distributional concerns of environmentally related taxes are important, but should be administered outside of the tax itself whenever possible.

5.1. Why taxes?

In an unregulated economy, environmental damage occurs because there is no market incentive for firms and households to not harm the environment – polluting entails no direct cost. The sheer complexity of environmental problems means that the victims of pollution (both in the present and in the future) are not able to work together to force polluters to pay.[2] This leaves most environmental problems unresolved, opening a door for government involvement.

The range of policy instruments available to governments is great. In the past, environmental policy was typically dominated by so-called "command-and-control" arrangements. These regulation-based instruments (standards, bans, etc.) are generally prescriptive and can be tightly targeted (for example, emissions limits and regulations on using specific technologies for specific industries). Over the past number of decades, there has been growing interest, especially among OECD countries, in using market-based instruments (such as taxes and tradable permits). Both of these approaches are typically used in combination with other environmental policies, such as information campaigns to help shift consumption patterns (*e.g.* the European Union's labelling scheme for passenger vehicles) and research and development (R&D) policies to spur new environmental innovations.

So why have taxes started to take a central role in countries' environmental policies? First, taxes on actual emissions (which are actually quite rare in practice) provide the greatest range of abatement options to polluters. Instead of focusing only on one type of pollution determinant (such as mandating a cleaner type of fuel), emission taxes open up a range of abatement options. Any type of abatement that would be stimulated with a command-and-control instrument would also be stimulated with a tax, along with every other action that can reduce emissions (energy efficiency, reduced output, cleaner production processes). This expansion in the range of abatement options can lead to firms and citizens searching out the lowest-cost options.

Second, well-designed taxes do not discriminate between sources of pollution. Many command-and-control approaches: i) are sector-specific (power generation, transportation, agriculture, etc.); or ii) apply the same standards on each firm (*e.g.* reduction of 90% of emissions from smokestacks). Because the market identifies the best abatement options, the information needed by governments for taxes is much lower than undertaking a sector-by-sector approach. In this regard, environmentally related taxation has two main economic benefits over most other potential instruments of environmental policy. First, taxation creates static efficiency – that is, the lowest individual cost abatement measures are undertaken first – meaning that a given environmental target is reached at the lowest possible cost for society. Second, taxation creates dynamic efficiency – the incentive to abate is present for all levels of emissions, even after significant abatement may have already occurred. This is contrasted with regulatory emission limits, for example, where, once the regulatory threshold has been met, there is no further incentive to abate.

Third, properly designed taxes can be highly transparent in that it is clear on what goods the taxes apply and which polluters (if applicable) are exempted. Critical to this is the tax base – that the base should be as close as possible to the actual pollutant (or a close proxy thereof). While some regulatory-based approaches come close to the same base as a tax, many do not. Moreover, non-tax based instruments may not clearly demonstrate the cost on each polluter, since the effect of the policy on an individual polluter is not as discernable or as comparable across polluters. The discretion provided to governments is an additional area of concern. With environmentally related taxation, the scope of the discretion of government agents to reduce the effects on certain industries or groups of individuals is limited, as reduced rates or exemptions are usually quite visible. With other approaches, such as sectoral-specific regulatory strategies or negotiated agreements, there exists the potential to favour some industries or constituencies in a less transparent manner.

Finally, taxes have positive effects on innovation. By increasing the cost of pollution, taxes provide incentives both to develop new innovations as well as incentives to adopt them. They encourage a greater range of innovation types as well. The development and use of these innovations lower the overall cost to society of addressing the environmental challenge.

Many times, these positive features position taxes as being better than alternative policy instruments. The tax system can also be used, however, to effectively subsidise environmentally beneficial goods or actions, such as "green" products through sales tax or VAT exemptions on energy-efficient appliances or favourable depreciation schedules for some capital investments. Instead of increasing the price of the dirty good, these tax expenditures attempt to lower the price of the cleaner good. There is a general hesitancy among economists to embrace subsidies (Metcalf, 2009a) because:

- First, tax expenditures and subsidies can induce additional consumption of environmentally harmful goods or services by lowering costs. By lowering the firms' average cost (or the after-tax cost to consumers), they may provide perverse incentives with respect to output/consumption and therefore pollution.[3]

- Second, subsidies are difficult to design to only induce new actions. The majority of subsidies also reward activities that would have been undertaken even without their presence – making much of the expense of the programme ineffective.

- Finally, in trying to promote actions that are not something environmentally bad, the targeting of the policy instrument is less precise. As there are thousands, if not millions, of things that lead to reduced environmentally harmful behaviour, precision in deciding what to subsidise can be difficult.

Additionally, governments can also use regulations such as emissions standards, renewable portfolio standards, technology prescriptions, and many others. A technology prescription which mandates that Y firms install Z abatement technology (typically based on the best available technology) has certain advantages: the results can be predicted with significant accuracy, costs are rather certain, and enforcement can be relatively simple. Yet, command-and-control instruments do not generally bring about static and dynamic efficiency. The cost of achieving the last unit of abatement will generally differ across polluters – meaning that total costs are not minimised. By outlining limits on individual polluters, there is no benefit to emitting less than the prescribed limit or adopting a better technology than that prescribed by regulation. As a result, incentives for innovation are more muted (and are zero when firms have reached compliance with the regulations). They are also typically more tightly targeted, leaving less manoeuvrability for abatement and innovation.

Monitoring costs play a large role in determining the optimal policy response. For example, fuels are a relatively easy tax base on which to levy carbon taxes, since the future pollution is determined by the quantity of fuel consumed.[4] However, in situations where monitoring costs are higher (due to the disconnection between easily taxed proxies and the emission characteristics of most pollutants), the case for using taxes weakens. In addition, taxes may not be nimble enough to properly address pollution that has different impacts in different locations, such as emissions causing local air pollution, or that cause problems only temporarily.

Finally, it should be noted that there can be less certainty of outcome with taxes than with other policy instruments (the importance of this factor varies both with the scale of the problems and with the presence of tipping points). Policy makers attempt to estimate the impact of taxes on pollution but are not able to fully assess the behavioural impacts. It may be necessary, therefore, to adjust tax rates as more information is revealed about the behavioural impacts. Other policy instruments, such a tradable permits, have a guaranteed outcome (through a set emissions cap) but with greater uncertainty over the resulting cost.

5.2. Making effective environmentally related taxation

The imposition of environmentally related taxation requires careful consideration of a number of factors. Badly designed taxes can have reduced, even negative, effects on environmental and economic performance. This section outlines the key considerations in determining how to optimally implement a "green" tax.

5.2.1. Defining a tax base

On what the tax is levied – its base – is one of the most critical factors in making effective policy. Simply put, the tax should be levied as directly as possible on the pollutant or action causing the environmental damage, as this stimulates abatement incentives for all possible abatement options. This occurs through: i) cleaner production processes, such as reducing fuel use per unit of output or reducing NO_x emissions per unit of fuel; ii) end-of-pipe abatement, such as measures to capture and neutralise emissions before they enter the environment; iii) completely new products, such as reduced-vapour paints; and iv) output reductions. Moving away from taxing the polluting activity itself generally reduces the available abatement options, such as when an environmentally related tax is levied on an intermediate good, like coal. In this case, a tax on coal which attempts to address sulphur emissions would only stimulate a subset of abatement options, such as reducing coal use or potentially finding coal with a lower sulphur content. Undertaking end-of-pipe measures and some cleaner production processes, which would have an effect on sulphur emissions, would provide no financial benefit to the firm. In other cases, such as motor vehicle fuel, the release of carbon into the atmosphere is highly correlated with fuel use and there are few end-of-pipe abatement options, making motor vehicle fuel a very efficient proxy for taxing CO_2 emissions.

An additional issue raised with levying taxes on intermediate goods is that the implicit tax on the pollutant is not necessarily transparent and can differ across fuels or activities. In a number of countries, so-called "carbon" taxes that are levied on fuels have implicit carbon tax rates that differ between coal, petrol, diesel and so forth, due to various political economy issues. This highly distortive approach can undermine the efficacy of carbon taxes by encouraging excessive abatement in specific sectors and/or fuels, potentially even encouraging switching to dirtier fuels. It also can undermine faith in the fairness and effectiveness of the environmental policy.

5.2.2. Setting the tax rate

The tax rate ought to be set to reflect society's value of the harm done by the pollutant as well as the need of governments to raise revenues. Doing so should fully account for the fact that polluters are not charged for their damage to, and overuse of, the environment in an unregulated economy. Some of these damages are relatively easy to measure – the damage of raw sewage releases on the harvest value of oysters or the damage done by acid rain on the productivity of forests for timber production. Where the damage is done to something that does not clearly have a market value, the process of valuing the damage can be much more difficult – what is the value of cleaner air, more biodiversity, or a less volatile climate? At the same time, account must be taken of the effect that environmental impacts have on the morbidity and mortality of humans.[5] Implicit in these analyses are calculations relating to the value of a human life (and quality aspects thereof). It is much easier, for example, when a specific environmental outcome is aimed for (such as 550 ppm CO_2e for climate change) and the tax rate can then be implicitly determined to achieve this target.[6]

At the same time, the tax bases on which environmentally related taxes are levied are typically associated with other issues beyond environmental ones. Local air pollutants from motor vehicles, for example, can cause respiratory problems for residents and the wasted time caused by traffic congestion has negative economic repercussions. These other outcomes suggest that the overall level of taxes on environmentally related bases (such as motor fuel) should be higher than simply the estimated value of the environmental damage to society. They should approximate the additive effect of all of these different externalities.

Governments also levy taxes simply to raise revenues to fund public spending. Many of the environmentally related taxes (motor fuel and motor vehicles are of note) are prime candidates for such taxation, since tax rates are unlikely to shift behaviour significantly – that is, they are inelastic in demand.[7] The use of hypothecated taxes (i.e. taxes levied to fund specific activities, such as taxes on fuel to support highway maintenance) also has to be considered, although such taxes are simply sub-optimal user fees.

Except for motor vehicles and motor vehicle fuels, the values of environmentally related taxes in OECD countries are typically quite low, in most instances being levied at a rate much below the value of the relevant damage. Therefore, only a few OECD economies are at risk of levelling environmentally related taxes that are too high. There is a tendency, however, to levy very high tax rates on some intermediate goods to pollution, such as motor vehicles (see Chapter 2). The disparity between tax rates in different jurisdictions can also be striking, such as Sweden's charge on NO_x emissions[8] that is set at EUR 4 150 per tonne compared to the level of Italy's at EUR 105 per tonne.

By contrast, other environmentally related policy instruments, such as consumer rebates, typically have a much higher implicit cost per unit of pollution avoided that can vastly exceed what an optimal tax would be. In an analysis of European countries, it was found that reducing the VAT rate on more energy-efficient appliances would shift consumption patterns away from more energy intensive models (European Commission, 2008). Extending countries' reduced VAT rates (which are generally focused on essentials) to energy-efficient refrigerators in the European Union, for example, would lead to a reduction in CO_2 emissions of 1.6 million tonnes over an average fifteen-year life. This would cost treasuries EUR 119 million in foregone revenues, implying a carbon price of EUR 73 per tonne CO_2. For freezers, the implicit carbon price is EUR 25 per tonne CO_2, while for washing machines the implicit carbon price is very high at EUR 167 per tonne CO_2.[9] It is critical to note, however, that any emissions reductions arising under an emissions trading system (as is the case with electricity in the European Union) are completely offset by emissions increases elsewhere, as long as the cap on emissions is fixed. The analysed tax rate reductions would thus not cause any reduction in CO_2 emissions, but governments would have significantly less revenue.

5.2.3. Providing consistent incentives

Ensuring that the environmentally related tax provides similar abatement incentives on every unit of pollution is important to ensuring that firms and households abate optimally. Homogenous taxes encourage abatement at the most efficient source. Nevertheless, there are considerations regarding the impact of such taxes on selected polluters, such as low-income households or pollution-intensive, trade-exposed businesses. For example, lower tax rates at low levels of consumption/pollution are sometimes put into place, such that marginal incentives are reduced for some but not others (for social reasons, for example).[10] Such differing incentive structures make the overall costs of meeting a given environmental target more costly since abatement falls disproportionately on some polluters. Governments should therefore try to implement taxes as broadly as possible.

5.2.4. Facilitating general policy predictability and credibility

Environmental policy, especially taxes, can affect pollution abatement through two processes: behavioural responses and structural responses. Behavioural responses can occur very much in the short-term in response to prices, taxes, and other stimuli. Firms can reduce output and consumers can find less polluting activities, such as carpooling or changes in room temperature. If the stimuli were reduced, economic agents could easily resume former activities without much cost or effort.

On the other hand, structural responses are quite different. These responses typically take longer and can require significant analysis and investment in actions. Whether families move from a petrol-fuelled vehicle to a hybrid, whether firms invest in technologies and revamp their productions processes or whether venture capital funds invest in start-up alternative energy firms crucially depend on their long-term views and assumptions. The long-term price factor is the overriding consideration in many of these decisions. In addition to the initial level of the tax, the predictability of the rate and the policy's credibility (i.e. whether it is likely to remain in place over the medium- to long-term) are fundamental to making informed decisions. Of course, lack of this predictability and credibility can have pronounced and negative impacts on abatement and innovation efforts. Japan's tax on SO_x emissions provides a sobering example of its effects on innovation (see Annex I).

While general predictability is important for long-term investment and abatement decisions, it is not to say that tax rates should never change. After having been set, tax rates should continually reflect a range of factors: *inter alia*, inflation and real economic growth (since most environmentally related taxation is in the form of excise taxes), citizens' changing preferences for environmental protection, and the effect of innovation on the cost of pollution abatement. The process of changing tax rates, however, relies on its transparency so that polluters understand the potential determinants and timing of future modifications. Denmark, for instance, has recently built such a feature into their system: excise taxes on environmentally related bases will now be automatically indexed to annual inflation, removing the need for *ad hoc* adjustments at typically infrequent intervals.

It is important to note that taxes are inherently stable, except when explicitly changed by policy makers. Tradable permits, alternatively, can have significantly more price volatility in return for certainty as regards to the environmental outcome. Nevertheless, well-functioning secondary markets should provide financial instruments to firms to hedge future price volatility. Price floors and ceilings in between which permit prices can freely float offer another mechanism to provide additional predictability.

5.3. Using the revenue generated

Unlike other environmental policy instruments, environmentally related taxation (and tradable permits) does have the possibility of raising revenues for governments. As seen in Chapter 2, the vast majority of environmentally related taxes do not raise significant revenues for governments and cannot be drawn upon as a major revenue source; only a handful of taxes and charges [CO_2 taxes and taxes on driving (fuel, vehicles and tolls)] generate the vast majority of revenues from environmental bases. Even so, environmentally related taxes account for approximately 5% of total tax revenues in OECD countries. Moreover, the intent of these taxes is to shrink the tax base which is contrasted against most other taxes which attempt to raise revenues while doing the least to distort tax bases. As revenues are generally small and will tend to decline as environmental performance is enhanced, governments should apply cautious revenue assumptions when incorporating environmentally related taxation into long-term fiscal consolidation.

Nevertheless, the increasing use of environmentally related taxes and auctioned tradable permits, particularly for issues such as climate change, is likely to have non-minimal effects on government revenues over the medium term. What then should governments do with this increased revenue? The notion of compensating those most affected by the environmental damage aligns with an inherent sense of justice and one of the intended purposes of externality-correcting taxes. Yet, measuring the individual impact of environmental damage from a range of pollutants is extremely difficult; many environmental issues also have public good aspects, suggesting that taxation should amend for loss of public goods and increased costs for hospitals, adaptation, etc. Many environmental issues also have significant intergenerational aspects (*e.g.* the CO_2 emissions of today are likely to have a large impact on citizens 200-300 years hence). Obviously, the "polluter-pays" principle would suggest that polluters are least deserving of compensation.

In the absence of the above, revenues should be treated as revenues from other tax sources. That is, they should not be hypothecated and should go into general government funds. Governments can then use the proceeds to augment general government spending in other areas, maintain spending levels, reduce debt or reduce other taxes. With governments' fiscal position only starting to recover from the financial crisis and the culmination of years of earlier deficits, the need for additional tax revenues may be strong. Raising taxes on environmentally related bases may be more politically acceptable than other forms of tax increases.

At one point, there was considerable interest in the potential of a "double dividend" of environmentally related taxation. That is, that the imposition of "green" taxes would yield environmental improvements – the first dividend. The resulting revenues could also be used to reduce the effects of existing distortions in the tax system (for example, by reducing personal and corporate tax rates) – the second dividend. As summarised by Metcalf (2009b), such ideas, although seductive, do not take account of the fact that the raising of environmentally related taxation may distort, for example, labour supply in the same way as consumption taxes.[11]

Given the presence of pre-existing distortions in the economy due to previous government policies, the imposition of environmentally related taxation in addition may accentuate these distortions, having adverse effects on economic growth. Using part of the revenues to reduce these distortions, such as by reducing personal and corporate tax rates, can help offset some of the unintended effects of environmentally related taxation within the economy, simultaneous with creating a more efficient tax code.

In a political economy context, a reduction of other taxes can also serve to achieve political support for the introduction of environmentally related taxes. In many instances, new environmentally related taxes are announced simultaneously with an associated reduction in other taxes as a means to ease acceptance. In the case of the United Kingdom and their Climate Change Levy, the levy was announced simultaneously with a 0.3 percentage point reduction in employers' social security contribution rates. In other countries, more direct approaches have seen cheques being sent to all households to accompany the "green" tax implementation, although such measures do not potentially address other distortions in the economy. Revenues can also be used to offset some of the more direct effects of environmentally related taxation, such as distributional aspects, as outlined in the following section.

5.4. Overcoming challenges to implementing environmentally related taxes

The policy recommendations outlined above paint a picture of a world where taxes can be levied relatively easily – policy makers have complete and solid information at their disposal, tax administration costs are low, and political economy issues (specifically, distributional and sector competitiveness issues) are largely non-existent. Yet, such conditions rarely exist in the real world. Policy makers must decide how to implement taxes in a second-best environment. The following describes such issues and techniques to overcome some of them.

5.4.1. Addressing distributional concerns

One of the largest sources of pollution (and therefore from which environmentally related taxation can raise the greatest amount of revenue) is fuel-based energy generation and use – being from carbon emissions or pollutants related to local air pollution. At the same time, energy is essential to households and can account for a significant part of the household budget. Increased taxes on combustion-related emissions can have significant impacts on those at the lower end of the income scale. Much the same is true of water use. While the two areas are quite broad, most other environmentally related tax bases form only a small proportion of the overall consumption bundle and are therefore unlikely to have significant distributional concerns.

Clearly, governments should not ignore the distributional impacts of environmentally related taxes. In recognition of this fact, a wide range of features have been built into environmentally related taxes to help soften the impact. Some taxes avoid the entire issue by exempting all households, such as with the UK's Climate Change Levy. Others try to target economically depressed regions, such as with a reduction of duties on natural gas for Southern Italy. Rates that progress with quantity are many times used for water and electricity as a means to provide reduced rates on "necessary" consumption with full rates on subsequent consumption.

Attempting to make taxes both address the environmental issue and address any potential adverse distributional concerns risks undermining the ability of the tax to do either. Such concessions typically result in some loss of the abatement incentive. Progressive block tariffs or reduced rates, for example, provide fewer incentives for environmentally beneficial action and go against the notion that the polluter should pay. Moreover, many of these features are in fact very poor measures of addressing income distribution, since those who are wealthy tend to use more fuel. Such "progressive" measures can sometimes be regressive.

Therefore, policy makers should be concerned not necessarily with the distributional impacts of specific policies and taxes, but with the redistributive aspects of overall governmental policy.[12] That is, measures to account for the potentially regressive nature of some environmentally related taxes may many times be better actioned through broader means, such as lowering personal income taxes, supplementing low-income supports or even providing "green cheques" to some or all citizens. Such measures can reduce the administrative complexity of the environmentally related tax (but may increase overall complexity) and build upon existing platforms to address income inequality while removing distortions to the design of the tax that can have negative economic and environmental impacts.

5.4.2. Recognising competitiveness issues

By seeking to protect the environment, environmentally related taxation is by definition intended to distort production decisions and have a disproportionate impact on polluters. In a closed economy, factors of production and consumer behaviour would be switched such that environmental outcomes were met. In the modern world, the concept of a closed economy – one with no trade – is an aberration. The ability to trade across borders implies that, for a wide range of goods and services, the factors of production are highly mobile as well. There is concern that high levels of environmentally related taxation that disproportionately fall on some sectors can encourage those businesses to relocate, while the goods and services can still be imported. Such issues can cause economic detriment with minimal environmental gain. This effect is sometimes referred to as "carbon leakage" in the climate change context.

By far, the most effective method to minimise potential carbon leakage is to co-ordinate environmental policies across countries. By expanding the reach of policies, potential areas for relocation are reduced and leakage diminishes quickly. Even where full co-ordination does not occur, it is important to recognise that a wide range of factors determine where firms locate: general tax rates, proximity to markets, business climate and access to talented labour are a few. Environmental policies are only one factor. In analysis done by the OECD (2009), it is estimated that, if the European Union were to act alone to cut 50% of 2005 emissions by 2050, carbon leakage would be 11.5%. With all Annex I countries of the Kyoto Protocol[13] acting to achieve this target (which notably excludes Brazil, India and China), leakage would only be 1.7% in 2050. International co-ordination, even if imperfect, is the optimal solution.

Although the EU ETS unifies multiple countries' climate policies for the largest emitters, there is some concern in other countries and for some currently excluded sectors about the impact of such taxation. In response, countries have undertaken a range of mitigation strategies to ease the sectoral competitiveness impacts of environmentally related taxation, recognising that any such measure violates the polluter-pays principle.

Beyond global pollutants, however, there is significantly less rationale for co-ordinated global action for other pollutants that are more local in nature, such as NO_x and SO_x. Since the optimal rate of taxation will likely differ between countries and even within different regions within a country (since, for example, the impacts of local pollution may vary with existing levels of local pollution, population densities and local climatic conditions), a co-ordinated mechanism would be unlikely to be so sensitive to these effects and would likely not mitigate the competitiveness concerns of industries which happen to be situated in regions where tax rates are, or should be, higher.

One of the least distortive means to address sectoral competitiveness issues is to potentially provide some lead-in time for affected firms to undertake mitigation measures. Since capital investments have a long productive life and cannot generally be replaced quickly, an environmentally related tax announced and implemented relatively quickly can penalise companies for historical decisions that result in current emissions. A lead-in period can provide firms time to significantly retool their operations and purchase new capital without being penalised for historical decisions. Yet, lead-in times should not be too great as adoption of off-the-shelf technologies can be quite quick: with the introduction of the NO_x charge in Sweden, firms adopting some form of emissions mitigation technology went from 7% to 62% in one year. An escalating tariff over a set time period can also ease the initial burden of a tax and leave additional financial flexibility for firms to invest in mitigation or R&D activities to minimise future payments. Credibility in the commitment to escalate rates towards the "standard" level is critical.

Countries have also taken to providing favourable mechanisms to businesses. In areas where revenues from environmentally related taxes are recycled to the affected firms (on a basis different from the collection), the marginal abatement incentive is generally maintained; yet, the average firm is little worse off from a cost and profit point. This means that the polluter pays principle is violated via such a mechanism – the price to consumers of pollution-intensive products is not increased. Only those goods of the same type that are relatively more pollution intensive are reduced. For those that are relatively less pollution intensive (yet still pollute significantly), the production costs are effectively subsidised.

Other measures have also been widely used. Rate reductions and exemptions for energy-intensive users simply shift some of the abatement burden to others or result in an inferior environmental outcome. Policy makers must keep in mind that measures to offset the full impact of environmentally related taxation on some firms or sectors provides an implicit subsidy to environmentally harmful activities and forces other sectors to undertake greater efforts or finance those subsidies through higher taxes. They can even shift consumption patterns away from less environmentally harmful activities (taxed regularly) towards pollution-intensive activities (taxed lightly).

Finally, one potential measure to address sectoral competitiveness issues and carbon leakage is the possibility of using the tariff system. So-called border adjustment taxes have been discussed as a means to compensate for products that are produced in exporting jurisdictions with weaker environmental policies than the importing country. Under such a system, it is suggested that tariffs would be levied on products to compensate for the economic impacts of the different environmental policies. Such a policy would then place domestic and imported goods on a level footing. While such policies have some intellectual appeal and may be compliant with trading rules of the World Trade Organization, real-world implementation issues make these a highly contentious topic. Because environmental policies within any given country are complex, encompass a wide range of policy tools and rely on existing economic structures, comparing them with an importing country's and then setting a compensating figure for the thousands of import codes poses challenges (potentially also differentiated by firm). It also risks aggravating international dialogue to liberalise trade. As co-ordination grows, these measures become significantly less important as carbon leakage drops precipitously.

5.4.3. Simplifying tax administration

The infrastructure, paperwork and human effort needed to administer and verify compliance with tax laws is substantial. As such, the administrative burden on both governments and taxpayers needs to be carefully assessed. In addition, the complexity of the system through large numbers of taxpayers or various exemptions can lead to evasion of taxes.

As discussed in preceding paragraphs, the ideal environmentally related tax is levied on the actual polluting activity. In some cases, tax administration can be stymied by the fact that there are numerous and diffuse pollution sources. Setting up monitoring systems, collecting data and administering taxes on such bases can prove overwhelming. New technological developments affecting both sophistication and reduced costs for implementation are making such possibilities more realistic, such as with the Netherlands' proposed road pricing scheme.

Nevertheless, the type of pollutant can play a large role in determining if pollution should be taxed at the source or whether there are opportunities to minimise the tax administration burden by levelling the point of tax incidence at higher levels in the supply chain where there are fewer potential taxpayers and fewer occasions for tax evasion. For pollutants where the type of activity that causes the pollutant does not affect the level of pollution, taxing intermediate goods may be much easier without compromising environmental outcomes. Carbon emissions, for example, have a direct correlation to the type of fuel used; the manner in which the fuel is combusted (for a given fuel consumption) does not affect CO_2 emissions, unless carbon capture and storage is used (which is unlikely for small and mobile sources of carbon, such as vehicles). Therefore taxing motor fuel at the refinery or wholesaler is much easier than monitoring the emissions from individual vehicles. For other pollutants, where the combustion process is integral to the level of emissions (for example, NO_x emissions), levying the tax at higher levels would significantly impair the ability to target environmental outcomes.

Moreover, the overlapping of multiple instruments on the same emissions can create duplicative compliance costs, in addition to the potentially harmful economic and environmental effects outlined in Section 5.5.

5.4.4. *Gaining trust and communicating a plan*

Despite the theoretical issues in favour of green taxes, past introductions have sometimes caused significant concern among citizens regarding their impact and the motivations for their use. As seen above, concerns about the distributive aspects and competitiveness concerns have placed significant pressure on policy makers. In addition, there has often been either public scepticism over the intentions of the tax (*e.g.* that tax rates are simply being increased and disguised as being green) or concerns over the economic impacts on such fundamental activities.

Coming into full swing in the mid-1990s, a number of European countries undertook significant "ecological tax reforms" to varying degrees of success. In all cases, the path to implementation was not smooth and there were significant barriers. Focus group assessments of ecological tax reform in Denmark (Klok, 2006), Germany (Beuermann and Santarius, 2006), the United Kingdom (Dresner, 2006) and France (Debroubaix and Lévèque, 2006), as well as in Ireland (Clinch and Dunne, 2006), where ecological tax reform did not take place, indicate that there are significant commonalities across countries.

First, there was a lack of knowledge about the overall scheme. Second, citizens were highly sceptical about governments using the funds to reduce other taxes and instead felt that ecological tax reform was a guise to generally increase taxes. It was also felt that the connection between the introduction (or augmentation) of environmental taxation and reduction in other taxes was not necessarily appropriate and that revenues should be used for environmental purposes. Such issues will likely continue to face governments in the future.

These findings suggest that pre-emptive mitigation measures can help smooth implementation of such policies. The utilisation of green tax reform commissions, led by respected and arms-length citizens, can help ensure that the policy prescriptions are perceived as credible and not as politically driven. Moreover, open, transparent, and adequate information campaigns can better inform citizens and businesses about the potential ramifications of shifts towards more environmentally friendly tax regimes.

5.5. Environmentally related taxes alone are not the answer

Despite the significant benefits associated with environmentally related taxation, taxes alone cannot always bring about the intended outcome. Issues and distortions within the economy may prevent optimal actions from occurring. In such circumstances, additional policy tools may be needed to provide an optimal instrument mix. Three examples are illustrated below.

First, consumers may be unaware of the environmental impacts of their purchases (and the long-term tax/price liability that they will face) with the market alone. This is generally true of a wide range of goods, with large household appliances being of particular note. Therefore, the imposition of a tax on energy may not induce changed behaviour or altered consumption patterns simply because consumers are not able to translate the effect of a tax into a demonstrable idea of the impact on utility bills. This information constraint can be overcome through, for example, government schemes that provide easy-to-understand and comparable information on energy consumption across models.

Second, incentives that are not fully realised can limit the scope for enhanced environmental performance. The classic example is of landlords and tenants with respect to energy/water efficiency and conservation. Tenants that pay utility bills have an incentive to minimise their energy use. Many of the most efficient ways to do so are the responsibility of the landlord: insulation, replacing aging windows, etc. If the landlord is not paying the energy bills, there are fewer incentives for investment in such items; for the tenant, the transitory nature of renting makes investments unlikely to be profitable. In such cases, taxes would not have the full effect as on owner-occupied housing; building codes may be more efficient.

Third, the role of innovation to deliver improved environmental outcomes at lower costs is critical. Environmentally related taxes can encourage the adoption and development of more market-ready innovations; however, the breakthrough technologies that will lead to fundamental environmental improvements are less likely to be developed under a tax-only regime. The long-term and more fundamental nature of such projects creates significantly uncertainty for investors and entails a high probability of failure. In such cases, taxes may need to be supplemented by targeted investments in basic R&D.

As outlined above, instrument mixes can play an important role provided that the instruments are mutually reinforcing and do not provide similar deterrents on the same environmentally harmful activity. Where instrument mixes do overlap, they can have either a negligible effect or can work to distort abatement and innovation decisions, leading to an overall less efficient environmental policy. In many OECD countries, multiple environmental policy instruments are used on the same pollutants. One of the most common is the use of carbon taxes and emission trading schemes. Using two instruments can still provide strong environmental outcomes when the instruments are on different sources, such as tradable permits on stationary emissions and carbon taxes on transportation. When the instruments perfectly overlap, the price of the tradable permit is exactly lowered by the tax rate.[14] Increasing tax rates therefore have no overall effect on emissions, except where they are high enough to form a *de facto* price floor. Where there is not perfect overlap between sectors, increasing tax rates can induce additional (and likely less efficient) abatement in some sectors compared to others. In this light, for example, the advent of the EU ETS system has encouraged the Danish government to abolish carbon taxes on emissions also covered by the EU ETS starting in 2010.

5.6. Conclusions

Environmentally related taxation has a significant role to play in addressing environmental challenges, especially compared to other instrument types. Taxes can be extremely effective, provided that they are properly designed, levied as close to the environmentally damaging pollutant or activity as possible and across all sources of pollution, and set at an adequate rate. The revenues generated can be used to help with fiscal consolidation or reduce other tax rates. At the same time, taxes may need to be combined with other instruments to obtain an overall environmental policy package. Administration costs or barriers may necessitate that proxies to environmentally harmful activities are targeted instead. Finally, the imposition of environmentally related taxes may exacerbate distributional or competitiveness concerns but solutions should be found outside of the tax itself. Therefore, environmentally related taxes should play a central role in countries' approaches to environmental policy but that taxes alone may not be able to adequately address all the environmental issue nor overcome some of the challenges to its implementation.

Notes

1. For a discussion of the similarities between taxes and tradable permits, see Box 3.4.

2. Theoretically, this would be the Coase theorem. See endnote 3 in Chapter 1 for more information.

3. The classic example is subsidies or tax reductions for energy-efficient appliances. The price reduction on an energy-efficient air conditioner, for example, encourages people to switch consumption away from energy-inefficient models. However, by lowering the price of air conditioners, it also encourages the use of air conditioners *versus* other goods in the economy (which may have less environmental impact).

4. However, new end-of-pipe technologies, such as carbon capture and storage, can make this connection less applicable.

5. For further information on a review of the literature on the economic value of morbidity impacts of pollution, as well as a meta-analysis of "Value of statistical life" estimates, refer to OECD (2010a) and (2010b).

6. There is no guarantee, however, that *a priori* the environmental target is set optimally.

7. Such tax bases can also be correlated with items that governments typically find difficult to tax (*e.g.* leisure), suggesting a higher tax rate than the environmental component alone. West and Williams (2007), for example, show that petrol use is correlated with leisure activities, which governments typically find hard to tax. In their scenario this feature suggests taxes near the current level in the US, not counting for the externality issues present. On the other hand, fuel use may have impacts on labour supply, especially where labour mobility, and therefore may be important to keep taxes low.

8. Sweden's charge on NO_x emissions is refunded, making the net impact of the tax less burdensome and the political economy aspects of implementation easier. The rate applied in Sweden was not set based on estimates of the marginal damage – and is about twice as high as the rate in the (un-refunded) Norwegian NO_x tax, which was based on estimates of the marginal damage.

9. Such figures are, of course, very country-specific, varying on countries' price levels, standard and reduced VAT rates, and the composition of electricity generation.

10. One way that this is done is through tax bands, used typically in consumer goods such as white goods and motor vehicles. While such tax structures can simplify the message to consumers, they can also reduce incentives for better environmental behaviour. Under these systems, where there

are different tax rates (typically a flat fee for each band) based on environmental performance, marginal abatement options become skewed. The tax provides increased incentives among tax bands but not within. Where tax bands are large, there can be considerably little incentive for movements within bands. A large family is unlikely to switch a purchasing decision from a minivan to a sub-compact but may be interested in a less-polluting minivan; taxes applying only a few bands may not provide such marginal incentives.

11. In this analysis, the optimal environmental tax would be divided by the marginal cost of funds. The marginal cost of funds is the effect on the economy of levying one unit of tax revenue. With the generally distorting nature of tax systems, this level is generally above one (that is, one unit of tax revenue withdrawn from the economy has a cost of more than one unit on the economy).

12. It must be noted that other instruments of environmental policy also have distributional aspects but they are generally less visible than those of taxes. Sutherland (2003), for example, shows that energy efficiency standards in the United States were regressive. Initial appliance costs increased but, since low-income households generally have a much higher discount rate than higher-income households, low-income households were disproportionally affected, even having negative impacts on their welfare.

13. Annex I countries include all members of the European Union (less Cyprus and Malta) plus Australia, Belarus, Canada, Croatia, Iceland, Japan, Lichtenstein, Monaco, New Zealand, Norway, the Russian Federation, Switzerland, Turkey, Ukraine and the United States.

14. This may be desirable where the permits have been distributed freely and the tax seeks to recover some of the windfall gains of polluting firms.

References

Beuermann, Christiane and Tilman Santarius (2006), "Ecological Tax Reform in Germany: Handling Two Hot Potatoes at the Same Time", *Energy Policy*, Vol. 34(8), pp. 917-929.

Clinch, J. Peter and Louise Dunne (2006), "Environmental Tax Reform: An Assessment of Social Responses in Ireland", *Energy Policy*, Vol. 34(8), pp. 950-959.

Debroubaix, José-Frédéric and François Lévéque (2006), "The Rise and Fall of French Ecological Tax Reform: Social Acceptability *versus* Political Feasibility in the Energy Tax Implementation Process", *Energy Policy*, Vol. 34(8), pp. 940-949.

Dresner, Simon, Tim Jackson and Nigel Gilbert (2006), "History and Social Responses to Environmental Tax Reform in the United Kingdom", *Energy Policy*, Vol. 34(8), pp. 930-939.

European Commission (2008), "The Use of Differential VAT Rates to Promote Changes in Consumption and Innovation", 25 June 2008, available at *http://ec.europa.eu/environment/enveco/taxation/pdf/vat_final.pdf*.

Klok, Jacob *et al.* (2006), "Ecological Tax Reform in Denmark: History and Social Acceptability", *Energy Policy*, Vol. 34(8), pp. 905-916.

Metcalf, Gilbert (2009a), "Tax Policies for Low-Carbon Technologies", *NBER Working Paper*, No. 15054.

Metcalf, Gilbert (2009b), "Environmental Taxation: What Have We Learned this Decade?" in Alan D. Viard (ed.), *Tax Policy Lessons from the 2000s*, American Enterprise Institute for Public Policy Research, Washington DC, available at *www.aei.org/docLib/9780844742786.pdf*.

OECD (2009), *The Economics of Climate Change Mitigation: Policies and Options for Global Action Beyond 2012*, OECD, Paris. *http://dx.doi.org/10.1787/9789264073616-en*.

OECD (2010a), *Valuing Lives Saved from Environmental, Transport and Health Policies: A Meta-Analysis of Stated Preference Studies*, available at *www.olis.oecd.org/olis/2008doc.nsf/linkto/env-epoc-wpnep(2008)10-final*.

OECD (2010b), *A Review of Recent Policy-Relevant Findings from the Environmental Health Literature*, OECD, Paris, available at *www.olis.oecd.org/olis/2009doc.nsf/LinkTo/NT00008EBA/$FILE/JT03278752.PDF*.

Sutherland, Ronald J. (2003), "The High Costs of Federal Energy Efficiency Standards for Residential Appliances", *Policy Analysis*, No. 504, Cato Institute: Washington DC, available at *www.cato.org/pub_display.php?pub_id=1362*.

West, Sarah E. and Roberton C. Williams III (2007), "Optimal Taxation and Cross-Price Effects on Labour Supply: Estimates of the Optimal Gas Tax", *Journal of Public Economics*, No. 91, pp. 593-617.

Case Studies

Case Studies

ANNEX A

Sweden's Charge on NO$_x$ Emissions

This case study outlines the tax on NO$_x$ emissions in Sweden, implemented in 1992. The tax rate is very high compared to other OECD jurisdictions but nearly the full amount is refunded to firms. Through a number of metrics, it was found that the tax did have significant impacts on innovation. Many of these were process innovations that made the firms' existing operations less pollution intensive, even for firms not adopting capital-based abatement strategies. This case study also includes a theoretical exposition of the impact of the recycling mechanism on the incentives for innovation.

Rationale for the environmental policy

Sweden implemented a charge on emissions of NO$_x$ emitted from all stationary combustion sources producing at least 50 MWh of useful energy per year, starting in 1992. The decision was part of an overall strategy to reduce overall NO$_x$ emissions in the country by 30% between 1980 and 1995. Quantitative emission limits had already been introduced in 1988 on an individual basis for stationary combustion plants; however, it soon became apparent that these measures alone would not be enough to attain the desired reductions. The NO$_x$ charge was introduced as a complementary instrument.

Design features

The charge came into effect on 1 January 1992 and initially about 200 plants were regulated (those with energy output greater than 50 MWh). Due to its effectiveness and falling monitoring costs, the charge was extended, first in 1996, to about 270 plants producing at least 40 MWh useful energy per year, and then from 1997 onwards to about 400 plants producing at least 25 MWh useful energy per year. Currently, all stationary combustion plants are subject to the NO$_x$ charge if they produce above the energy output threshold and belong to any of the following sectors: power and heat production, chemical industry, waste incineration, metal manufacturing, pulp and paper, food and wood industry. Exempt from the charge due to concerns about unfeasibly high costs are, for example, the cement and lime industry, coke production, the mining industry, refineries, blast-furnaces, the glass and insulation material industry, wood board production, and the processing of biofuel.

The NO_x charge was given a unique design. Plants pay a fixed charge per kg NO_x emitted and the revenues are entirely (except for an administration fee of less than 1% withheld by the regulator) refunded to the paying plants in relation to their respective fraction of total useful energy produced. The design encourages abatement among plants for attaining the lowest NO_x emissions per amount of useful energy produced relative to other plants. The result is that firms having an emissions intensity at the average of all other firms will pay no net tax; relatively cleaner plants will receive a net refund while dirtier plants will pay a net tax.

There were a number of reasons for the Swedish Environmental Protection Agency (SEPA) to use a refundable charge. First, continuous monitoring of NO_x emissions was considered important due to the complex formation of NO_x throughout the combustion process; however, it entails high monitoring costs (making it feasible only to target large combustion plants). Therefore, it was a way to counteract the effects of distorted competitiveness between the large regulated and the smaller unregulated plants. Second, refunding helped to avoid strong political resistance from emitters and thereby facilitated a charge level high enough to attain significant effects on emissions.[1]

Environmental effectiveness

The NO_x charge has provided significant environmental benefits since its introduction. The first panel of Figure A.1 shows how NO_x emissions from regulated plants have been decoupled from increases in energy production. The second panel of Figure A.1 shows the development of NO_x emissions per unit of useful energy produced (i.e. emission intensity) for regulated plants. Overall emission intensity among regulated plants fell by 50% between 1992 and 2007. Larger plants have managed to reduce average emission intensities to 194 kg NO_x per GWh in 2007, which is less than the average of 330 kg NO_x per GWh achieved by plants producing 25-50 MWh useful energy per year. This is probably a

Figure A.1. **Effectiveness of Swedish charge on NO_x emissions**

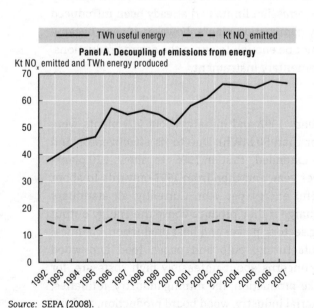

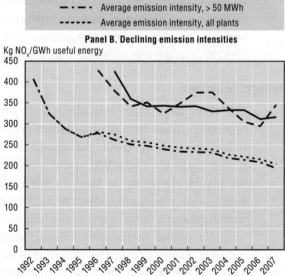

Source: SEPA (2008).

StatLink ᴍᴴᴴ http://dx.doi.org/10.1787/888932317502

result of large producers being able to exploit economies of scale, but also a consequence of the nature of the available NO_x abatement technology, which is characterised by indivisibility and high costs for the most effective types of technology.

It should be noted that the NO_x charge level of SEK 40 per kg NO_x was kept constant in nominal terms between 1992 and 2006, leading to an effective depreciation of around 25% in real terms. Such a cut in the incentive effect of the charge may have contributed to the levelling off of the fall in emission intensities that can be observed in later years. After 2006, the tax was increased to SEK 50 per kg NO_x.

Effects on innovation

From a technology point of view, the introduction of the charge created a strong incentive for the immediate adoption of existing abatement technologies. As seen in Table A.1, there is a significant jump in firms utilising established technologies, as rates of technology usage go from 7% to 62% in the first year alone. These comprise both post-combustion technologies [such as selective catalytic reduction (SCR) and selective non-catalytic reduction (SNCR)] and combustion technologies, such as trimming.

Table A.1. **Adoption of NO_x mitigation technology in Sweden**
Plants regulated by the Swedish NO_x charge, 1992-2007

| | Output threshold (MWh per year) | Number of regulated plants | Fraction of plants with NO_x technology installed | | | | | |
| | | | Plants with NO_x mitigation (%) | Post-combustion technology | | Combustion technology | | Flue gas condensation (%) |
				SCR (%)	SNCR (%)	Trimming (%)	Other (%)	
1992	50	182	7	1	3	0	3	3
1993	50	190	62	3	21	18	30	4
1994	50	203	68	5	26	21	36	4
1995	50	210	72	5	30	22	40	4
1996	40	274	69	5	25	22	40	19
1997	25	371	60	3	22	17	39	19
1998	25	374	62	3	23	19	39	21
1999	25	375	65	3	24	20	43	23
2000	25	364	69	4	26	21	47	26
2001	25	393	67	3	25	20	47	30
2002	25	393	71	4	26	20	50	33
2003	25	414	70	5	26	20	48	32
2004	25	405	70	4	28	19	49	34
2005	25	411	69	5	30	18	47	34
2006	25	427	72	6	32	19	47	34
2007	25	415	71	6	33	18	46	34

Source: SEPA (2008).

StatLink ⬛⬛ http://dx.doi.org/10.1787/888932318110

In addition to technology adoption, the Swedish charge induced innovation. Three methods were used to ascertain the innovation effects: patent data analysis, emission intensity analysis and marginal abatement cost curves.

Patent data analysis

Counting the number of patent applications filed for NO_x mitigation technologies can give an indication of changes in the incentives for developing this type of technology. It should, however, be stressed that the number of patent applications is not a direct measure of innovation levels, since the relative importance of different patents is highly variable and a single patent may be more important in terms of NO_x abatement than dozens of others. Furthermore, not all granted patents are brought into use and only innovations to which exclusive rights can be clearly defined are possible to protect through patents. As many innovations in NO_x mitigation technology take place through small alterations in the combustion process, without additional installations of physical equipment, the analysis of patent data is limited in its scope to indicate incentives to develop NO_x mitigation technology. Moreover, patent levels for such specific innovations in small countries can lead to very low levels of patenting.

Table A.2 shows that Sweden has been quite active in NO_x technology development, ranking among the top four countries of patents per million inhabitants. What is most striking is the significant increase in patenting in Sweden during the 1988-93 period – exactly when the charge was being discussed and implemented. Two different hypotheses could explain this phenomenon. First is that the introduction of a charge of a high magnitude spurs incentives to engage in R&D in NO_x abatement technology. Alternatively is that the decision to set a high charge level was made possible by an existence of effective Swedish NO_x abatement technology. This is a political economy argument that suggests lobbying, or at least interaction, between the innovating firms and the decision makers. Conclusive evidence of either hypothesis is not available and would require a much more detailed analysis of each individual patent.

Table A.2. NO_x patent applications across countries

Innovations in NO_x technologies by inventor country

	Number of patents 1970-2006 by country of residence of inventor			Average number of patents per year measured per million inhabitants			
	Total	of which: Combustion technology (%)	of which: Post-combustion technology (%)	1970-2006	1970-87	1988-93	1994-2006
Austria	20.3	27	73	0.071	0.062	0.147	0.047
Australia	1	0	100	0.001	0	0	0.004
Belgium	4	0	100	0.011	0.008	0.017	0.011
Canada	14.7	20	80	0.014	0.007	0.022	0.020
Czech Republic	2	0	100	0.005	0	0	0.015
Denmark	10.5	19	81	0.055	0.049	0.194	0
Finland	15.6	19	81	0.083	0	0.144	0.146
France	54.8	35	65	0.026	0.015	0.032	0.039
Germany	353	28	72	0.120	0.131	0.164	0.085
Italy	20.5	41	59	0.010	0	0.023	0.012
Japan	289	11	89	0.066	0.072	0.063	0.060
Korea	9.3	14	86	0.005	0	0.004	0.014
Netherlands	12.5	40	60	0.023	0.014	0.033	0.030
Norway	6	75	25	0.037	0	0.040	0.086
Russian Federation (incl. USSR)	5	20	80	0.001	0.000	0.001	0.002
Spain	2.2	0	100	0.001	0	0.004	0.002
Sweden	24.3	47	53	0.076	0.033	0.223	0.067
Switzerland	58.5	69	31	0.232	0.138	0.587	0.197
United Kingdom	47	24	76	0.022	0.017	0.021	0.029
United States	269.6	33	67	0.028	0.020	0.049	0.029
Other countries	10.1	30	70	n.a.	n.a.	n.a.	n.a.
World	1 230	27	73	n.a.	n.a.	n.a.	n.a.

Source: Worldwide Patent Database (2009).

StatLink ⬛🔗 http://dx.doi.org/10.1787/888932318129

It is important to keep in mind that the patent values are quite low – as seen in Table A.2, over the 36-year period 1970-2006, only 24.3 patents can be attributed to Sweden, less than one per year. Moreover, emissions from plants regulated by the Swedish NO_x charge are not significant from an international perspective. In fact, they only make up less than 1% of total emissions from stationary sources (power plants and industrial boilers) in the 19 European Union member states that have ratified the Gothenburg Protocol and thereby committed to NO_x emission reductions. Thus, if mitigation technology developed in Sweden is primarily intended for an international market, the introduction of the NO_x charge is unlikely to affect invention activity levels. If, however, inventions are primarily driven by the specific needs of the domestic market, invention activity levels can be affected by the charge. It is possible that inventions first intended for the regulated Swedish market with its high abatement incentives, spill over and become adopted on the broader international market.

Emission intensity analysis

Table A.3 presents the results of the largest power generators with respect to annual changes in emission intensities. From an innovation and technology development perspective, this is interesting because of the moderate, continuous declines in average emission intensities that can be observed from 1997 onwards in both pre-mitigation (firms which did not make capital installations to address post-combustion emissions) and post-mitigation plants (firms that did). In 1997, the large plants had been regulated by the NO_x charge for five years and plant engineers should have had enough time to adopt and try out existing technology to find the most efficient NO_x emission intensity level for their individual plant. If it is assumed that this is the case,[2] explanations other than investments in existing mitigation

Table A.3. **Plants subject to the NO_x tax: Descriptive statistics**

Sample of pre-mitigation and post-mitigation plants, larger than 50 MWh

	Pre-mitigation plants > 50 MWh				Post-mitigation plants > 50 MWh			
	Number of plants	Weighted average emission intensity	Annual change in emission intensity (%)	TWh useful energy produced	Number of plants	Weighted average emission intensity	Annual change in emission intensity (%)	TWh useful energy produced
1992	168	402	..	34.5	12	438	..	2.7
1993	72	345	−14	13.2	117	309	−29	27.7
1994	68	294	−15	12.9	131	279	−10	31.9
1995	75	279	−5	12.2	133	260	−7	34.1
1996	92	327	17	13.6	154	260	0	41.6
1997	86	298	−9	14.0	146	242	−7	35.6
1998	93	301	1	13.1	153	229	−5	38.4
1999	97	289	−4	16.1	145	221	−3	33.8
2000	70	277	−4	10.7	165	225	2	35.8
2001	74	260	−6	11.9	177	221	−2	40.5
2002	82	258	−1	13.1	189	221	0	43.1
2003	89	255	−1	13.7	198	219	−1	47.3
2004	85	252	−1	13.6	189	204	−7	46.8
2005	85	242	−4	14.1	192	200	−2	45.3
2006	81	249	3	12.0	200	193	−3	49.1
2007	79	234	−6	12.2	191	181	−6	48.3
Average 1997-2007			−2.9				−3.2	

Source: SEPA (2008).

StatLink 🔗 http://dx.doi.org/10.1787/888932318148

technology need to be found to explain why emission intensities for this group of plants continue to fall and, in particular, why they continue to fall both for plants that report to have undertaken mitigation measures and for plants that report no NO_x mitigation measures. Three main explanations are presented:

● Plants improved their performance without investing in new equipment, *e.g.* by learning better to control NO_x formation, by optimising the various parameters in the combustion process given the boundaries of the existing physical technology, or by changing routines and firm organisation. Such changes in the non-physical mitigation technology show up as a fall in emission intensity in both sets of plants.

● Plants improved the efficiency of physical mitigation installations: by adopting mitigation technologies at a later point in time, they were able to attain lower emission intensities than those having invested at the beginning of the period.

● The realisation of the full mitigation potential of an investment in physical mitigation equipment may not have been immediate, but may have required testing and learning that took several years before working optimally.

The first two explanations are effects of innovations both in physical mitigation technology and non-physical mitigation technology. The last explanation is a mere effect of that it may take more than a year of phasing in and testing before an investment in existing technology becomes fully efficient. If this effect can be separated out, the residual would be the effect on emission intensity that (with some plausibility) can be referred to as effect of innovations in mitigation technology.

Figure A.2 shows the annual adjustment in emission intensity levels following an installation in NO_x abatement. The analysed sample includes those plants that have only reported one installation during the period 1992-2007 and the installation should be SCR,

Figure A.2. **Changes in NO_x emission intensities**

Annual change in emission intensity level following a NO_x mitigation installation

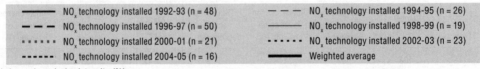

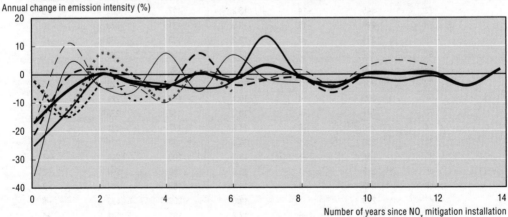

Annual change in emission intensity (%)

Number of years since NO_x mitigation installation

Note: Only plants that have made investments in SCR, SNCR or combustion technology at one occasion in time included (n = 216, *i.e.* 50% of plants > 50 MWh).

Source: SEPA (2008).

StatLink http://dx.doi.org/10.1787/888932317521

SNCR or installations in physical combustion technology. The adjustment is relatively rapid. On average, emission intensities drop by 17% in the first year and 6% in the second year after installation of a NO_x mitigation technology. After the first two years, the average annual change revolves around zero with an average annual drop of 0.9%. Thus, the phase-in of a new technology, including testing and learning how to use it optimally, appears to take one to two years. After the phase-in period, additional gains from optimising the existing technology are limited and slow and may well be the effects of innovations in non-physical mitigation technology like trimming.

Therefore, the continuous fall in average emission intensity that can be observed for large plants from 1997 onwards in both the pre-mitigation and post-mitigation group of plants cannot be explained by long adjustment periods that drag on for many years before the phasing in and testing of installations in physical mitigation technology are completed. Instead, much of the annual decline in emission intensity of 2.9% in pre-mitigation plants and 3.2% in post-mitigation plants is likely to come from improved knowledge about how existing technology should be run more efficiently and adoption of innovated mitigation equipment. For pre-mitigation plants, the entire improvement in emission efficiency can be linked to innovations in non-physical mitigation technology. For post-mitigation plants, the continuous decline of –3.2% per year after 1997 is partly (i.e. by –0.9% per year) explained by improved knowledge about how to operate existing SCR, SNCR and combustion technology installations more efficiently and partly by adoption of innovated physical mitigation technology.

In the analysis above, it is not possible to visualise the evolution of emission intensity in individual plants. However, of interest is whether it is typically the same plants that improve their performance or whether emission intensity varies strongly from one year to the next for the same plant. Figure A.3 plots the average emission intensity of the plants in 2006-07 against the average emission intensity in 1992-93 for a set of 137 large plants that were regulated by the NO_x charge in both periods. The dots situated to the right of the 45-degree line ($e_{2006-07} = e_{1992-93}$) have lowered emission intensity levels between the two periods. As expected, a majority of plants (76%) is in this category. Only a few units have significantly worsened their emissions in relation to output between the two periods.

Roughly half of the plants reduced emission intensity by up to 50%. Another third cut emission intensity by more than 50%, while four plants cut them by more than 75%. Two of these are oil fuelled plants that have installed SCR technology, while the other two have made major shifts from fossil to bio fuel. Every single plant with really high emission intensity in 1992-93 (> 600 kg NO_x per GWh) improved its performance, although their emission intensity levels in 2006-07 are still high relative plants starting from lower initial levels. This indicates a large spread between individual plants in the best performance levels that are technically attainable.

Increases in emission intensity were experienced by 24% of plants, but the increases were small – only for eight plants (i.e. 6%) did it exceed 50%. Of the 33 plants that had worsened the performance, nine of them had started from already low levels (< 250 kg NO_x per GWh) in 1992-93 and made slight increases (< 10%) in emission intensity.

Twenty-four plants remain that started from levels above 250 kg NO_x per GWh in 1992-93 and still worsened emissions per output in 2006-07. Seven of these plants did not report any installations of NO_x mitigation technology during the period 1992-2007, which may partly explain why these plants did not improve. For the other plants, the main reason for worsening performance appears to have been fuel switches from fossil fuels or

Figure A.3. **NO_x emission intensities at individual plants**
2006-07 relative to 1992-93

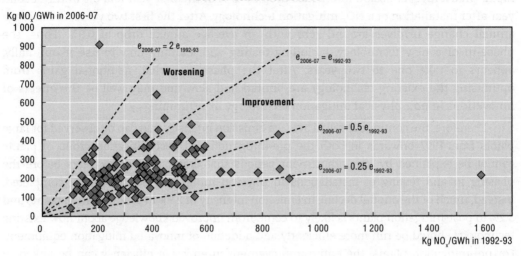

Source: SEPA (2008).

StatLink ⫸ http://dx.doi.org/10.1787/888932317540

pure biofuels to less pure biofuels such as unsorted municipal waste, recycled wood, fat waste, unsorted residual products from forestry, and black liquor from pulp-and-paper production. Such fuels have higher nitrogen content and switches are generally driven by economic factors unrelated to the NO_x charge. For instance, some may have reacted to the rising costs of fossil fuels and emitting carbon. In some cases, they were using "alternative" biofuels that meet climate goals but are still significant sources of local pollutants like NO_x. In some cases, access to waste such as bark and other by-products was plentiful and their use as fuel was promoted by other policy initiatives.

Marginal abatement cost curves

The final indicator to investigate innovation impacts is the use of marginal abatement cost curves. If abatement cost savings for given emission intensity levels can be used as indicator for the occurrence of innovations in abatement technology, one could measure the incidence of innovations by measuring changes in abatement costs for given emission intensity levels over time. This, however, requires detailed information about actual investment and operation costs of abatement technologies from firms having actually installed the technologies. Systematic collection of this kind of abatement cost data is very rare.

The results of a survey of 114 plants regulated in 1992-96 provide a nice basis. Estimations were performed for three industrial sectors: energy, pulp-and-paper, and chemical and food. Innovation effects were measured as downward shifts of the marginal abatement cost curve from one year to the next. The energy sector had been most active in abatement during 1990-96 and only for this sector was it possible to find statistically significant evidence for falling marginal abatement costs over time. Compared to year 1996, marginal abatement costs were significantly higher for the same level of emission intensity in years 1991, 1992, and 1994. The predicted marginal abatement cost functions for these years are presented in Figure A.4. These show, for example, how the emission intensity attainable at zero abatement cost (i.e. the efficient abatement level without regulation) moves from 557 kg per GWh in 1991 to about 300 kg per GWh in 1996.

Figure A.4. **Declining marginal NO$_x$ abatement cost curves**

For 55 plants in the energy sector regulated by the Swedish NO$_x$ charge, 1992-96

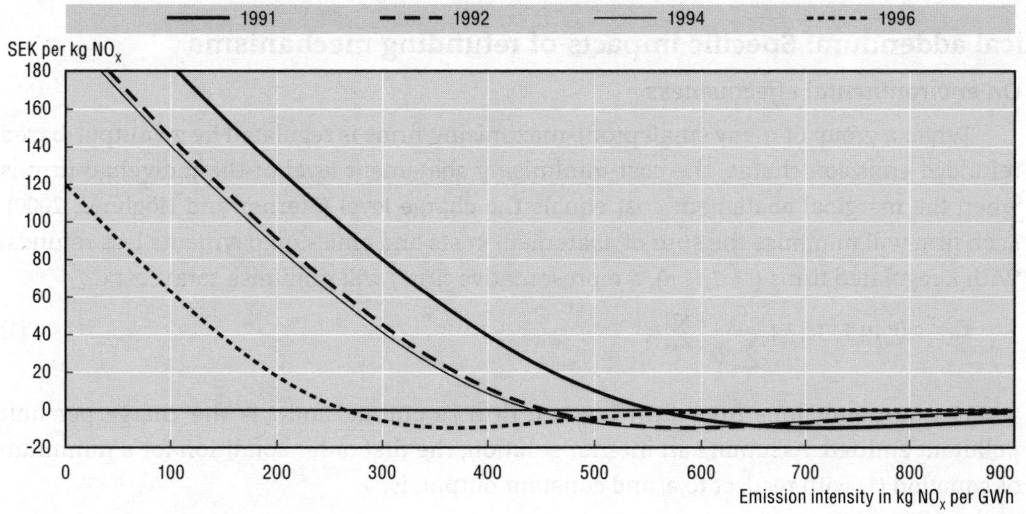

Source: Höglund-Isaksson (2005).

StatLink ⟐ http://dx.doi.org/10.1787/888932317559

This shift is likely to come from the adoption of innovations in abatement technology, which has made it possible to produce energy with less NO$_x$ emissions without increasing costs. To a large extent, the effects occur because of trimming activities. The introduction of the charge revealed opportunities to pick "low-hanging fruit" in abatement. Some of these opportunities existed also before the introduction of the NO$_x$ charge, but the charge, with its requirement to monitor NO$_x$ emissions continuously, made it possible for firms to discover and develop them to attain even lower emission intensity levels.

For the other two sectors, pulp and paper and chemical and food, parameters measuring shifts in marginal abatement costs over time were not found significantly different from zero and could accordingly not show any evidence of innovation effects.

Conclusions

This case study has clearly shown that taxes are an important driver for innovation. The tax rates on emissions in Sweden were particularly high compared to other countries – likely achieved because of the refunding mechanism. Finding the linkages, however, required a range of approaches. In addition to patent data, analysis of marginal cost curves and emission intensities were central to highlighting the impacts. It is interesting to note as well that ongoing emissions reductions by firms occurred both for firms adopting (capital-based) abatement technologies and those not doing so, indicating that a significant amount of the abatement reduction was driven by cleaner production innovation, such as learning how to better optimise the existing capital stock.

This exposition also showed how the design of the tax – the refunding mechanism in particular – influenced the level of innovation (and by whom it may be undertaken). The greater the level of market concentration, the lower is the incentive for innovation, given the reduced refund payment associated with increased abatement. This finding can be extrapolated to the innovation impacts from collective investments in innovation under such a refunding mechanism.

For more information on the Swedish NO_x charge, the full version of the case study (OECD, 2009) is available at *www.olis.oecd.org/olis/2009doc.nsf/linkto/com-env-epoc-ctpa-cfa(2009)8-final*.

Technical addendum: Specific impacts of refunding mechanisms

On environmental effectiveness

When a group of many small profit-maximising firms is regulated by an output-based refunded emission charge, the cost-minimising abatement level of the individual firm is when the marginal abatement cost equals the charge level (Sterner and Höglund, 2000). Each firm will minimise the sum of abatement costs and emission payments less refunds. With n regulated firms ($i = 1,...,n$), a representative firm j will minimise total cost C_j:

$$C_j = c_j(e_j, q_j) + te_j - t\frac{q_j}{\sum_i q_i} * \sum_i e_i \tag{1}$$

where e_j are emissions from firm j, q_j are firm j's output, and t is the charge per unit pollutant emitted. Assuming an interior solution, the first order condition for a minimum of equation (1) with respect to e_j and constant output, is:

$$-\frac{\partial c_j}{\partial e_j} = t * \left(1 - \frac{q_j}{\sum_i q_i}\right) \tag{2}$$

With many small regulated firms, each firm's contribution to total regulated output becomes very small, i.e. $\frac{q_j}{\sum_i q_i} \to 0$, and the optimal abatement level is found when marginal abatement cost approximately equals the charge level. Thus, in terms of effectiveness in emission reductions, a refunded charge is equivalent to a conventional emission tax without refunding. In the case of the Swedish NO_x charge, the largest fraction of total output ever produced by a single owner in one year has been 12%.

On innovation incentives

Now allow for the possibility of innovations in abatement technology and that an innovation takes place in one of the regulated firms denoted firm j (Höglund, 2000). After adoption, firm j supplies the innovation to all other regulated firms $i = 1,...,n-1$ at the royalty price, P. Firm j has an exclusive right to the innovation and the right is protected through a patent. Other firms are supposed not to be able to imitate the innovation and are accordingly not able to acquire any of its usefulness without paying the patent royalty. Firm j is therefore a monopolist in the market for innovation and is able to set a profit-maximising royalty price. The demand-side of the innovation market consists of many, small and non-co-operative regulated firms, where a single firm cannot affect the adoption decision of other firms in any way.

Variables for abatement technology (k_j) for firm j, as well as R&D costs (D_j), and revenues from royalty payments (R_j) from m non-innovating regulated firms adopting the innovation are introduced. The royalty price (P_m) will correspond to the reservation price of the last firm adopting the innovation, i.e. the reservation price of firm m. Output is assumed constant throughout the analysis.

The innovated technology affects firm costs both directly and indirectly. Directly, by affecting abatement costs, R&D costs or royalty revenues and, indirectly, by reducing tax costs as the optimal emission level is reduced to meet a downward shift in the marginal cost curve with respect to emissions. To find an interior solution, the following properties are assumed for the relevant interval of the cost curve. Both emission level and production cost are supposed to be decreasing at a constant or increasing rate in k_j, i.e. $\partial e_i/\partial k_j < 0$, $\partial^2 e_i/\partial k_j^2 \geq 0$, $\partial c_i/\partial k_j < 0$, and $\partial^2 c_i/\partial k_j^2 \geq 0$. Thus, the cost-saving from adopting an innovation increases at a decreasing or constant rate with improved innovation level.

Suppose that the innovating firm j has enough information about the adopting firms to set a profit-maximising royalty price, which maximises royalty revenues (R_j):

$$R_j(k_j) = m(k_j)P_m(k_j) \qquad (3)$$

where $\partial R_j/\partial k_j > 0$ and $\partial^2 R_j/\partial k_j^2 \leq 0$.

Firm j will choose an innovation level which minimises the following total cost function:

$$C_j = c_j\left(e_j(k_j), q_j, k_j\right) + D_j(k_j) - R_j(k_j) + te_j(k_j) - t\frac{q_j}{Q}\sum_{i=1}^{n} e_i(k_j) \qquad (4)$$

By setting the first derivative of equation (4) with respect to changes in technology k_j equal to zero, the following condition for a minimum is obtained:

$$\frac{dC_j}{dk_j} = \frac{\partial c_j}{\partial k_j} + \left(\frac{\partial c_j}{\partial e_j} + t\left(1 - \frac{q_j}{Q}\right)\right)\frac{\partial e_j}{\partial k_j} + \frac{\partial D_j}{\partial k_j} - \frac{\partial R_j}{\partial k_j} - t\frac{q_j}{Q}\sum_{\substack{i=1, \\ i\neq j}}^{n}\frac{\partial e_i}{\partial k_j} = 0 \qquad (5)$$

where $\dfrac{\partial R_j}{\partial k_j} = P_m\dfrac{\partial m}{\partial k_j} + m\dfrac{\partial P_m}{\partial k_j}$ and $\left(\dfrac{\partial c_j}{\partial e_j} + t\left(1 - \dfrac{q_j}{Q}\right)\right) = 0$.

Alternatively, the latter condition can be shown by applying the envelope theorem. The change in the total cost function when adjusting emissions (e_j) in an optimal way is equal to the change in the total cost function when emissions are not adjusted. From this follows that $\left(\dfrac{\partial c_j}{\partial e_j} + t\left(1 - \dfrac{q_j}{Q}\right)\right) = 0$. Note that this does not imply that the indirect effect always has to be zero. It only implies that the sum of the direct and indirect effects is equal to the direct effect when emissions are unchanged. By rearranging the resulting terms, the condition for an optimal level of innovation for firm j is obtained:

$$\frac{\partial D_j}{\partial k_j} = -\frac{\partial c_j}{\partial k_j} + \frac{\partial R_j}{\partial k_j} + t\frac{q_j}{Q}\sum_{\substack{i=1, \\ i\neq j}}^{m}\frac{\partial e_i}{\partial k_j} \qquad (6)$$

where $Q = \sum_{i=1}^{n} q_i$ and $\partial D_j/\partial k_j > 0$ and $\partial^2 D_j/\partial k_j^2 \leq 0$.

Equation (6) equates the marginal cost of innovation with the marginal benefit of innovation for firm j, where the latter can be decomposed into three different terms. The first term is the cost effect, which expresses the magnitude of the marginal effect on production cost, e.g. in terms of reduced abatement costs or in terms of reduced tax costs as emissions are reduced, or in terms of effects on both. The second term is the royalty revenue effect, which reflects the marginal revenue from royalty sales to other regulated firms adopting the innovated technology. The third and last term is the marginal effect on the refund from reduced overall emissions when other regulated firms adopt the innovation. Note that the marginal refund effect is not infinitely small even if $q_j/Q \rightarrow 0$, since also a very small output share is approximately constant for changes in the technology k_j. Instead, the marginal effect on the refund depends on the marginal change in the overall emission level, which cannot be assumed to be infinitely small.

If a conventional emission tax, set to the same level, had been used instead, firm j would be minimising the total cost in equation (4) less the last refund term. The corresponding condition for an optimal R&D level is accordingly:

$$\left[\frac{\partial D_j}{\partial k_j}\right]^{Tax} = -\frac{\partial c_j}{\partial k_j} + \frac{\partial R_j}{\partial k_j} \tag{7}$$

Comparing the condition for an optimal R&D level under a refunded charge (equation 6) with the condition under a conventional emission tax (equation 7), the difference in marginal R&D cost (i.e. marginal spending on R&D) is caused by the refund term in equation (6). It is, however, less straightforward to compare equilibrium levels of marginal spending on R&D between the two regimes, since the marginal effects on costs and royalty revenues are likely to differ between innovation levels. A comparison requires further restrictions.[3] With approximately constant marginal effects on production costs and revenues from royalty sales, firm j is willing to invest in R&D to a lower marginal cost when using a refunded emission charge than when using a corresponding conventional emission tax. The discrepancy is approximately equal to the marginal effect on the emission refund.

The intuitive explanation is that with an emission charge with output-based refunding, a regulated firm's willingness to share innovations with other regulated plants is hampered by the refund, since a spread of the innovation to other regulated firms will reduce firm j's own refund. By keeping the innovation to itself, the innovating firm is able to improve its relative position within the charge system, thereby increasing its net refund. With a conventional emission tax, there are no gains[4] to be made from reducing a firm's emission intensity relative other regulated firms.

A special case, which is of interest to mention because it has relevance for NO_x abatement, is when the royalty price for an innovation is zero. This may for example occur when a regulated firm through experience accumulates knowledge, which improves the environmental effectiveness of the firm but is too indistinct to protect through a patent. Compared with a tax, refunding restricts any spread of knowledge among regulated firms and particularly knowledge about emission reducing innovations that cannot be protected through a patent, i.e. often the small and simple, but sometimes effective, measures. This may have been important in the case of the Swedish NO_x charge, where extensive emission reductions were attained at a low or even zero cost through trimming activities.

Firms outside the regulated group of firms may develop and supply new and improved abatement technologies to the regulated firms. Innovation incentives then depend on the general demand for innovated technology. Is the demand for a given innovation the same under a refunded charge as under an equivalent conventional emission tax? It appears that this generally holds when the demand-side of the innovation market consists of many small and non-co-operating regulated firms.

When calculating the profit-maximising price, the monopolist innovator will take into consideration the cost of innovation and the expected number of royalties sold. The price will correspond to the reservation price of the last firm adopting the innovation. The reservation price will, in turn, correspond to the additional profit the last adopting firm makes from adopting the innovated technology ($k = 1$) compared with not adopting it ($k = 0$). The total cost function of the last adopting firm m is:

$$C_m^{k=1} = c_m^{k=1}\left(e_m^{k=1}, q_m\right) + P_m^{k=1} + te_m^{k=1} - t\frac{q_m}{Q}\left(\sum_{i=1}^{m} e_i^{k=1} + \sum_{i=m+1}^{n} e_i^{k=0}\right) \tag{8}$$

With a refunded charge, a new innovation adopted by some of the regulated firms affects the cost of firms *not* adopting it by reducing the refund as the innovation deteriorates the firm's environmental effectiveness relative to the adopting firms. In its decision between adoption and non-adoption, the last adopting firm therefore compares the cost of adoption with the cost of non-adoption:

$$C_m^{k=0} = c_m^{k=0}\left(e_m^{k=0}, q_m\right) + te_m^{k=0} - t\frac{q_m}{Q}\left(\sum_{i=1}^{m-1} e_i^{k=1} + \sum_{i=m}^{n} e_i^{k=0}\right) \tag{9}$$

The reservation price of the last adopting firm is accordingly:

$$P_m = C_m^{k=0} - C_m^{k=1} = \Delta c_m + t\Delta e_m\left(1 - \frac{q_m}{Q}\right) \tag{10}$$

With all firms being small, the effect of the last firm's adoption decision on the same firm's refund can be taken to be very small. Hence, the reservation price of the last adopting firm for a given innovation will be approximately the same as under an equivalent conventional emission tax, namely:

$$P_m^{Tax} = C_m^{k=0} - C_m^{k=1} = \Delta c_m + t\Delta e_m \tag{11}$$

Note that the resulting reservation price holds only when the regulated group of firms consists of many firms that are small in relative size and not co-operating. In the special case when regulated firms co-operate and act as one entity and bargain over the price in a situation where either all regulated firms adopt the innovation or none, incentives to adopt are likely to be considerably weakened. If all firms adopt and the innovation is equally effective (in terms of effects on emissions) for all firms, the change in net refund is zero. Incentives to invest in improved technology are therefore the same as in the completely unregulated case. The assumption of many non-co-operating firms in the market for innovations is accordingly crucial for the result that the reservation price (and demand) for a given innovation is approximately the same under a refunded emission charge as under an equivalent emission tax.

Notes

1. The Swedish charge, for example, is many times that of the French charge.

2. This may not be completely the case. In a sample of 114 plants regulated by the NO_x charge in 1992-96, about half of the plants comply (or over-comply) with the charge in 1996, while the residual half of plants do not attain an efficient investment level in abatement, *i.e.* where the marginal abatement cost equals the unit charge.

3. An assumption that appears plausible is that $\partial^2 c_j / \partial k_j^2 < 0$ and $\partial^2 R_j / \partial k_j^2 > 0$ for low levels of k_j and $\partial^2 c_j / \partial k_j^2 > 0$ and $\partial^2 R_j / \partial k_j^2 < 0$ for high levels of k_j. Cost-savings from adopting innovations are then assumed to increase at an increasing rate for low levels of innovation and at a decreasing rate when higher levels of innovation are reached. Under these assumptions it is difficult to speculate on the direction of the difference in the level of $\left(-\partial c_j / \partial k_j + \partial R_j / \partial k_j\right)$ between a refunded charge and a tax. Still, if the difference in optimal k_j-level between the regimes is not too extreme, a plausible assumption seems to be that the main effect on differences in marginal spending on R&D comes from the refund term and not from differences in the sum of the marginal cost-saving and the marginal revenue.

4. If regulated firms compete on the same market for final output, sharing knowledge for free about how to reduce emission tax payments, could potentially change relative production costs and the competitiveness of the firm in the output market. Since this indirect effect would be the same under a refunded charge as an emission tax, it does not affect the findings and does not enter into the analysis.

References

Höglund, L. (2000), "Essays on Environmental Regulation with Applications to Sweden", Ph.D. thesis, Department of Economics, Göteborg University, Sweden.

Höglund-Isaksson, L. (2005), "Abatement Costs in Response to the Swedish Charge on Nitrogen Oxide Emissions", *Journal of Environmental Economics and Management*, No. 50, pp. 102-120

OECD (2009), *Innovation Effects of the Swedish NO$_x$ Charge*, OECD, Paris, available at *www.olis.oecd.org/olis/ 2009doc.nsf/linkto/com-env-epoc-ctpa-cfa(2009)8-final.*

SEPA (2008), *Database of Information from Annual Surveys of Plants Regulated by the Swedish NO$_x$ Charge*, data used by kind permission of the Swedish Environmental Protection Agency, Östersund, Sweden.

Sterner, T. and L. Höglund (2000), "Output-based Refunding of Emission Payments: Theory, Distribution of Costs and International Experience", Discussion Paper, No. 00-29, Resources for the Future, Washington DC.

Worldwide Patent Database (2009), *http://ep.espacenet.com/*, European Patent Office, Vienna.

ANNEX B

Water Pricing in Israel

This case study explores water pricing in Israel in light of the constant pressures over water resources in this semi-arid region. It first looks at the differentiated approaches across industrial, agricultural and household uses, highlighting the fact that pricing reflects use, type of water and varies on quantity. The Israeli experience in conserving water is clearly a success and has been very innovative. A multitude of factors have contributed to this, including water pricing structures, government information campaigns and governments investments in water technologies.

Rationale for the environmental policy

Water scarcity and water-related environmental threats beset Israel, making it a unique experience for OECD countries to study. Israel is in a semi-arid region with an uneven distribution of its water resources.* It was decided early in its establishment to develop regions that were also remote from water sources. "Blooming the desert" was perhaps one of the initial driving forces for the Israeli economy and the National Water Carrier was built to bring water from the north to the south. Settlements, food security and agricultural development have put further pressure on water resources. Increasing population growth and a large inflow of immigration have created an additional burden on the already overexploited and environmentally degraded resources. The result is that fresh water levels are low and existing water resources have been degraded. Many policy instruments have been used to address these issues. This study focuses on the pricing schemes for the use of water in a variety of sectors.

Design features

To address the scarcity of water in Israel by encouraging reduced consumption and recycling of water, strong pricing signals have been placed into Israel's water policy. There are differentiated rates for the agricultural, industrial, and household and tourism sectors.

---

* The statistical data for Israel are supplied by and under the responsibility of the relevant Israeli authorities. The use of such data by the OECD is without prejudice to the status of the Golan Heights, East Jerusalem and Israeli settlements in the West Bank under the terms of international law.

Like all sectors, there is progressive pricing for water use for agriculture, based on the level of quota held by individual farmers. Over the ten year period 1995-2005, real prices for water increased substantially, as outlined in Table B.1. In addition to the prices for fresh water outlined below, agricultural users can be offered the use of marginal, recycled water and saline water for use in their operations, which are priced at a significant discount to the use of fresh water.

Table B.1. **Agricultural prices for fresh water in Israel**

USD per m^3 at 2005 prices

Level	1995	2005	Increase (%)
A	0.165	0.282	70.9
B	0.199	0.335	68.3
C	0.267	0.441	65.2
Mean	0.196	0.330	68.3

Source: OECD (2009).

StatLink ᴹˢᵖ http://dx.doi.org/10.1787/888932318167

For agricultural users, the price steps to which quantities apply are determined by farm-specific quotas. Availability of water beyond allocated quota is not guaranteed, but the "quotas" are not constraints. Farms can, in most cases, use more than their quota, but a higher price is paid for over-quota use and a lower price is paid if use is sufficiently less than quota. Since each farmer is free to adjust use within these intervals, each farmer's marginal price bracket tends to reflect the true marginal value of water on that farm (unless the quota is not fully used). Most importantly, individual quotas serve to differentiate water prices among users because they determine the levels where rate steps occur.

Increasing water scarcity and price inequities have led to questions regarding agricultural water subsidisation and social efficiency of the agricultural sector under its present structure. The drought of the early 1990s highlighted the potential for allocation of water away from agriculture. Largely because of consecutive years of drought in 1990 and 1991, the real price of water to agriculture was increased and the quota was reduced as a means of dealing with the temporary shortage. Some 47% increase in agricultural water prices occurred from July 1990 to May 1992 for use levels at 80-100% of quota, suggesting a substantial reduction in the indirect agricultural subsidy. Recently, water quotas were cut by at least 40%.

Industrial users also have individual water quotas and pay a higher price for above-quota use. Industrial quotas are set on an individual basis according to production norms. Firms can submit petitions for increased quota when businesses expand. Industry paid approximately the same average prices as agriculture from 1966 until May 1994, but has paid roughly 35% more than agriculture since. Currently, industrial water prices are close to the gate price paid by municipalities.

Water for household users is delivered by municipalities or by local water consortiums who buy at established prices or extract water locally, paying the government Extraction Levy, and sell at much higher prices to residents. These rates more than cover the costs of local water delivery. Water consumption is metered and users face increasing block-rate pricing. All households face the same block-rate schedule. Domestic consumers pay for water according to three increasing block rates: the first level covers the first eight cubic metres per month per family of up to four people, the second level covers an additional seven cubic metres per month and the third level reflects any consumption per month thereafter. Families with more than four members are entitled to apply for an additional twenty cubic metres per month at a reduced price. The average rate is USD 1.02 per m^3, where the third level is approximately double than the first level, as seen in Table B.2.

Table B.2. **Domestic water prices in Israel**

ILS per cubic metre at nominal prices

Consumption level	2004	2005	2006	2007	2008	% change, 2004-08
Level C: For consumption above 15 m^3 per month	6.132	6.648	6.471	6.695	7.648	24.7
Level B: From 8 m^3 to 15 m^3 per month	4.342	4.779	4.651	4.811	5.495	26.6
Level A: The first 8 m^3 per month	3.042	3.521	3.329	3.444	3.934	29.3

Source: OECD (2009).

StatLink ⧉ http://dx.doi.org/10.1787/888932318186

It should be noted, of course, that water pricing is but one facet of Israeli water policy. Like other OECD economies, water policy is made up of many interrelated issues: policies regarding abstraction and supply, water transportation and distribution, wastewater policies and policies aimed at reducing water demand. All of these factors have impacts on the demand for and innovation incentives of water pricing and teasing out the effectiveness (innovation and environmental) can be difficult.

Environmental effectiveness

Agriculture has historically used around 70% of Israeli water, but its share has been decreasing since the mid-1980s. In recent years, the agricultural sector has relied more on recycled and saline water sources for irrigation, accounting for about 50% of total water demand for irrigation. This process is a result of a massive effort not only in converting to drip irrigation, but also in moving towards more appropriate crops, removing water-intensive trees and replanting with water-saving types, training farmers through educational programmes and launching awareness campaigns.

Interestingly, decreased agricultural potable water use has not been accompanied by a decrease in the overall value of agricultural output, as outlined in Figure B.1. For example, between 2000 and 2005, the fruits sector was exposed to an average 35% cut in water quotas while increasing its production by 42%. Whether agricultural demand for water will continue to decline depends both on opportunities to expand use of currently available irrigation technologies and on discovery of new irrigation technologies and new sources of recycled or saline water, such as in the case of citrus, where the majority of the plantations are now been irrigated using reclaimed water or, in the case of aquaculture, using saline water.

In fact, absolute agricultural water use has declined even as a share of policy-imposed water use quotas. Farm water quotas were reduced in 1991 as a result of drought, but water use did not increase accordingly when quotas were again increased. Beyond the continuous increase in efficiency in the use of each unit of water, this reduced use relative to quota is explained by changes in the agricultural water pricing structure, and by the fact that price of water in agriculture rose 100% over the last decade.

Changes in recent years in water used in the agricultural sector indicate that farms do respond to changes in price. For example, an increase of 11.7% in water prices resulted in a 2.4% increase in quantity demanded in 2003 relative to previous year. In 2005, an increase of 12.4% in water prices created a greater impact and reduced demand by 2.3% relative to

Figure B.1. **Agricultural output value per unit of irrigation water**

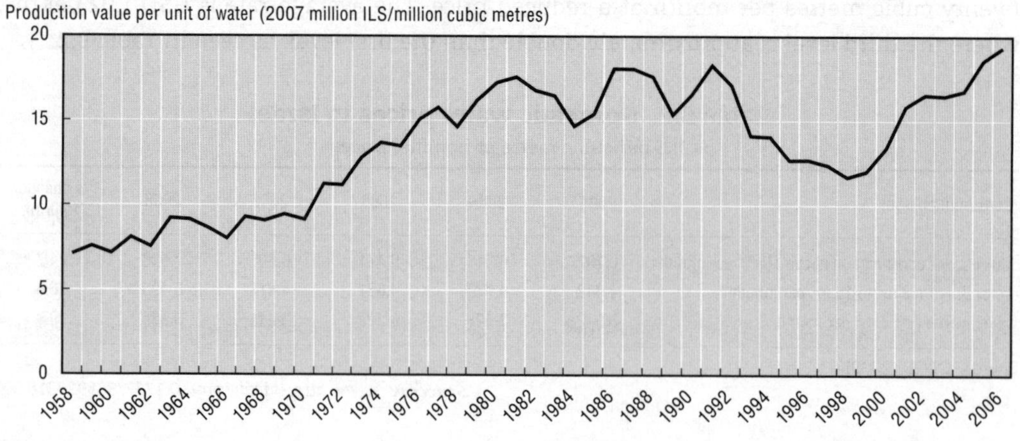

Production value per unit of water (2007 million ILS/million cubic metres)

Source: OECD (2009).

StatLink ᕫᕫᕫ http://dx.doi.org/10.1787/888932317578

previous year. This price increase kept farms at a 74.5% usage rate of the total allocated quotas for 2005. Total value of water as a fraction of total inputs to agricultural production was 7.9% in 2003, rising to 8.9% in 2005, increasing the significance of water in farmers' budgets and hence creating greater motivation for water saving.

Many farms that were able to adjust to the progressive pricing schedule attained a lower water price bracket by reducing use relative to quota. The decline in national agricultural water use as a share of quota, from 89% in 1990 to 70% in 1992, suggests that many farmers moved to lower price brackets. Thus, the marginal water price (averaged among all farmers) increased less than the 47% average increase in the price schedule.

To overcome the increase in water scarcity, substantial public investment was made in highly efficient irrigation technology, concurrent with decreasing quotas and the introduction of a progressive water pricing schedule. Computerised sprinklers and drip irrigation systems have led to increasing efficiency of water use in agriculture. Water-saving technology has evidently caused a decline in agricultural water demand.

Many of these gains have been supported by public investments. For example, specific government investments targeted at agriculture include aiding in the removal of marginal plantations and the planting water-saving trees, such as olive and almond trees, as well as the utilisation of water-saving technologies, such as drip irrigation. These measures are in addition to programmes to expand the availability of recycled and saline water and other government initiatives to reduce water consumption.

Industrial water use increased about 3.5% per year from 1960 to 1980, 1.7% per year from 1980 to 2000 and decreased 7.4% from 2002 to 2004, perhaps in anticipation of price increases and due to an economic slowdown. About 22% of the water consumed by industry comes from saline and marginal sources. Despite the gradual slowdown in demand growth until 2000, and the absolute decline in demand since 2002, industrial product value per unit of water use has increased steadily and future industrial water consumption is expected to increase roughly in proportion to population, corrected by the decline achieved due to improved efficiency in industrial production processes that use water. Stringent environmental regulations related to the quality of industrial effluents impose on the polluting industry the responsibility to treat industrial sewage on the factory site prior to leaving the plant and reaching public sewage facilities. The treatment cost and related operations, along with the purchasing cost of water and sewage levies, imply a loss in potential profit and hence motivate the industry to conserve water, develop water-saving production processes, and increase the use of recycled and marginal water in industrial operations.

Household consumption of water in Israel has been growing at roughly 2.5% per year. About 80% of this growth is due to population growth, with the rest attributed to income growth. Increased demand due to population growth is predicted to cause serious water shortages. Water demand from the sector has increased tremendously during the years. For example, from 1970 to 1980, it increased by 56%, from 1980 to 1990 by 28.5%, from 1990 to 2000 by 37.4% and, from 2000 to 2005, the increase has relatively stabilised and was only 8%. Per capita domestic water demand reflects a rise in the standard of living. In 1970, demand per capita was 79.3 m^3, 94 m^3 in 1980, 100 m^3 in 1990, and since it has relatively stabilised to 102.32 m^3 in 2005.

Domestic users are not generally influenced by water prices, and demand remains relatively inelastic to water price increases. Laws and ordinances, such as limiting irrigation of private gardens to specific months and metering quantities used, prohibition on washing cars with pipes, use of dual-flushing toilets, water-saving devices for faucets and shower heads, etc., are in place, but rarely enforced unless a year of drought has been officially announced. National water-saving campaigns have been proven to be effective in lowering consumption for the duration of the campaign. The 2000-01 water-saving media campaign was successful in reducing domestic consumption by 6% using a budget of about USD 2.3 million. In 2008, the national water saving campaign had a downward impact on water consumption of 3.3% relative to 2007. However, once the campaign was over, domestic consumption began to rise again, as seen in Figure B.2. This suggests that water-saving campaigns must focus on tools and methods that would cause long-lasting water saving (i.e. education and technology).

Figure B.2. **Impact of the national water saving campaigns**

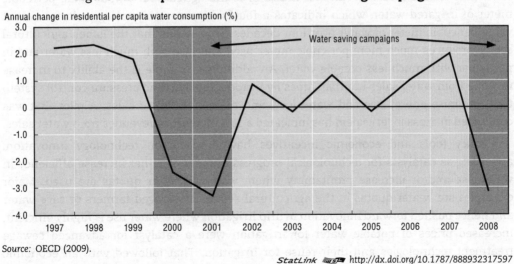

Annual change in residential per capita water consumption (%)

Source: OECD (2009).

StatLink ⟪ᵜᵴ⟫ http://dx.doi.org/10.1787/888932317597

It is important to stress that a significant saving in the domestic sector has the potential in delaying costly investment in desalination plants. For example, a 5% decrease in domestic water use is comparable to a desalination plant with a production capacity of 35 million m^3 per year, such as a plant that is currently under construction.

In addition, there were effects of the water policy on the water firms themselves. On average, water lost due to leakages in local municipalities reaches 10%. Municipalities are subject to fines once unbilled water quantities exceed 12% of total water consumed by the town. Since water lost in the system is also a waste of income to municipalities, they make an effort to fix leakages. Despite that, many municipalities fail in managing and maintaining their water infrastructure in good shape.

A common assessment is that the new Urban Water Corporations, which are driven by for-profit motivation, would increase efficiency in water use within urban areas (*e.g.* by fixing leaking infrastructures, etc.). By the end of 2008, fourteen such corporations functioned in Israel, serving twenty municipalities and 35% of the urban population. The remaining urban water consumption within 170 towns was still supplied by municipalities. Water losses reported by the Urban Water Corporations reveal higher figures than reported prior to the corporations' establishment. This may suggest that the business motivation of the water corporations pushes them to measure losses more accurately, in order to fix infrastructure and avoid losing water and hence money. For example, in one city, the water loss was estimated at 14% prior to the establishment of the corporation and a year after it was estimated at 24%. After two years, the corporation had already reduced water loss to 19.5%.

Effects on innovation

Various measures can be established to indicate technological innovation. Such measures can include growth in exports, research and development funding as a share of GDP, water saving and leakages/loss in water networks.

Water loss in Israeli municipalities has declined dramatically in recent years, to a national average of 10% in 2007 of the total water consumed in the municipalities (in comparison to a European average of around 25%). Leakage detection technologies contribute to this measure. Another indicator is the agricultural output value per cubic meter of irrigated water, which indicates a fourfold increase in real agricultural output value per cubic meter of water over four decades. This means that the Israeli agricultural sector produces much more per cubic meter used, but also with much less water and in particular with much less potable water. An additional example is the ability to increase revenue from water sales in urban areas by introducing dynamic pressure control system that minimise energy use and water loss, and maximises water sales. A pilot that was conducted in areas in Jerusalem has indicated a 10% increase in revenues from water sales.

Policy tools and economic incentives have impacts on technology innovation, appearing as catalysts for technological progress in order to either increase efficiency in water use and/or increase profitability where water prices or quotas are used. Major examples are: water quotas in the agricultural sector encouraged farmers to save water and hence pushed forward innovation in drip irrigation where water use is highly efficient. Increased prices of potable water for irrigation were a catalyst for advanced sewage treatment technologies and their reuse for irrigation. That followed with an economic incentive in the form of lower prices for treated water for irrigation. Stringent environmental policy for sewage dumping also contributed to a range of technological developments in water treatment technologies. High prices for industrial and domestic water have contributed to water-saving devices for domestic use and for domestic and public irrigation. Water loss fines for municipalities at a level above 12% created incentives for the development of water loss detection and dynamic water pressure equipment. Economic incentives and support in private initiatives in improving water quality in closed drinking wells also brought improvement in low-scale water treatment technologies.

From time to time, due to cyclical droughts, especially when droughts have lasted a few consecutive years, or when the economy experienced a dramatic increase in consumption (for example, due to a large immigration influx in the early to mid-1990s), the administrative and economic systems reacted with discrete changes in prices and/or quotas. Periods of water droughts in late 1980s pushed forward the establishment of a three-tier quota regime in the agricultural sector where quotas began to be sold in a progressive rate. Farms adjusted by adopting water saving technologies and related farm practices to reduce water consumption. In years when the country experienced quantities of renewable water sources close to a multi-year average level, water prices were only adjusted according to the consumer price index. In the years characterised by hydrological shortages or sharp increases in consumption, one could notice an increase in the motivation to find technological solutions, either pushing innovation or simply adopting technology that previously was not economically feasible. For example, the recent five consecutive years of drought led to a significant increase in water prices. In 2009, an additional "surplus use" fee has been imposed on domestic uses, to discourage excessive water consumption. During these years, one could observe establishment of many water technology start-ups and also implementation of technologies at all scales – from home water-saving devices to accurate reading of water meters to establishment of new desalination plants. Also, stricter environmental enforcement activities and litigation in the area of urban and industrial sewage disposal have increased innovation and adoption of sewage treatment technologies in a multitude of ways.

While being unique and dynamic, Israel's market is small and has limited opportunity in local growth. In addition, although improving in recent years, the market lacks awareness of the worldwide potential in the government and private sectors. There are inefficiencies in government financial support in the industry and not many venture capital funds are willing to carry large R&D. Lack of finance to build beta-site plants also delays entrance to foreign markets.

Nevertheless, policies have had a large impact on the Israeli water sector. As of 2007, 270 water-technology companies operated in Israel, employing almost 8 000 people. About 60 companies among the 270 were start-up companies, established after 2001, and were involved in R&D. In addition, exports of the water technology sector grew from USD 700 million in 2005 to some USD 850 million in 2006, a 21% increase. In 2007, exports were estimated at around USD 1 100 million, a 28% increase on the year previous.

Water technologies relating to water demand, such as water efficient irrigation technology, were estimated at USD 300 million in 2007, 30% growth per year, produced by three major Israeli companies. Another technology area that is growing quickly and is oriented to domestic water use is monitoring and water metres. On the water supply side, some 50 companies associated with conveyance systems, valves, etc. have employed around 3 000 employees and generated USD 430 million in 2007. Desalination firms are operating on a larger scale in Israel in recent years following policy support of sea-water desalination production. Previously, these firms operated mostly abroad. The area of wastewater technologies attracts start-ups and some 60% of the start-ups in water technologies are in this area.

Conclusions

The Israeli case study clearly highlights the power of prices to induce change behaviour among water uses, with the shifts seen between types of water used by agriculture being a clear example and the efficiency of agriculture with respect to water use per unit of output. Prices also stimulated wide adoption of innovation, such as with new irrigation equipment or new water-saving techniques. At the same time, the contemporaneous impact of government efforts to find innovative means to secure fresh water supplies (such as through desalination plants) further extended innovation in this area. For such reasons, providing clear linkages between water pricing and innovation creation is somewhat more difficult.

For more information on water policy in Israel, the full version of the case study (OECD, 2009) is available at *www.olis.oecd.org/olis/2008doc.nsf/linkto/com-env-epoc-ctpa-cfa-rd(2008)36-final*.

Reference

OECD (2009), *The Influence of Regulation and Economic Policy in the Water Sector on the Level of Technology Innovation in the Sector and its Contribution to the Environment: The Case of the State of Israel*, OECD, Paris, available at *www.olis.oecd.org/olis/2008doc.nsf/linkto/com-env-epoc-ctpa-cfa-rd(2008)36-final*.

ANNEX C

Cross-country Fuel Taxes and Vehicle Emission Standards

This case study looks at the effect of emissions regulations, fuel efficiency standards, petrol prices and petrol taxes on innovation in the motor vehicle industry, focusing on the United States, Germany and Japan. The study finds that regulations on emission standards have generally induced innovation in related areas (for example, nitrous oxide emission regulation and innovations in engine design). The effects of petrol prices and petrol taxes on patenting are not as straightforward. Fuel taxes (which can be predicted) had an impact on innovations related to fuel efficiency, whereas petrol prices and fuel efficiency standards did not. However, further analysis of the interplay of taxes and prices highlight some of the empirical issues that result from analysing the innovation impacts of taxation.

Rationale for the environmental policy

By the combustion of fuel, motor vehicle use causes a wide range of environmental issues, compounded by the scale of motor vehicle use across the globe: smog, acid rain, climate change, and others. Many instruments have been used by governments to tackle these various challenges: fuel taxes, regulatory standards on specific pollutants, taxes on vehicles and driving, and fuel efficiency standards. This study focuses on fuel taxes and regulatory standards (both for specific pollutants and for fuel efficiency).

On the one hand, environmental outcomes are clearly top-of-mind with the use of regulatory approaches. These approaches have set out upper limits of pollution intensities (or fuel efficiency) in order to bring about significant reductions in emissions levels. On the other, the rationale for fuel taxes is less clear. These instruments have historically been implemented because they provide a relatively stable base on which to levy taxes and therefore provide a revenue stream for governments. Although not necessarily intended to have an environmental impact in the early years, increased taxes can impact the quantity of fuel used and types of fuel purchased by drivers. Over the last few decades, fuel taxes have been seen as instrument to achieve environmental goals, such as with differential taxation on leaded and unleaded fuels.

Design features

Fuel taxes

Fuel taxes are used in every OECD country and generally provide a significant revenue stream for governments. The development of diesel excises over time is presented in Figure C.1; the trends are quite similar for unleaded petrol. Remarkable differences exist between the countries, in particular between the United States, Japan and Germany. At face value, the variation appears quite similar, in particular because (real) excise rates in the United States were generally constant over time. There seems to be some convergence for European Union member states, due to harmonisation efforts and the implementation of a minimum diesel excise rate within the European Union. Both Japan and the United States had relatively low levels until 1985, whereas Germany rapidly lowered their rates to almost similar levels in this year. Since then, Germany increased levels gradually over time, in particularly after 2000 and Japan more or less followed this pattern though at considerably lower levels.

Figure C.1. **Excise tax rates on diesel in select OECD countries**

Tax rates per litre in real 2000 USD

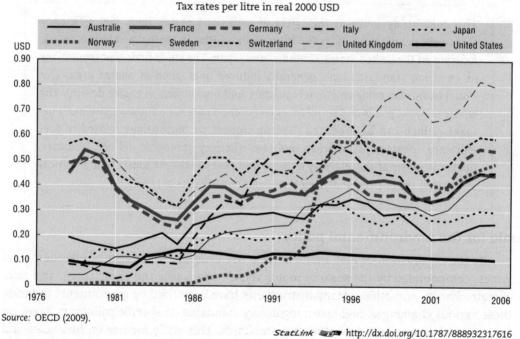

Source: OECD (2009).

StatLink 🔗 http://dx.doi.org/10.1787/888932317616

Tailpipe standards

The United States, European Union and Japan have all introduced increasingly stringent tailpipe standards on car exhaust for CO, HC and NO_x and PM. Figure C.2 provides an example of the development over time of HC and NO_x standards in the United States, European Union and Japan. Some interesting observations of the pattern of the regulations can be made:

- US regulations were introduced rather early. Restrictions became more stringent in the 1970s for both petrol- and diesel-driven cars, but remained rather generous since this initial initiative. Overall restrictions have always been much more lenient than those in Japan with the exception of regulation for HC.

Figure C.2. **Regulatory tailpipe limits for petrol-driven vehicles**

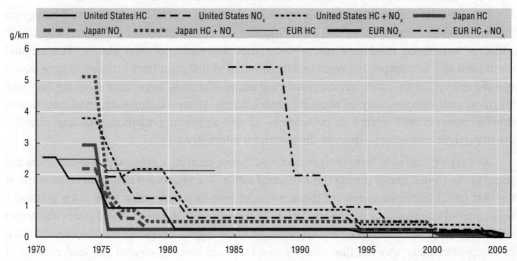

Source: OECD (2009).

StatLink ⟨⟩ http://dx.doi.org/10.1787/888932317635

- Japan introduced regulations for CO, HC and NO_x somewhat later than the United States, but these regulations have been particularly strict from the outset. Only regulation for diesel cars has been more lenient, probably because the share of diesel cars was also very small throughout the sample period.

- The European Union was typically late and rather lenient for most exhaust gases from the very beginning, probably due to its initial limited regulatory power. Since the introduction of the Euro I standard in 1992, the standard-setting process in the European Union has rapidly caught up with, and subsequently sometimes even appears to outrun, the stringency of regulations in the United States under Euro III. Although care should be taken in comparisons based on absolute standards, the differences in level seem to have become much smaller over time and Japan's regulations tend to remain the strictest for the three exhaust gases considered.

- The difference in regulation between petrol- and diesel-driven cars can be substantial, particularly in the European Union, where diesel cars obtained a substantial market share rather early. CO standards became even stricter for diesel cars compared to petrol cars starting in 1996. In the United States, where diesel cars make up only a small share of the passenger fleet, no such differences exist for CO; Japan has similarly equal standards. As to the regulation of HC and NO_x, substantial differences can be observed. Particularly in the European Union and Japan, standards have always been considerably more stringent for petrol-driven cars.

- Regulation of particulate matter (PM) is rather recent. Here, regulation started only in 1990 with the European Union leading. Indeed, the share of diesel driven cars rapidly increased in the 1980s, particularly in Germany with its relatively (compared to petrol) low diesel tax. When in Japan the share of diesel gradually increased as well, regulations were also tightened. The European Union typically took the lead with their Euro I-III standards in the 1990s. Since 2000, further restrictions could be observed in all areas.

Fuel quality regulations

Regulation of fuel quality is mainly related to the quality of the combustion technology on the one hand and emissions of CO, HCs, NO_x and PM on the other. In particular, anti-knock additives have been used to improve detonation resistance of fuel blends. The original motivation was to improve the combustion potential of fuel (and thus increase engine power and durability). In the past, various lead-containing additives were used because this was the most cost-effective way of boosting octane levels. However, environmental and health considerations of lead-related air pollutants – as well as the incompatibility of lead with the use of catalytic converters – spurred the search for alternatives.

As a result, lead standards were introduced, hence creating a gradual phase-out of leaded petrol in the United States during the 1970s and 1980s. The phase-out of lead in Japan – one of the first OECD countries to reduce the amount of lead in petrol – also took place gradually. Japan started its phase-out during the 1970s; by the early 1980s, only 1-2% of petrol contained lead. The production and use of leaded petrol has now been fully eliminated in Japan. Finally, in Europe, Germany was the first country to adopt standards to control the lead content of petrol. In 1981, the European Union set a standard of 0.4 grams of lead per litre, which lagged almost a decade behind the German law. As of October 1989, all European Union member states had to offer unleaded petrol, with a maximum of 0.15 grams of lead per litre. The 1998 Aarhus Treaty required the use of only unleaded petrol by 2005.

Policies aimed directly at improving fuel efficiency

Mandatory fuel efficiency requirements, which typically apply to the average of a fleet of cars with specified weights, are exceptional across the world. In fact, the only example is the application of the Corporate Average Fuel Economy (CAFE) standards in the United States introduced in 1978. After an initial increase in stringency, the gradual tightening was shortly relaxed after 1984 when it was quite stringent. Since 1989, however, the standard has never been changed. In contrast, voluntary schemes have been applied much more often in OECD countries, such as Germany and Japan. Recently several countries have negotiated with car manufacturers and importers to further improve fuel efficiency in order to reduce car-related greenhouse gases like CO_2.

Policies in combination

It is important to note that the relationships regarding the formation of different pollutants and other factors (fuel efficiency, power, etc.) are complex, as is suggested by Figure C.3. The figure suggests that maximum power is obtained for a slightly rich mixture (less air to fuel), while maximum fuel economy occurs with slightly lean mixture. During the period before emissions regulations were introduced, cars were thus designed to run on richer mixtures for better power and performance.

However, a rich air-fuel mixture leads to production of relatively large amounts of CO and unburned HC emissions, since there is not enough oxygen for complete combustion. A lean mixture helps reduce CO and HC emissions – unless the mixture becomes so lean that misfiring occurs. Hence, after the first regulations of CO and HC emissions were introduced in the 1960s in the US, the initial response of manufacturers was to redesign cars to run on a less rich mixture (introduction of air-to-fuel ratio devices). The introduction of catalytic converters, which have their own exacting specifications for efficiency, further presents issues of optimality across the range of pollution issues.

Figure C.3. **Engine calibration and emission levels**

Effect of air-fuel ratio on emissions, power, and fuel economy (petrol engines)

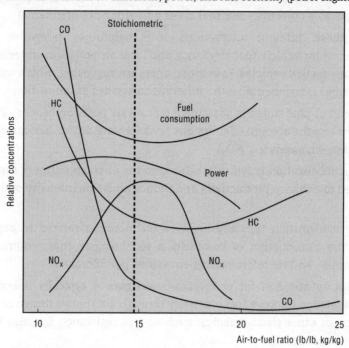

Source: Masters and Ela (2008).

Innovation impacts of environmental measures

Of interest are patents as an observable output-indicator of R&D activities related to innovation within the automobile sector, as a result of taxes, regulations and other forces. The assumption is that environmental policy – whether this is through a standard or a specific tax – signals to (new) producers that it is beneficial to be engaged in dedicated R&D to meet the requirements of the standard or to reduce tax payments. If this is indeed the case, one would expect a rise in R&D activity specifically dedicated to the invention of new technologies (products) or the improvement of existing ones addressing the concern as signalled by the regulatory device.

New technologies can be expected from regulations and taxes that address major pollutants emitted by motor vehicles: carbon monoxide (CO), hydrocarbons (HCs), nitrogen oxides (NO_x), particulate matter (PM), lead, sulphur dioxide (SO_2) and volatile organic compounds (VOCs). In the automobile sector, the relevant new technologies or products would involve not only changes in petrol and diesel engines of cars, but also cars driven by entirely new engines, as well as changes in the design of the cars to increase fuel efficiency. The effects of these policies on different emissions can be complicated and there are interactions between policies targeted on different pollutants. Several aspects need to be considered:

- Pollutant-by-pollutant regulation can induce engineering trade-offs and hence may lead to perverse effects (*e.g.* emission standards for NO_x may actually increase fuel consumption, and thus CO_2 emissions).

- Type of policy instrument generally differs by emission – emission standards (CO, HC, NO_x, PM) *versus* fuel taxes (CO_2 indirectly, sulphur and lead in some cases).

- The inter-relationship between different variables of interest, such as the additive effects of pre-tax fuel prices and fuel taxes, and the joint use of policy measures to achieve comparable objectives (*e.g.* fuel taxes and efficiency standards).

Given all these different interactions, it is helpful to categorise the potential inventions relevant for vehicle fuel efficiency and local air pollution emissions abatement for conventionally fuelled vehicles. Four broad areas are suggested, which will help identify the effect of various instruments on the different categories of innovation:

- First, typical end-of-pipe emission abatement for cars are post-combustion (after-treatment) devices that reduce the amount of emissions per kilometre driven, like catalytic converters, lowering tailpipe emissions (*e.g.* NO_x).

- Second, input substitution is typically related to the characteristics of the fuels and the additives used to enhance productivity and reduce emission intensity of the combustion process.

- Third, factor substitution typically involves technologies related to engine redesign, *e.g.* through the introduction of combustion technologies that require less fuel per kilometre driven – and therefore reduce emissions per kilometre.

- Fourth, output substitution for petroleum-based cars is typically linked to measures primarily designed to improve fuel efficiency through alternative design of cars, like their aerodynamics, or other characteristics, such as tyre resistance, but also substitution of materials to decrease weight.

Like in other areas of environmental innovation, the most important of the major car-producing countries are Japan, Germany and the United States for the specific areas of innovation that are being investigated. Together these countries account for roughly 89% of the overall number of patents, with Japan filing by far the largest number of patents with its contribution of almost half of the overall number of counts (47.2%), followed by Germany (28.3%) and the United States (13.7%).

The evolution of the number of patent applications in Japan, US and Germany for the period 1965-2005 is shown in Figure C.4. Hardly any innovative activity is present in the first part of the period. Apart from a spike around 1975 in Japan, patenting activity increases steadily from the early 1970s. After an initial rise of patenting activity in the 1970s, there is more or less stabilisation until 1995 when another five-year take-off period can be observed, in particular in Germany. Overall, patent activity grew steadily in these countries until almost the end of the sample period, and this trend was particularly prominent and early for Japan and Germany.

In order to describe when innovation in each technological category occurred, Figure C.5 plots the number of patent applications of each group for the period 1965-2005. In particular, the largest technological subfield, input combustion, shows an upsurge both in the 1970s and again between 1995 and 2000, as well as a sharp relative decline since 2002. Patenting of tailpipe technologies ("emissions") shows a remarkably steady increase over time, with only a sharp increase in the years preceding 1975 and 1998. To a great extent, the evolution of patenting in the domain of emissions-related technologies is similar to the pattern for input combustion; however, it is always at a considerably lower absolute level of patent applications. Patents for technologies that directly reduce fuel consumption through an improvement in aerodynamics or rolling resistance tend to increase steadily in the 1980s, with a clear peak

Figure C.4. **Patent applications for relevant vehicle technologies**

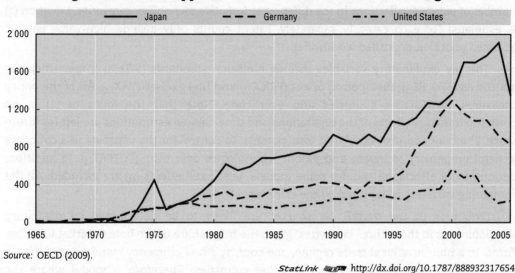

Source: OECD (2009).

StatLink http://dx.doi.org/10.1787/888932317654

Figure C.5. **Patent applications for the four technological categories**

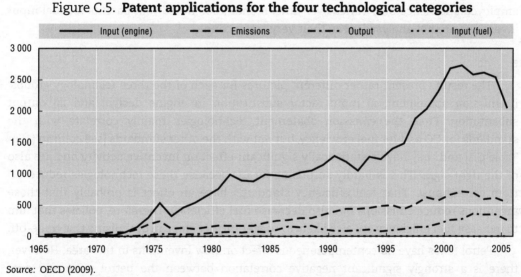

Source: OECD (2009).

StatLink http://dx.doi.org/10.1787/888932317673

in 1986-88, and reveal again a sharp boost in the years before 2002. Then, as with the other technological domains, the degree of patenting goes down again. Finally, for patenting related to input fuel technologies, hardly any activity seems to be occurring for the period 1965-2005.

The model

The empirical model to investigate the effect of public policy (standards, taxes) and other determinants on inventive activity in the main automotive technology classes takes the following form:

$$ENVPAT_{i,t} = \beta_1 STD\_X_{i,t} + \beta_2 STD\_FE_{i,t} + \beta_3 PRICE_{i,t} + \beta_4 TAX_{i,t} + \beta_5 R\&D_{i,t} + \beta_6 TOTPAT_{i,t} + \alpha_i + \gamma_t + \varepsilon_{i,t} \qquad (1)$$

where i indexes country and t indexes year. The dependent variable is measured by the number of patent applications in the different automotive technology categories. Equation (1) is estimated for each category separately. Patent counts only include high-value patents (claimed priorities, deposited worldwide).

The key explanatory variables include emission standards ($STD\_X_{i,t}$), fuel efficiency standards ($STD\_FE_{i,t}$), fuel (petrol) prices ($PRICE_{i,t}$) and fuel excises ($TAX_{i,t}$). All of the policy measures vary across countries and over time. Note that the focus of (1) is on contemporaneous effects of the regulations and time-related estimations are left for future work. The major control variable is total patents, to control for the variation in a country's general propensity to invent and patent technologies over time ($TOTPAT_{i,t}$). In addition, country fixed effects (α_i) and, for some models, year fixed effects (γ_t) are included. All the remaining variation is captured by the error term ($\varepsilon_{i,t}$).

Dynamics in the overall car market are likely to be determined by regulatory developments in these three countries, given the huge share in the home market for these firms. In a non-autarkical trade regime, one country's fuel efficiency standard might have repercussions for inventive activity in other countries. Therefore, a model where the variable STD_FE_t represents the lowest efficiency standard in any of the three countries is employed and hence only varies over time. It should be noted that patents for fuel input inventions are not analysed given their very small count.

Results

The results present rather different pictures for each of the three technology groups: i) emission abatement; ii) input factor substitution in engine design; and iii) output substitution. First, the emission abatement technologies mainly correlate with the standards for CO and for fuel efficiency, but not with the other standards [see column (1) in Table C.1] and they have a statistically significant effect on inventive activity and are also of the right sign.* This is hardly surprising for CO because these technologies reduce CO from car exhaust. That fuel efficiency standards have an effect is probably that these inventions reduce emissions but also decrease fuel efficiency. Therefore, policies that aim to increase fuel efficiency are also likely to trigger further steps in optimising this trade-off.

Petrol taxes have no contemporaneous effect on new inventions in this area. However, there is a strongly significant negative correlation between the petrol price and new inventions. This negative correlation exists across all specifications for emission abatement technologies with the exception of adding time fixed effects [see column (3) in Table C.1]. Adding time fixed effects to the standard model, however, lowers the explanatory power of equation (1), suggesting model (1) as the base model. An explanation for the negative correlation is that rising (or falling) petrol prices are unlikely to have a contemporaneous effect on inventions. Oil price spikes are usually unexpected and the first reaction by consumers is to reduce consumption of fuel by driving less and buying more fuel efficient cars from the existing stock of car models. This demand side reaction already reduces emissions on its own and therefore signals to inventors less pressure for inventing new technologies that control emissions.

---

* Note that the fuel efficiency standard is measured in litres of fuel per 100 kilometres driven; hence, the expected sign of this variable is negative. For the standards, the measurement is km/g; hence, the expected sign is positive.

Table C.1. **Empirical results: Emission abatement technologies**

	(1)	(2)	(3)
Standard CO	9.30***	8.33***	9.54***
	(2.84)	(2.94)	(2.75)
Standard HC	−0.78	−0.70	−0.95
	(0.61)	(0.63)	(0.71)
Standard NO$_x$	1.60	2.70	−2.93
	(4.07)	(4.24)	(5.27)
Standard PM	−0.38	−0.65	−0.13
	(0.75)	(0.80)	(0.97)
Standard FE	−3.00***		−0.52
	(1.13)		(1.28)
Standard FE (low)		−0.09	
		(1.25)	
Petrol tax	5.67	39.42	−209.04***
	(60.76)	(61.60)	(68.90)
Petrol price	−67.01***	−96.46***	101.16***
	(22.21)	(22.35)	(36.41)
Time fixed effects	No	No	Yes
Adjusted R^2	0.76	0.74	0.65

Note: All regressions include a control for total patents and country fixed effects, and were performed with OLS. They also each have 108 observations and three groupings. P-values in parentheses, based on robust standard errors.
* $p < 0.05$.
** $p < 0.01$.
*** $p < 0.001$.
Source: OECD (2009).

StatLink ᴍᴤᴘ http://dx.doi.org/10.1787/888932318205

These basic findings are robust to the exclusion of correlated standards such as the NO$_x$ standard. However, there is no evidence for the hypothesis that inventors of emission abatement technologies are responsive to the strictest worldwide contemporaneous fuel efficiency standards [see column (2) in Table C.1]. Although the other effects are hardly affected, the strongly significant negative effect of local regulation disappears. This suggests that inventors of new technology are mainly driven by local policy measures, just as has been observed for SO$_2$ and NO$_x$ abatement technologies for electric power plants in other studies.

The results for the most important technology group in terms of counts, the input technology category, are quite different [see column (1) in Table C.2]. Clearly CO has no effect on the overall number of patent counts for the underlying technologies, whereas NO$_x$ reflects a strongly positive effect in this case. CO and NO$_x$ standards appear to have a complementary effect on this type of invention because CO becomes significant if this model is re-estimated without the somewhat problematic NO$_x$ standard. Somewhat surprisingly, however, are the results for both HC and PM, as both standards appear to reduce contemporaneous inventive activity. Looking more carefully in the original data of Germany and Japan, it appears that this type of inventive activity peaked at the end of the 1990s, which is several years before further restrictions were introduced, in particular Euro IV in the European Union. This also fits observations that Euro IV regulations created pressure on the automobile industry to find new ways to reduce the main pollutants from car exhausts jointly, particularly also for diesel cars. This explains why the standards for HC and PM appear to have had even a negative impact, because they were tightened before and particularly after the main inventive period.

Table C.2. **Empirical results: Input (improved engine design) technologies**

	(1)	(2)	(3)
Standard CO	−11.58	−15.60*	−9.24
	(8.96)	(8.78)	(7.41)
Standard HC	−8.83***	−8.13***	−8.36***
	(1.91)	(1.88)	(1.91)
Standard NO$_x$	57.05***	62.90***	40.12***
	(12.83)	(12.63)	(14.12)
Standard PM	−6.25***	−8.13***	−5.54***
	(2.36)	(2.38)	(2.60)
Standard FE	−4.49		0.59
	(3.55)		(0.86)
Standard FE (low)		8.72**	
		(3.72)	
Petrol tax	456.34**	491.81***	−223.65
	(191.37)	(191.37)	(185.62)
Petrol price	−78.35	−196.32***	468.72***
	(69.96)	(66.64)	(98.10)
Time fixed effects	No	No	Yes
Adjusted R^2	0.90	0.90	0.89

Note: All regressions include a control for total patents and country fixed effects, and were performed with OLS. They also each have 108 observations and three groupings. P-values in parentheses, based on robust standard errors.

* $p < 0.05$.
** $p < 0.01$.
*** $p < 0.001$.

Source: OECD (2009).

StatLink ⟐ http://dx.doi.org/10.1787/888932318224

The strong positive effect of the petrol tax on engine redesign technologies is also remarkable. This effect is statistically even stronger if the model is re-estimated with the lowest fuel efficiency standards [column (2) in Table C.2] or without the (insignificant) standard for fuel efficiency (not included). However, this result fails to pass several robustness checks, including adding time fixed effects [see model (3) in Table C.2]. Somewhat surprisingly, the signs of both tax and petrol price switch, whereas only the petrol price remains significant if time fixed effects are allowed. Again this specification is robust to both inclusion or exclusion of different variables in specification (1) including petrol tax and price individually. As such, this result is not robust enough to state that increasing petrol taxes induce innovations in car engine technologies.

A final set of estimations looks at the main drivers of the output technologies, mainly fuel efficiency improvement technologies. One would typically expect fuel efficiency standards to be the most important driver here. However, neither these measures nor a positive contemporaneous effect by fuel market prices seem to have had an effect at all [see Table C.3, column (1)]. The most important driver, however, is petrol taxes. The positive effect for taxes is confirmed by other specifications, including one with the lowest fuel efficiency standard [model (2)], and a model without NO$_x$ standards which controls for potential multicollinearity with other standards [model (3)]. In addition, adding time fixed effects does not change this strong correlation [models (4) and (5)]. So, increasing petrol taxes induces inventors strongly to invest in new technologies, in particular in inventions that reduce fuel use per kilometre driven directly.

Table C.3. **Empirical results: Output technologies**

	(1)	(2)	(3)	(4)	(5)
Standard CO	−2.78**	−2.99**	3.52***	−1.64	−1.40
	(1.29)	(1.27)	(0.93)	(1.20)	(1.16)
Standard HC	−0.97***	−0.91***	−0.14	−0.56*	−0.56*
	(0.28)	(0.27)	(0.28)	(0.31)	(0.30)
Standard NO$_x$	11.57***	11.96***		6.40***	5.85***
	(1.85)	(1.83)		(2.30)	(2.20)
Standard PM	−1.60***	−1.75***	−0.06	−1.31***	−1.24***
	(0.34)	(0.34)	(0.28)	(0.42)	(0.41)
Standard FE	0.29		−0.04*	0.18	
	(0.51)		(0.60)	(0.56)	
Standard FE (low)		1.06*			
		(0.54)			
Petrol tax	108.05***	103.00***	106.34***	88.27***	73.40***
	(27.62)	(26.55)	(32.54)	(30.10)	(24.52)
Petrol price	−32.34***	−38.25***	−17.93	−13.87	
	(10.10)	(9.63)	(11.58)	(15.91)	
Time fixed effects	No	No	No	Yes	Yes
Adjusted R^2	0.66	0.65	0.63	0.80	0.80

Note: All regressions include a control for total patents and country fixed effects, and were performed with OLS. They also each have 108 observations and three groupings. P-values in parentheses, based on robust standard errors.
* $p < 0.05$.
** $p < 0.01$.
*** $p < 0.001$.
Source: OECD (2009).

StatLink ⛓ http://dx.doi.org/10.1787/888932318243

Estimating the sensitivity of patenting of output technologies for the tightening of emission standards produces similar results compared to the patenting of engine redesign technologies at first sight. In this case, however, the results are quite sensitive to multicollinearity problems caused by the inclusion or exclusion of the NO$_x$ standard. Without this standard, the estimations produce a very simple and intuitive story [see model (3)]. Not only are the other emission standards no longer significant (including those with negative signs), but also the fuel efficiency standard and the CO standard have the expected signs. Also the negative effect from the real petrol price disappears in that case. All of these results do not fundamentally change if time fixed effects are controlled for.

Conclusions

Important regulatory interventions by governments in Germany, Japan and the United States have induced serious inventions in the car market. Specifically, key findings from this case study include:

● In inducing innovation, regulatory pressure (including taxes) is much more important than changing net-of-tax petrol prices. This is particularly true for contemporaneous innovations, since inventors may react slowly when they are taken by surprise (rising oil prices are notoriously difficult to predict and, therefore, anticipate).

● There is some evidence that standards, in particular for CO and to a lesser extent NO$_x$, strongly correlate with inventions in the main technology groups distinguished in this paper, emission abatement ("emission"), engine redesign ("input") and fuel efficiency ("output") technologies.

● Petrol taxes seem to have had an impact, in particular on the technologies that increase fuel efficiency. This may be due to the fact that such taxes can be anticipated by innovators and automobile manufacturers may be able to gain market share by selling consumers vehicles that reduces fuel use (because of rising excise taxes on motor fuel).

● Somewhat remarkable is the limited effect observed for fuel efficiency standards, particularly for inventions in fuel efficiency and engine redesign technologies. For emission abatement technologies, an effect is observed but only from local policies, including negotiated agreements.

Clearly these conclusions are conditional on further work that should be undertaken. The simplest and clearest observation is that the estimation methodology used so far should be subject to further refinement, like the use of count data methods and the inclusion of other countries. Potentially more important, however, is that new and convincing hypotheses could be built on a deeper analysis of how regulation and technologies are related. Both the technologies involved, as well as the regulatory interventions, have many relevant dimensions that sometimes, but not always, are closely linked, such as the serious technical trade-offs in controlling pollution. There are also likely effects from inventions, which are mainly limited to specific countries, but easily cross borders as embodied technologies in new models. Finally, there is the area regarding how regulators interact and respond to autonomous or regulation-driven changes in the car market. For instance, the growing number of diesel cars in Germany forced the regulators to respond by increasing exhaust regulation, in particular PM, but also seems to have been the result of its own fuel tax policy, with petrol taxes increasing more compared to diesel taxes.

For more information on fuel taxes and emission standards, the full version of the case study (OECD, 2009) is available at *www.olis.oecd.org/olis/2008doc.nsf/linkto/com-env-epoc-ctpa-cfa(2008)32-final*.

References

Masters, Gilbert and Wendell Ela (2008), *Introduction to Environmental Engineering and Science*, Third Edition, Prentice Hall, Upper Saddle River, NJ.

OECD (2009), *Fuel Taxes, Motor Vehicle Emission Standards and Patents Related to the Fuel Efficiency and Emissions of Motor Vehicles*, OECD, Paris, available at *www.olis.oecd.org/olis/2008doc.nsf/linkto/com-env-epoc-ctpa-cfa(2008)32-final*.

ANNEX D

Switzerland's Tax on Volatile Organic Compounds

> This case study looks at the innovation impacts of Switzerland's tax on volatile organic compounds. Introduced in 2000, the tax covers all emissions of VOCs in Switzerland – both in the production and in the consumption of products containing them. Focusing on three industries, the case study found that innovation did take place by firms. The vast majority of it, however, consisted of incremental innovations and homemade solutions that were not patented. Also highlighted were the barriers that individual firms face to innovating, such as capital equipment sourced from a large manufacturer. The tax on VOCs appears to have also led to significant environmental improvement.

Rationale for the environmental policy

Volatile organic compounds (VOCs) comprise a wide variety of chemicals, characterised by their ability to vaporise quickly and their non-aqueous nature. Although the broad definition generally includes substances such as methane, hydrocarbons and ozone-depleting substances, focus is usually placed on a more limited definition of substances relating to solvents (alcohols, acetone, benzene, etc.). They can be found in various products like paints, varnishes and some detergents and are used in many industries for cleaning purposes, including metal fabrication and dry cleaning. Released into the atmosphere, they interact with nitrous oxides to form high concentrations of ozone at low altitude (summer smog). They are also known to have negative human health effects for exposed workers.

Design of the instrument

The enabling legislation for the VOC tax entered into force 1 January 1998, and the tax was levied from 1 January 2000, with a rate of CHF 2 per kg. The tax was increased, as planned, to CHF 3 per kg at the beginning of 2003. The tax does not apply to all products classed as VOCs, partly because of the excessive administrative burden for customs clearance. Therefore, there is both a "positive list of substances" (*e.g.* benzene, butanes, ethers) that are VOCs themselves as well as a "positive list of products" (*e.g.* solvents, colorants, paints, perfumes, beauty products) for products containing VOCs that are subject to the tax.

As emissions are difficult to measure within a given firm, VOCs are taxed on entry into production and on importation into Switzerland. Imported products containing VOCs are taxed on importation according to the quantity of VOCs they contain. Products manufactured in Switzerland are taxed indirectly through the tax already levied when VOC substances are purchased. The VOCs remain liable for the tax if they escape into the environment or if they are sold (transferred) to Swiss consumers. However, VOCs exported as substances or in products not liable to the tax are exempt because they are not released into the environment in Switzerland. To account for all these imports, exports and uses of VOCs, firms are required to keep a VOC balance sheet.

Exemptions likewise apply to VOCs in products whose VOC content does not exceed 3% and to VOCs in products not included in the positive list. In addition, firms that have taken measures on a stationary installation and reduced emissions significantly below stipulated limit values can be exempt from the tax. These limits refer to levels that are 30% lower than the maximum limit (since 31 December 2003) and 50% lower (since 31 December 2008).

The direct effect of the tax is to increase the cost of making products with a VOC content of more than 3%. If the tax is passed on, products intended for the domestic market become more expensive to buy. In that respect, Swiss and foreign products are treated alike in tax terms. Exemption from the tax for exported products helps to keep Swiss products with a VOC content of more than 3% competitive on export markets. That is no longer the case if production costs in Switzerland increase because VOCs that escape into the environment during production are taxed. Under these circumstances, Swiss products made using taxed VOCs are at a disadvantage in Switzerland in comparison with substitution products, and in other countries in comparison with competing untaxed products.

On the domestic market, the increase in the relative price of products that are more expensive to produce on account of the tax discourages consumption of such environmentally harmful goods and services. Thus, final and intermediate consumers are encouraged to shun products whose manufacture is a source of emissions in favour of cheaper and potentially less harmful substitution products (if they exist). Firms can react in two ways, depending on whether the problem lies with the production process or the product:

- They can reduce emissions by changing the production process. Firms may be expected to "innovate" if their current and future (discounted) direct and indirect costs are lower than the tax they would otherwise have to pay. Firms that use small quantities of VOCs thus have little incentive to innovate in order to further reduce VOC emissions.
- They can reduce or eliminate the VOCs contained in their products (or cut the concentration of VOCs to less than 3% by volume), as long as that does not significantly alter their quality or end use. However, they are unlikely to do so if the tax represents only a small fraction of the product's value.

Revenue rose from CHF 67 million in 2000 to a peak of over CHF 140 million in 2005, falling back to CHF 126.7 million in 2006 and 2007. It is estimated that the figure will level off at CHF 125 million annually over 2008-10. The tax, which is redistributed to the population, represents only 0.3% of federal revenue and 0.1% of all public authority revenue.

Environmental impacts of the tax

Emissions of VOCs liable to the tax fell significantly between 2001 and 2004, having already declined between 1998 and 2001. Table D.1 shows the estimated reduction for the most polluting industries; the reduction for all industries since 1998 is estimated to be around a third.

Table D.1. **Largest VOC reductions by industry**

Industries liable to the tax	Change 1998-2001		Change 2001-04	
	Tonnes	%	Tonnes	%
Industry, crafts and households	−9 700	−12	−17 200	−25
Paint applications	−3 100	−13	−11 000	−54
Printing	−1 800	−16	−4 900	−51
Metal cleaning	−700	−18	−1 100	−34
Wood protection applications	−270	−15	−730	−48
Emissions of solvents, miscellaneous	−200	−11	−500	−29
Hairdressing salons	40	5	−480	−59

Source: OFEV (2007).

StatLink http://dx.doi.org/10.1787/888932318262

Three activities – printing, metal cleaning/degreasing and paintmaking – were chosen as industries for case studies for the rest of the study. Printing was chosen because of its large emissions (4 179 tonnes in 2004), the fact that several studies had already looked at the Swiss print industry, and that the industry had also organised itself in the effort to reduce VOC emissions (making the industry easy to approach and the existence of technical documentation). A generally high level of emissions and a relatively large emission factor also spoke in favour of metal cleaning (2 065 tonnes in 2004). Cleaning as part of the metalworking process is an important activity in several industries, such as automobile parts, clockmaking, medical equipment, electrical engineering and machine construction. Finally, an activity with relatively low VOC emissions that hitherto has been the subject of little, if any, analysis was chosen: paintmaking (448 tonnes in 2004). Although the industry is relatively homogeneous, it has a wide range of applications (construction, wood, etc.).

VOCs are used differently among the three industries. In printmaking, VOCs can be found in the inks, colours, and toners (to help the finished product be crisp and of high quality), as well as being used to clean the printing machines. In paintmaking, VOCs are also used to clean the machines and equipment in addition to the fact that (generally) VOCs are a constituent part of the final product (paint, varnishes, lacquers) to help with drying. Metal cutting is somewhat different. To prevent oxidisation of metals after fabrication, greases are usually applied to protect them. VOCs are then used later on to remove the grease, providing a non-aqueous solution that limits oxidation.

A sample of firms in each industry was interviewed. The group should include, for each VOC-emitting activity, at least one small, one medium and one large firm, one firm belonging to a foreign group and one independent firm. About one third of the firms contacted finally granted an interview. Some, especially smaller firms, did not feel particularly concerned by the tax, either because they use negligible quantities of VOCs or were not motivated to take part in the survey, or because they regarded the tax as an "aberration" but were not inclined to say any more on the subject. Cantonal experts were questioned in parallel to evaluate the strength of the incentive to innovate generated by cantonal air protection services in managing the VOC tax and to see the innovation behaviour through the eyes of cantonal experts.

Generally, the VOC tax did cause changes in the three sectors analysed: activities generating VOC emissions were scaled back by adapting production processes (printing, paintmaking, metal cutting) or by putting new products on the market (paints). In the first case, this mainly involved raising awareness about and making less use of products that emit VOCs, washing and cleaning products and solvents, and replacing them with water-based products. In the second case, it involved new products like water-based and solvent-free paints and paints containing few solvents (*e.g.* high solid paints). In particular, in the metal cutting industry, the VOC tax was a decisive financial incentive for changing the way they cleaned parts. Distilling plants, benzene jars (to capture VOCs) and replacement cleaning products alone, which did not require large-scale investment, enabled them to reduce their emissions by an appreciable amount, perhaps by around 10-20%.

These emission declines occur for a wide range of reasons, with the VOC tax being a significant driver. However, the amount of the tax sometimes seems rather small, or even negligible, in relation to the product cost or price or sales, particularly for large firms. However, other factors were also quite important in accounting for reductions in VOC emissions. Increases in employee health due to improved air quality are important. Greater awareness among employees about the use of VOCs in production processes, greater know-how and the rising price of alcohol (a substitute) in relation to water are additional factors. In the printing industry, the advance of digital printing may also explain some of the changes in the production process. Growing environmental awareness among customers seems to favour a move towards more environmentally friendly production processes and products. In paintmaking, customers are demanding less solvent use in products (because of the smell). Finally, the existence of other regulations is encouraging demand for reduced-VOC products (such as EU directives on paint that have a large impact on Swiss paint exports).

Yet, there are limits to emissions reductions. In printing, for reasons of quality and perhaps also price, some printers – depending on their type of output (high-quality work, for example) – still appear to prefer VOC-based products. All the printers interviewed agreed that a minimum level of alcohol is still necessary in printing to guarantee high quality and to keep presses productive. Progress has been made in cleaning products and detergents, but again it often seems difficult to do away entirely with products containing VOCs, especially for productivity reasons. Another difficulty is that manufacturers of printing presses often advise against using VOC-free products, even on the latest machines, with possible consequences for the warranty. For paints used in construction, which are the most important products for the firms in the survey, climatic conditions (low temperatures and humidity) often mean that it is not possible to use only water-based products. With respect to metal cutting, there is still scope for further reductions by replacing VOC products but it is acknowledged that the use of non-VOC substitutes and the necessary changes to procedures are tricky, partly because the quality of degreasing and drying is often not (yet) guaranteed and partly because parts made of steel or iron begin to rust on contact with water, an irreversible process.

Finally, it seems that abolishing the tax would not reverse changes to manufacturing processes or products. The health and safety advantages of making less use of products containing VOCs are undeniable. In addition, environmental awareness within firms has increased, partly as a result of the VOC tax, and they are increasingly starting to play the "green" card.

Innovation impacts of the instrument

Printing

At first sight, most of the changes that have taken place are based on the introduction of existing formulae or technologies that reduce VOC consumption and emissions throughout the production process. In the printing industry, less use is made of isopropylic alcohol in production processes, both in printing itself and in cleaning products (used to clean the rollers in offset printing). The printing presses are often the same and it is up to the printers to find the right dose of alcohol and seeking a technical solution to minimise VOC use by varying the printing technique, the VOC content of colours and the water quality.

Apart from making less use of products containing VOCs when cleaning the equipment after each print run, the challenge consists in reducing the alcohol content responsible for reducing the surface tension of the water in the ink on contact with the print medium. Thus, printers are increasingly moving towards zero alcohol use in ink and colours, even though the goal is still difficult to achieve (technically and financially) for the same level of quality. One problem in this context also lies in Swiss firms' lack of influence on foreign manufacturers of printing machines, who often advise against the use of VOC-free inks and colours. In contrast, where colours are concerned, producers seem more inclined to listen.

Efforts have been made to reduce VOCs in the products used to clean the rollers, but none of the firms in the study has been able to entirely eliminate products containing VOCs. However, the brand new press installed in one firm in 2008 can be cleaned with alcohol-free products. The same firm also uses an osmosis device to soften its water, which also cuts alcohol consumption and hence VOC emissions.

Thus, the changes observed in the printing industry can mostly be characterised as process innovation, since they concern machines, developed by the manufacturers, and production inputs: less and less isopropylic alcohol is used in the production process. Changes can also be observed among staff: printing with little or no alcohol is becoming an integral part of printers' know-how and there is a growing awareness of the need to use VOCs sparingly.

Testing new machines to achieve low-VOC production (in this case, less use of isopropylic alcohol) is often expensive for firms. The main problem lies in the quality of the finished product, which is difficult to maintain while using less alcohol. However, tests carried out by individual firms lead to changes in production processes that can be qualified as innovations. Most printers seem to belong to the category of firms that adopt and take up innovations. None of the firms interviewed had an R&D unit, reflecting the general situation in the industry.

Paintmaking

Changes in the paintmaking industry tend to involve the introduction of processes that make less use of VOCs. The use of solvents in manufacturing processes has often been greatly reduced (e.g. in acrylic varnishes) or entirely replaced by water-based products. In addition, low-VOC or VOC-free products are increasingly used during production, especially to clean tanks. For example, one firm has introduced a solvent-free tank cleaning system that cost CHF 450 000 but has enabled the firm to reduce its VOC emissions by 30 000 kilograms.

Similarly, in 2007, one firm bought a new cleaning device for CHF 300 000 in order to clean tanks used for products that do not contain VOCs. Other benefits of this measure include less risk of accident, less risk to health and a reduction in smells. Changes in manufacturing processes here also affect the end product. Some products can therefore be classed as new (*e.g.* high solid and aqueous varnishes), thus representing technological product innovations. This was the case for four of the seven firms interviewed, without counting innovations by the parent of one firm outside Switzerland. As long ago as the 1980s, the Swiss paint and varnish industry had already set itself the goal of reducing or even completely eliminating VOC-based solvents. The introduction of existing exhaust air-purifying technologies or, in the case of one firm, of end-of-pipe scrubbing represents a change that can be qualified as innovation. An industry-wide recycling scheme for customers is another major change.

Metal cutting

VOC emissions can be reduced by two changes in the production process:

- Replacing VOCs (traditionally used to degrease metal parts without residue by evaporation and, in much smaller quantities, to clean machines) with water-based detergents or bacterial systems.

- Using VOCs only in closed recipients and devices so that they can no longer escape into the atmosphere, including used product recycling. Emissions can be further reduced by changing working practices involving VOCs and recycling used substances.

Replacement means changing degreasing processes, although more needs to be done to identify and select detergents suited to the types and materials of manufactured parts, usually by repeated on-site testing. In many cases, the replacement products and procedures are not (yet) entirely satisfactory, and what works for one firm does not necessarily work for all. Not all the firms interviewed systematically co-operate with others in the same industry: each one has its own "recipes" and firms neither co-operate on research nor share their experience. Switching to detergents sometimes involves relatively substantial investment, like buying a detergent-based washer instead of or in addition to existing VOC-based equipment.

Closed-circuit degreasing devices are now standard in the metal cutting industry: the firms interviewed made the change before the tax was introduced or during the early years (2000-01). Production equipment benefits from technical advances made by manufacturers and suppliers, who are at least partially in tune with the environmental demands of clients and politicians. Greatly encouraged by the tax, lidded benzene jars are used extensively for regular controls of manufactured parts. However, quality control staff often do not close the lid after dipping the parts, since the operation may be repeated dozens or hundreds of times a day.

Only two of the firms interviewed have research and development activity *per se*. R&D focuses on improvements to existing equipment, the construction of specific inspection devices and greater efforts to optimise production processes. The biggest firm in the sample has joined forces with a manufacturer to develop a prototype benzene jar with an automatic lid that would be entirely airtight, to prevent evaporation, and use a shower system to degrease parts rather than having to dip them into the liquid by hand.

Unsurprisingly, the relatively large number of changes announced by firms in the industry relate entirely to technological process innovations. Changes are driven by the acquisition of new equipment incorporating technological advances and by the introduction of new procedures. All the firms have learnt and invested in new degreasing techniques after extensive on-site testing.

The tax seems to have significantly reduced VOC emissions for metal cutters. Three categories of firms can be distinguished:

- The three large firms in the sample, which have been dealing with environmental issues for twenty years or more and included the aim of reducing VOC emissions in their "strategic" concerns when the VOC tax was introduced, have brought in the most up-to-date production technologies and have adopted cutting-edge technologies in other matters like water and waste treatment.

- Two firms became aware of the environmental problem of VOCs (and of other issues like workplace health and safety) when they had to start paying the tax and took measures relatively quickly.

- Two firms that have taken what they describe as "restrictive" measures to reduce emissions and the amount of tax payable, without being convinced of the administrative or technical efficacy of the tax.

Cantons' viewpoint

The cantons consulted put forward a large number of examples of technological product and process innovation according to their dominant industries. Above all, they mentioned product improvements in paints, colours and solvents. Most of the cantons that took part in the study also noticed a clear reduction in the VOC content of cleaning products and detergents. But the downside of these successes is often the risk of a reduction in quality that in some cases cannot be tolerated, as in metal cleaning, for example. Given the current state of technology and existing substitution products, it is difficult for these activities to eliminate VOCs altogether.

Two types of process innovation have been introduced. The first, end-of-pipe innovations were generally introduced by big firms in large-scale installations before the VOC tax came into effect (e.g. an incinerator at a cigarette maker, a biological washer in a chemicals and pharmaceuticals plant). The second type of innovation concerns continuous improvement of the production process (printing).

Factors explaining the differences in firms' innovation behaviour

Several factors seem to explain the differences in firms' innovation behaviour. The most important seem to be the firm's products (e.g. book or newspaper printing, interior or exterior paints, oil paints and the impossibility of eliminating VOCs from some products), customer demand (customers may be more or less environmentally demanding), the size of the firm (smaller firms seem to have to make more effort to innovate), the existence of an R&D unit (generally the case with paintmakers) and, finally, the firm's own attitude towards the environment (integrated environmental strategy).

Many medium-sized and family firms take advantage of the need to renew technologically and economically obsolescent equipment to reduce VOCs. However, considerable thought and consideration is given to the present and future financial impacts of investment. For the smallest firms in the sample, some innovations were not made, or were made only partially, because of the cost (problem of funding), except in the metal cutting industry; for others, some innovations like the installation of filters or the capture of VOC emissions are simply not financially viable. The financial obstacle is correlated to the size of the firm and whether or not it belongs to a group (national or international).

At canton level, the determinants of VOC innovation are quite variable. The perception of them may depend on the structure of the canton's economy and the presence of activities that generate VOC emissions. The factor most often mentioned is workplace health and safety, though other factors include the supplier and market demand, green credentials, the VOC tax, competition and international standards. Two cantons emphasised the importance of the tax as a factor favouring innovation.

All the cantons agreed that the frequency and mode of innovation depend to a very large extent on financial resources. Investing in green innovation is often too expensive for small businesses, while some big firms simply do not see any interest in it. Firms very often innovate of their own accord or follow innovations developed by suppliers. One canton noted that from 2009, small businesses will be able to get together to declare their emissions and obtain reimbursement, giving them a better basis for co-operation. Another canton with a dominant pharmaceutical industry said that many firms do R&D and develop innovations for other firms.

Few firms have suffered economic difficulties on account of the tax. No firm has moved, changed business or totally ceased production. But many, especially small firms, have not taken any measure that could be qualified as innovation because the tax is so small. Cantonal administrations also mention the problems of the quality of substitution products or technologically modified products.

All the cantons consulted found that there has been a definite improvement in products and processes in certain industries but that reducing VOCs does not seem to have been the main driver of innovation. Fewer than half the cantons interviewed had noticed any real change of behaviour in favour of the environment, and only one said that a small number of firms had innovated solely from a concern for the environment.

Ongoing incentive for emission reductions/innovation among industries

In the printing industry, the process of continuously adapting existing technologies certainly facilitates innovation and change. In order to remain competitive, especially in terms of printing speed, printers change their presses relatively often. This feature of the industry helps to explain what appears to be a very dynamic process. The costs of reducing VOC consumption generated by the various changes made are often negligible because the technology would be replaced in any case. Consequently, firms are not in a position to put a figure on the cost. In contrast, firms underline the regular effort that needs to be made to reduce VOC levels or keep them low. In compensation, they can sometimes reduce production costs because they use less alcohol. For these different reasons, the VOC tax can therefore act as an incentive to change or innovate, and environment-related technological aspects (in this case less use of VOCs) are included in the considerations driving change.

Innovation in the paint and varnish making industry also seems to be a dynamic process, but the link with the VOC tax sometimes seems less obvious. For the reasons already mentioned (health and safety, cost, quality, concern for the environment), and for reasons of compliance with EU regulations, for example, firms are continuing to reduce the use of products containing VOCs. However, for reasons of quality and climate (low temperatures, humidity) and according to usage (interior or exterior), it does not seem possible at present to eliminate solvents altogether.

Five of the eight metal cutters interviewed had no coherent environmental strategy. With the introduction of the tax they discovered a problem with VOCs which, while known, had not required any response or action on their part. While some may regret the heavy administrative burden, particularly of completing the VOC balance sheet, and see the tax as an additional factor undermining their competitiveness, they all recognise that something had to be done, even if only for the health and welfare of their employees. Measures have been taken continuously at a relatively rapid pace, creating an impetus for innovation, though not without taking profitability into account. Other measures to reduce VOC emissions are still possible, especially by using substitute products, though there is limited scope for substitution in a significant proportion of cleaning and degreasing activities.

If the tax were to be abolished, none of the firms interviewed would turn back the page. They seem increasingly to be playing the "green" card, which also increasingly corresponds to what customers expect. Low-VOC products could become a marketing plus, as is the case for other pollutants like CO_2, except perhaps in the metal cutting industry. Health and safety considerations in the production process also mitigate any return to the previous situation.

However, end-of-pipe measures often seem too expensive in the industries under consideration. The biggest printing firm interviewed, with over 400 employees, decided not to take end-of-pipe measures on cost grounds. One paintmaker with a catalytic/thermal air purifier regrets the high cost of the equipment and the associated energy consumption. The firm, which draws up a VOC balance sheet, thinks that abolishing the VOC tax could even encourage some firms to stop using their end-of-pipe equipment.

Conclusion

The Swiss tax has managed to cut VOC emissions and use by 20-50% in five to eight years in the firms interviewed and generated greater awareness of the environmental and other problems of VOCs (workplace health and safety). This is despite the fact that firms often regard the tax and the VOC balance sheet as a (heavy) administrative burden. Many firms, especially the smaller ones, are also ill-informed about how the tax works (VOC balance sheet, possibilities of exemption, etc.). There is also considerable displeasure about how the tax revenue is used; one proposal is that it should be used for projects to reduce VOCs.

Nevertheless, the VOC incentive tax seems to have had a positive effect on innovation. The relatively small amount of the tax means that it is not a major factor in making products more expensive. Although the firms interviewed have not so far really won customers because of their commitment to reduce VOC emissions, the innovations themselves are often profitable to the firms and neutral in terms of productivity.

In the three industries under consideration, the effects of innovation were highlighted. In printing and metal cutting, for example, changes have taken place in production processes that can be classified as innovation. In printing, this mainly concerns less use of products containing VOCs and of isopropylic alcohol; in metal cutting, it involves more efficient processes that reduce VOC emissions, including the use of substitutes. But as preserving end-product quality is vital, it is often not possible to entirely eliminate VOCs or products containing VOCs. In paintmaking, innovation has taken place not only in production processes (less use of products containing VOCs, especially solvents) but also in products (water-based products and products containing less solvent). In the industry, the tax seems to have been one factor among others (health, safety, EU directives, etc.) encouraging innovation.

The timing of the innovations is also of interest. Some changes had already been made before the tax was introduced in 2000, others spanned the entire review period from 2000 to 2008. A minority of firms had planned in advance for the future changes to reduce VOC emissions or the use of products containing VOCs.

Potential for further reduction still exists, especially in processes, because product quality requirements often mean that no more can be done at the current time. However, although reducing emissions and the amount of tax payable continues to be a concern, the effect seems to be dissipating. Some firms have indicated avenues of innovation in the near future.

- *Paintmaking*: The main changes relate to processes (cleaning with water, closed-circuit systems, solvent recycling) and products, *i.e.* the development of low-VOC, VOC-free and water-based products (mineral paints, high solid paints and varnishes).

- *Printing*: The chief concern is to continue reducing the use of isopropylic alcohol in printing processes while maintaining a given level of quality. Other possibilities include using low-VOC or VOC-free cleaning products, detergents and cleaning systems, and using osmosis water treatment systems.

- *Metal cutting*: The options for reducing VOC emissions are known; mostly they involve replacing the VOC with detergents for cleaning and degreasing parts and recycling used substances. One firm is trying to improve benzene jars with a shower system; another, with suppliers, is working on the use of detergents for parts liable to rust on contact with water. For these firms, however, the main concern is to optimise processes in order to reduce VOC emissions, with a limited reduction in the amount of tax payable.

For more information on VOCs in Switzerland, the full version of the case study (OECD, 2009) is available at *www.olis.oecd.org/olis/2008doc.nsf/linkto/com-env-epoc-ctpa-cfa(2008)35-final*.

References

OECD (2009), *Effects of the VOC Incentive Tax on Innovation in Switzerland: Case Studies in the Printing, Printmaking and Metal Cutting Industries*, OECD, Paris, available at *www.olis.oecd.org/olis/2008doc.nsf/linkto/com-env-epoc-ctpa-cfa(2008)35-final*.

Office fédéral de l'environnement (OFEV) (2007), *Flüchtige Organische Verbindungen (VOC), Anthropogene VOC-Emissionen Schweiz 1998, 2001 und 2004*, Internet publication, 26 February 2007.

ANNEX E

R&D and Environmental Investments Tax Credits in Spain

This case study examines the impacts of two tax credits in Spain: one for R&D investments and the other for investments in assets that relate to the environment. The tax credit for environmental investments did not seem to induce innovation, partly due to the fact that the tax credit could be triggered for investments needed to comply with existing environmental policies. On the other hand, the R&D tax credit seemed to support environmental innovation, given the number of firms that made use of the environmental investments deduction after having used the R&D deduction.

Rationale for environmental policy

Like all OECD economies, Spain faces a range of environmental challenges and its recent strong economic performance – GDP has grown 44% between 1995 and 2005 – has resulted in the need for stronger environmental policies. For example, greenhouse gas emissions increased by 52% between 1990 and 2005, mainly due to higher overall economic activity, higher energy intensity and due to emissions in the transport sector, which increased 78% over the same period. NO_x and ammonia emissions also increased, while emissions of SO_2, CO and VOCs were reduced. As regards municipal solid waste, generation per capita rose by 62% between 1990 and 2004.

Higher pressures have also been exerted on the use of natural resources. Water is one of the most critical resources in Spain. Agriculture accounts for more than 75% of total water consumption, although it almost stabilised in absolute terms during the period 1997–2004. On the other hand, water used for public supply (households, non-agricultural economic activities and municipal uses) increased by 31% between 1996 and 2004, without any progress being made in terms of reducing water use intensity. For all these reasons, policies in Spain have been enacted to help address a wide range of environmental issues.

Design of the instruments

This case study focuses on two policies, both tax schemes regulated within the Spanish Corporate Income Tax. The first – the R&D and technological innovation (R&D&I) tax credit – aims at stimulating expenditures on research and development for a wide variety of issues within the economy, including those related to the environment.

The base is 30% of the expenses on R&D incurred during a given year. Where these expenses exceed the average of the previous two years, a base of 50% is applied to the excess, making this tax credit both volume-based and incremental. In addition, an extra 20% tax credit applies for expenses on qualified researchers assigned exclusively to R&D activities, and for expenses related to projects entrusted to universities, public research bodies or innovation and technological centres. There is also a tax credit for 10% of the investments on fixed or intangible assets for R&D activities and for expenses on technological innovation incurred a given year. The base for this tax credit includes expenditures such as those on industrial design, engineering related to production processes, acquisition of advanced technology in the form of patents or licenses, obtaining the ISO 9000 and analogous certificates. The tax credit is 15% in the case of expenses related to projects entrusted to universities and other agencies, public research bodies, or innovation and technological centres. R&D and technological innovation expenses incurred abroad may qualify for this tax credit, provided that the main R&D activity takes place in Spain and the expenses incurred abroad do not exceed 25% of the total. There are also global limits to deductions.

With the aim to increase legal security for firms and to encourage them to make use of the R&D&I tax credit, firms may also voluntarily request reasoned reports from the government, which state the compliance of the proposed activity with the scientific and technological criteria required to qualify for the tax credit. The number of applications for reasoned reports for the R&D&I credit has increased steadily since being introduced, along with the number of reports issued (see Table E.1).

Table E.1. **Use of reasoned reports in Spain**

	2004 (FY 2003)	2005 (FY 2004)	2006 (FY 2005)	2007 (FY 2006)
Number of applications	298	561	905	1 215
Number of reports issued	252	496	696	–

Source: Ministerio de Industria, Turismo y Comercio (2007) and Gutiérrez (2008).

StatLink ⬛🔗 http://dx.doi.org/10.1787/888932318281

At the same time, according to a survey conducted in 2006 among manufacturing industries, awareness of the instrument is varied: 82.4% of companies with more than 200 workers were aware of the existence of this tax credit, whereas the percentage fell to 49.5% for those between 10 and 200 workers.

The second is designed to foster investments for environmental protection. The tax credit is 10% of the total investment in tangible assets devoted to environmental protection consisting of installations used to: i) avoid air pollution from industrial facilities; ii) prevent pollution of surface, underground and sea water; iii) reduce, recover or adequately treat industrial waste; and iv) generate renewable energy from selected processes. The tax credit is 12% in the case of purchases of new land-based means of transportation for commercial or industrial use. In order for these investments to qualify for the tax credit, they have to go beyond what is legally required, and must be included in programmes or agreements with the corresponding environmental authorities, who subsequently have to issue a certificate validating the investment.

Over the period 2000-05, the average value of the R&D&I tax credit was about EUR 70 000 (EUR 16 000 for environmental investments) and the percentage of companies making use of the tax credit broadly grows as their size increases, along with the average size of the tax credits (see Figure E.1). These two factors combine to make it possible for larger companies to capture a high share of the R&D&I tax credit (*e.g.* in 2005, 93.3% of all the deducted amount benefited companies with a net turnover higher than EUR 10 million, whereas they represented only 1.9% of all declarations and 72.9% of the total net corporate tax payable).

Figure E.1. **R&D&I and Environmental Investments tax credit use by firm size**
Percentage of firms by turnover (million EUR)

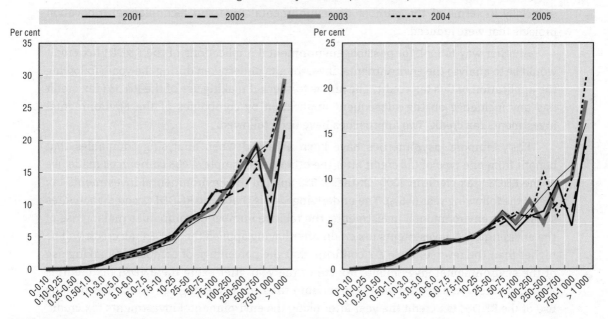

Source: Ministerio de Economía y Hacienda (2004, 2005, 2006, 2007, 2008).

StatLink http://dx.doi.org/10.1787/888932317692

The Environmental Investments (EI) tax credit is used by a limited number of companies (0.4% of all declarations in 2005) but its scope is significant. According to estimates in 2005, private companies invested EUR 1 033 million on environmental protection. Therefore, the tax credit represented 8.7% of total investment in environmental protection declared by private companies. Considering that the tax credit is approximately 10% of the quantity invested, this means that most of the investment on environmental protection undertaken in Spain was supported by this tax credit.

The legal condition that "investments have to go beyond what is legally required" implies that investments regulated are (in principle) undertaken on a voluntary basis. However, this tax credit also benefits in practice investments aimed to comply with the existing legislation (and that would be undertaken anyway). Of the EUR 384 million in eligible investments under this tax credit in 2005, over 60% was directed at air pollution, with about 20% at each of water and waste.

In 2006, a progressive phasing out of the two tax credits was agreed upon. The R&D&I deduction was to disappear in January 2012, whereas the other one will disappear in January 2011. In March 2009, however, it was agreed to not abolish the R&D&I tax credit.

Environmental impacts of the R&D&I tax credit

In fiscal year 2005, 7% of projects supported with R&D&I reasoned reports were specifically oriented to environmental protection and accounted for 4.6% of all validated expenses. These figures should be compared to the relative importance of the environmental sector in the Spanish economy. A rough comparison can be made with Spanish private industrial companies having invested EUR 1 033 million in environmental protection, being 4.9% of their total investment in machinery and equipment. Whereas the proportion of investments in environmentally motivated projects supported by reasoned reports is about the same, the number of environmentally motivated projects is significantly larger in relative terms. This suggests possible positive environmental consequences of the R&D&I tax credit, especially in terms of the number of environmental R&D and technological innovation projects that were induced.

Another way to look for possible environmental implications of the R&D&I tax credit would be to analyse the environmental investments undertaken during the years following the application of this tax credit. The aim is to check if making use of the R&D&I tax credit has any incidence on the subsequent application by companies of the environmental investments tax credit. Two approaches have been followed.

First, temporal asymmetries have been analysed when analysing companies that benefit from one type of tax credit after the other. For example, 4 408 companies made use of the R&D&I tax credit in 2000. Of these, 333 applied the environmental investments tax credit the following year (7.6%). The underlying idea is that if the R&D&I tax credit activates subsequent environmental investments, the relative number of companies making use of the environmental investments tax credit after having applied the R&D&I tax credit will be higher than the reverse. Table E.2 confirms that the percentage of companies making use of the environmental investments tax credit the year after using the R&D&I tax credit is always greater (on a year-to-year comparison) than the percentage of companies making use of the R&D&I tax credit the year after using the environmental investments tax credit.

Table E.2. **Sequential impact of tax credits**

	2001	2002	2003	2004	2005
Percentage of companies applying or generating the environmental investments tax credit the year after applying or generating the R&D&I tax credit	7.6	7.3	8.5	7.8	8.0
Percentage of companies applying or generating the R&D&I tax credit the year after applying or generating the environmental investments tax credit	7.2	6.8	7.8	7.8	7.3

Source: OECD (2008).

StatLink ⬛⬛⬛ http://dx.doi.org/10.1787/888932318300

A second approach tests if companies behave differently after having made use of the R&D&I tax credit, as compared to when no such tax credit had been used. Thus, Table E.3 presents the 34 071 companies that generated or applied any of the two tax credits during 2001-05 according to whether or not they made use of the R&D&I tax credit in the initial year of the two sub-periods (2000-02 and 2003-05).

Table E.3. **R&D&I tax credits and tax credit use**

Companies that benefited from either two tax credits, according to whether they made use or not of the R&D&I tax credit in years 2000 and 2003

Use of the R&D&I tax credit[1]		Number of companies using either R&D&I or EI credit
2000	2003	
Yes	Yes	2 015
Yes	No	2 393
No	Yes	3 941
No	No	25 722
Total		**34 071**

1. Either application or generation of the tax credit is considered.
Source: OECD (2008).

StatLink ᴍ᷒ᴦ᷅ http://dx.doi.org/10.1787/888932318319

From these four groups, Table E.4 focuses only on the two that can be used to compare the net effect of the use of the R&D&I tax credit on subsequent application of the environmental tax credit, *i.e.: i*) the group of companies that made use of the R&D&I tax credit in 2000 and did not apply it in 2003; and *ii*) the group that did not make use of the R&D&I tax credit in 2000 but did so in 2003. This change allows for the interpretation of firm behaviour in terms of the environmental investments tax credit in the two subsequent two-year periods (*i.e.* 2001-02 and 2004-05). For each of the two groups of companies and for each of the two periods, Table E.4 presents information on the number of companies making use of the environmental investments tax credit in any of the two years, as well as on the amount of the tax credit.

Table E.4. **Impact of R&D&I tax credit on use of EI credit**

Number of companies	Use of the R&D&I tax credit[1]		Companies making use the environmental investments tax credit[2]			Generated environmental investments tax credit (million EUR)			Applied environmental investments tax credit (million EUR)		
	2000	2003	2001-02	2004-05	Δ 04-05/01-02 (%)	2001-02	2004-05	Δ 04-05/01-02 (%)	2001-02	2004-05	Δ 04-05/01-02 (%)
2 393	Yes	No	192	136	−29.2	17.9	14.5	−19.1	4.8	3.8	−20.3
3 941	No	Yes	338	395	16.9	57.4	83.8	46.1	18.6	26.7	43.7

1. Either application or generation of the tax credit is considered.
2. Any of the two years.
Source: OECD (2008).

StatLink ᴍ᷒ᴦ᷅ http://dx.doi.org/10.1787/888932318338

The results from these two independent approaches confirm that R&D and technological innovation projects that benefit from the corresponding tax credit lead in subsequent years to a small, but significant, amount of additional environmental investments that can be observed by an increase in the environmental investments tax credit. This links with the perception that some environmental investments that benefit from the corresponding tax credit have their origin in previous R&D and technological innovation projects that benefited from R&D&I tax credits. This can be illustrated using a similar methodology. Again, two three-year consecutive sub-periods are analysed within the 2001-05 period. In this case, companies were classified according to whether they made use or not of the environmental investments tax credit in the last year of the two sub-periods (2002 and 2005) (Table E.5) in order to analyse what happened during the previous two years regarding application of the R&D&I tax credit.

Table E.5. **Environmental Investments tax credits and tax credit use**

Companies that benefited from either two tax credits, according to whether they made use
or not of the environmental investments tax credit in years 2002 and 2005

Use of the environmental investments tax credit[1]		Number of companies using either R&D&I or EI credit
2002	2005	
Yes	Yes	1 853
Yes	No	3 951
No	Yes	4 989
No	No	23 278
Total		**34 071**

1. Either application or generation of the tax credit is considered.
Source: OECD (2008).

StatLink http://dx.doi.org/10.1787/888932318357

From the four groups above, results in Table E.6 refer only to those two that can be used to compare the net effect of the previous application of the R&D&I tax credit on the use of the environmental investments tax credit, *i.e.:* i) the group of companies that made use of the environmental investments tax credit in 2002 and did not apply it in 2005; and ii) the group that did not make use of it in 2002 but did so in 2005. It shows that the use of the R&D&I tax credit is relatively higher before the use of the environmental investments tax credit than when this tax credit is not used.

Table E.6. **Impact of environmental investments tax credit in use of R&D&I tax credit**

Number of companies	Use of the environmental investments tax credit[1]		Companies making use the R&D&I tax credit[2]			Generated R&D&I tax credit (million EUR)			Applied R&D&I tax credit (million EUR)		
	2002	2005	2000-01	2003-04	Δ 03-04/00-01 (%)	2000-01	2003-04	Δ 03-04/00-01 (%)	2000-01	2003-04	Δ 03-04/00-01 (%)
3 951	Yes	No	283	305	7.8	90.7	107.8	18.8	48.1	22.3	−53.6
4 989	No	Yes	245	362	47.8	95.1	155.0	62.9	21.4	60.2	181.2

1. Either application or generation of the tax credit is considered.
2. Any of the two years.
Source: OECD (2008).

StatLink http://dx.doi.org/10.1787/888932318376

In summary, this suggests that R&D and technological innovation projects that benefit from the corresponding tax credit lead in subsequent years to additional tax credits on environmental investments, which derive from additional environmental investments. This has been proved for the 2000-05 period. To reinforce this conclusion, for the same period it has also been proved that the use of environmental investments tax credit tends to be preceded by a higher application of R&D&I tax credits. However, there is debate about the effectiveness of fiscal incentives on R&D. In the case of the Spanish R&D&I tax credit, the tax credit may act only as an incentive for companies that are already undertaking research. Other limitations in the design and use of the tax credit may also undermine its possible beneficial environmental effects, including: regulations governing the tax credit have been modified repeatedly; insufficient awareness; lack of clarity and practicality of the legal definitions; and, uncertainty regarding application after 2011.

Innovation impacts of environmental investment tax credit

To study possible innovation impacts, the details of companies making use of both tax credits are useful. Although the number of companies making use of both tax credits simultaneously rose during the 2000-05 period, their relative presence was quite limited in percentage terms (only around 4% of the companies applying any of the two tax credits during that period), and this variable remained quite stable. However, they accounted for a very significant proportion of total amounts deducted, particularly in the case of the environmental investments tax credit. This indicates that the average tax credit (either R&D&I or environmental investments) is higher for companies making use of both tax credits simultaneously, as seen in Table E.7.

Table E.7. **Characteristics of tax credit use**

Value in euros	2000	2001	2002	2003	2004	2005
R&D&I tax credit						
Number[1]	4 408	5 767	5 585	5 956	6 037	6 045
Generated	562 666 120	1 070 207 317	657 094 753	773 828 103	881 520 933	934 942 943
Applied	185 566 986	220 256 602	204 860 450	251 088 783	299 880 114	348 084 993
Pending	377 099 134	849 950 716	452 234 304	522 739 320	581 640 819	586 857 950
Average tax credit generated	127 647	185 574	117 653	129 924	146 020	154 664
Environmental investments tax credit (EI)						
Number[1]	4 594	6 218	5 804	6 107	6 396	6 842
Generated	207 963 080	187 176 047	160 204 069	171 557 840	186 638 055	219 979 982
Applied	58 086 821	61 188 366	56 652 641	55 625 625	89 599 204	89 391 208
Pending	149 876 259	125 987 681	103 551 428	115 932 216	97 038 851	130 588 774
Average tax credit generated	45 268	30 102	27 602	28 092	29 180	32 151
Both tax credits						
Number[1]	323	468	498	502	488	496
Generated (R&D&I)	85 611 136	117 589 518	118 046 576	185 810 955	166 023 172	198 275 167
Applied (R&D&I)	33 402 507	65 949 366	55 727 463	68 418 295	76 548 040	87 630 444
Pending (R&D&I)	52 208 629	51 640 152	62 319 113	117 392 660	89 475 132	110 644 722
Average tax credit generated (R&D&I)	265 050	251 260	237 041	370 141	340 211	399 748
Generated (EI)	31 688 138	39 693 774	31 954 965	38 974 226	57 978 517	76 908 038
Applied (EI)	20 851 741	24 318 229	11 321 804	16 036 477	38 634 037	33 474 812
Pending (EI)	10 836 397	15 375 545	20 633 160	22 937 749	19 344 480	43 433 226
Average tax credit generated (EI)	98 106	84 816	64 167	77 638	118 808	155 057
Companies making use of any of the two tax credits	8 679	11 517	10 891	11 561	11 945	12 391
% of companies making use of both tax credits	3.7	4.1	4.6	4.3	4.1	4.0
R&D&I, % of total applied tax credit that benefits companies making use of both tax credits	18.0	29.9	27.2	27.2	25.5	25.2
EI, % of total applied tax credit that benefits companies making use of both tax credits	35.9	39.7	20.0	28.8	43.1	37.4

1. Number of companies having generated or applied the tax credit.
Source: OECD (2008).

StatLink ᓗᑭᔊ http://dx.doi.org/10.1787/888932318395

Although between 2000 and 2005 the R&D&I and the environmental investments tax credits were used (generated or applied) on 33 798 and 35 961 occasions, respectively, only 34 071 different companies benefited from any of the two tax credits during this period. This means that the use of the tax credits is concentrated in some companies. Actually,

14 921 companies made use of these tax credits on more than one occasion during this five-year period. This also implies that having made use of either of the two tax credits increases the probability of making use of them again.

In 2005, the Spanish autonomous communities were asked to classify environmental investments validated for the tax deduction between "end-of-pipe" and "cleaner production" solutions, the former accounting for 67.8% of the entire invested amount. This has traditionally been the dominant approach to address environmental impacts. However, cleaner production solutions are gaining importance among environmental investment decisions. In practice the tax credit on environmental investments still benefits investments that simply aim to comply with the environmental legislation. The motivation of these investments seems more to be the need to fulfil the environmental legislation rather than the tax credit itself. This may explain why, despite the fact that the environmental investments tax credit can be deemed as a flexible policy instrument, its results in terms of promotion of cleaner production investments are lower than could have been expected.

If end-of-pipe investments are associated more with a reaction to environmental obligations and cleaner production is more associated with an initiative aiming at cost savings, it seems reasonable to suppose that the latter may entail more research and technological innovation. Therefore, the present weight of end-of-pipe technologies in the investments applying the environmental investments tax credit suggests a limited incidence of this tax credit in terms of innovation.

Given the specific focus of the environmental investments tax credit on air pollution, water pollution and industrial wastes, it is interesting to analyse patenting activity in these areas in order to detect possible positive innovation consequences of this tax credit. As shown in Figure E.2, the absolute number of environmental technology patents in the areas of air, water and waste has been growing for the last few decades in Spain. However, the growth rate in the number of these patents was similar before the introduction of the environmental investments tax credit (1997). The trend followed by the number of patents in these three areas has not been significantly different from that followed by the total number

Figure E.2. **Patent applications in Spain and EU15**

Patents in the areas of air, water and waste

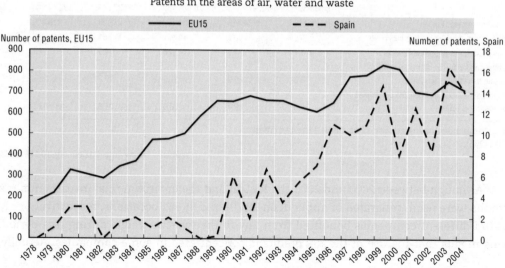

Source: OECD (2008).

StatLink http://dx.doi.org/10.1787/888932317711

of patents, which indicates that the evolution in the number of patents in the three analysed areas is parallel to the general growth of innovation activities in the country. The starting point of Spain, back in the 1970s, was very low, both as regards innovation activities and the application of environmental policies.

The relative number of Spanish patents in the areas of air, water and waste are 4.7%, 84.7% and 10.5%, respectively, for the 1997-2004 period. These percentages are very different from the relative importance of these investments in the tax credit, which indicates that investments on air pollution clearly dominate within those applying the environmental investments tax credit. This poor correspondence between the composition of environmental investments that benefit from the tax credit and innovation in these areas (measured by the number of patents) suggests again a low incidence on innovation of the tax credit on environmental investments.

Limitations on the configuration and use of the tax credit on environmental investments may also undermine its possible beneficial effects on R&D and technological innovation. Some of the present limitations are: lack of harmonisation among Autonomous Communities; possible misuse because Regional Administrations validate while the Central Administration pays; tax credits for investments that would have taken place regardless; and, progressive loss of intensity of the tax credit, due to phasing out.

Conclusions

This case study has focused on the analysis of two tax credits regulated within the Spanish Corporate Income Tax: the R&D and technological innovation tax credit and the environmental investments tax credit. Only a small percentage of tax returns include the tax credits analysed, particularly in the case of the R&D&I tax credit. Among companies making use of these tax credits, a high percentage of the amounts deducted benefits a very limited number of large companies, which usually undertake larger and more costly projects. It has also been found that having made use of either of the two tax credits in the past increases the probability of using them again. The scope of the tax credit is particularly significant in the case of the environmental investments tax credit, which has been estimated to benefit a majority of the investments on environmental protection undertaken in Spain.

As regards the effects of these two tax credits, it is difficult to know what expenditures would have also taken place without the analysed incentives, and it was not the aim of this case study to analyse their effectiveness. Several studies analysed the tax credit on R&D and technological innovation in the past and generally concluded that it is effective in stimulating such expenses. No analyses exist on the effectiveness of the environmental investments tax credit. This case study focused on the analysis of the environmental effects of the R&D&I tax credit and on the effects on innovation of the environmental investments tax credit.

Although the two tax credits are largely independent, evidence has been found regarding positive environmental consequences of the R&D&I tax credit. For the period analysed (2000-05), it was found that the percentage of companies making use of the environmental investments tax credit in the year after applying the R&D&I tax credit was systematically greater than the percentage of companies making use of the R&D&I tax credit the year after applying the environmental investments tax credit. For the same period, using another methodology, it was also found that the application of the R&D&I tax credit increased the application of the environmental investments tax credit (i.e. additional environmental investments being induced) in subsequent years. This conclusion is also supported by the fact that the proportion of environmentally motivated projects among projects supported by reasoned reports is higher than the proportion of environmental expenditures among total investment in machinery and equipment expenditures by Spanish industrial companies.

However, it seems that the relative presence of environmentally related R&D&I projects taking benefit of this tax credit is similar to that in other public measures to support R&D&I at the national level. In both cases, these percentages are higher than the percentage of internal R&D expenditures dedicated to environmental control and protection by private companies, which suggests a positive environmental bias in the innovation activities that receive public financial support.

As regards the environmental investments tax credit, several results suggest that it has no or very low impact on activating innovation. On the one hand, expenditures on cleaner production (as opposed to end-of-pipe expenditures) among investments are clearly lower than in average environmental investments. The present weight of end-of-pipe technologies in the investments taking benefit of the environmental investments tax credit suggests a limited incidence of this tax credit in terms of innovation, since environmental innovations are more often identified with cleaner production.

Moreover, the evolution in the number of patent applications in the areas of air, water and waste did not change significantly with the introduction of the environmental investments tax credit. Furthermore, the relative number of Spanish patents in the areas of air, water and waste is very different from the relative importance of these investments within the amounts deducted. This poor correspondence again suggests a low incidence on innovation of the tax credit on environmental investments. In any case, and despite the estimated positive environmental consequences of the R&D&I tax credit, it seems that if environmental R&D and technological innovation is to be fostered, stimulating R&D and technological innovation and environmental investments separately is not the best option. Several programmes specifically addressing environmental R&D and technological innovation could be reinforced or new ones could be created.

Some of the limitations of the two tax credits examined in this case study are associated with a changing legal framework (this has a cost in terms of stability and predictability), unawareness by companies on their existence (which leads to limited use), complexity and bureaucracy (which leads to higher administrative costs), uncertainty about their future (due to progressive phasing out), and legal uncertainty regarding possible tax audits. Some of these limitations could be overcome and this could lead to a greater use and effectiveness of these two tax credits. Some other limitations are their lack of flexibility (changing a law is required to modify their intensity, as opposed for example to the flexibility of subsidy programmes) or the fact that a positive tax payable is necessary in order to benefit from the tax credits (they can be deferred, but only for a limited number of years). In relation to this, it may happen that two companies that undertake the same activities benefit from the tax credit to a different extent.

Tax credits (and other subsidies as well) may be economically justified in some cases, for example when positive externalities appear. No subsidies or tax credits should be granted to actions that are compulsory to undertake, even though the EI tax credit continues to consider as eligible investments those that simply aim to comply with the existing environmental legislation. This is not according to the polluter-pays principle and not only results in inefficient public expenditure, but also hampers a potential side-effect of this tax credit on innovation. Since environmental standards are often established considering the capabilities of the best available technologies, innovation required to just fulfil the legislation is relatively moderate. If only measures going beyond legal obligations were eligible for the environmental tax credit, investments making use of the tax credit would essentially pursue cost savings, which tend to favour clean production and, therefore, innovation.

Yet, there are positive aspects of the tax credit scheme that could be used to support its continuation. For example, tax credits constitute a form of public support that distorts the market the least, since companies decide whether to make use or not of the tax credit, and this is automatically granted if the application qualifies. Tax credits also entail (at least in principle) less administrative costs than subsidies, both for public administrations and for companies.

If the environmental investments tax credit were to remain, other reforms could be considered, such as the possible inclusion of other areas eligible for the environmental investments (e.g. efficient use of raw materials or others that also would increase the weight of cleaner production expenditures) or a more explicit support to investments in the service sector or to logistics. Also a limitation, or at least a more restrictive application, of the tax credit on the purchase of new land-based means of transportation should be considered, particularly since the vehicle registration tax was reformed in 2008 to foster cleaner vehicles, and since there are governmental programmes aimed at substituting older vehicles that partially overlap with this tax credit. However, if deeper reforms are considered, they should be ideally framed in a broader ecological reform of the Spanish tax system.

For more information on the two tax credits in Spain, the full version of the case study (OECD, 2008) is available at *www.olis.oecd.org/olis/2008doc.nsf/linkto/com-env-epoc-ctpa-cfa(2008)38-final*.

References

Gutiérrez Monzonís, G. (Ministerio de Industria, Turismo y Comercio) (2008), *Evolución del sistema de informes motivados 2004-2008*, Conference held at Consejo Asesor de Aidit, 16 June 2008, Barcelona.

Ministerio de Economía y Hacienda (2004), *El impuesto sobre sociedades en 2001. Análisis de los datos estadísticos del ejercicio* (The Corporate Income Tax in 2001. Analysis of the Statistical Data of the Fiscal Year), Ministerio de Economía y Hacienda, Madrid, *http://documentacion.meh.es/doc/C9/Estadisticas/An%C3%A1lisis%20IS%2001.pdf*.

Ministerio de Economía y Hacienda (2005), *El impuesto sobre sociedades en 2002. Análisis de los datos estadísticos del ejercicio* (The Corporate Income Tax in 2002. Analysis of the Statistical Data of the Fiscal Year), Ministerio de Economía y Hacienda, Madrid, *http://documentacion.meh.es/doc/C9/Estadisticas/Analisis_IS_02.pdf*.

Ministerio de Economía y Hacienda (2006), *El impuesto sobre sociedades en 2003. Análisis de los datos estadísticos del ejercicio* (The Corporate Income Tax in 2003. Analysis of the Statistical Data of the Fiscal Year), Ministerio de Economía y Hacienda, Madrid, *http://documentacion.meh.es/doc/C9/Estadisticas/Imp.Soc-2003.pdf*.

Ministerio de Economía y Hacienda (2007), *El impuesto sobre sociedades en 2004. Análisis de los datos estadísticos del ejercicio* (The Corporate Income Tax in 2004. Analysis of the Statistical Data of the Fiscal Year), Ministerio de Economía y Hacienda, Madrid, *http://documentacion.meh.es/doc/C14/C1/Tributos/IMPTO%20SOCIEDADES%202004.pdf*.

Ministerio de Economía y Hacienda (2008), *Recaudación y estadísticas del sistema tributario español 1996-2006* (Tax Collection and Statistics of the Spanish Fiscal System 1996-2006), Ministerio de Economía y Hacienda, Madrid, *http://documentacion.meh.es/doc/C14/C1/Tributos/Recaudaci %C3%B3n%20y%20 Estad%C3%ADstica%201996-2006.%20Internet.pdf*.

Ministerio de Industria, Turismo y Comercio (2007), *Informes motivados para deducciones fiscales por actividades de I+D e innovación tecnológica. Informe Solicitudes 2006 (Ejercicio Fiscal 2005)* [Reasoned Reports for Fiscal Deductions for R&D and Technological Innovation Activities. Report on 2006 Applications (Fiscal Year 2005)], Ministerio de Industria, Turismo y Comercio, Madrid, *www.eqa.org/ boletin/oct07/imgboletin/pdf/I+D.pdf*.

OECD (2008), *Taxation, Innovation, and the Environment – The Spanish Case*, OECD, Paris, available at *www.olis.oecd.org/olis/2008doc.nsf/linkto/com-env-epoc-ctpa-cfa(2008)38-final*.

ANNEX F

Korea's Emission Trading System for NO_x and SO_x

This case study explores Korea's policies towards NO_x and SO_x emissions. A cap-and-trade system was implemented in 2008 and was expanded the following year. Although too early to investigate the environmental and innovation impacts, this instrument builds on previous policies targeted towards the same pollutants. These policies bought about increased patenting but mainly in end-of-pipe technologies.

Rationale for the environmental policy

Air pollution in Korea is a major problem, such that atmospheric conditions in the metropolitan region in Korea are some of the worst among OECD nations. Emissions of NO_x, SO_x and other pollutants are quite high and are a by-product of fuel combustion. NO_x contributes to the formation of ground-level ozone and smog, while SO_x contributes to the formation of acid rain. These undesired consequences of industrialisation have both environmental and non-environmental effects (such as that on human health). The goal is to improve air quality of the capital region to the level of other OECD nations by 2014.

Design of the policy instrument

The legal framework took effect in January 2005, the basic plan was established in November 2005, and the State Implementation Plan was established in January 2007. Key policy tools include implementing a cap-and-trade system for emissions from large factories, providing low-emission vehicles and strengthening emission standards for diesel-fuelled cars. The cap-and-trade programme targeting factories became effective in July 2007.

The target year of the basic plan is 2014 and target air pollutants are NO_x, SO_x, PM_{10} and VOCs. Specifically, the targets for 2014 are 22 parts per billion for NO_2 and 40 µg per m^3 for PM_{10}. Figure F.1 shows the goals regarding NO_2 and PM_{10} by 2014 compared to ambient levels in 2003.

One of the most interesting features of the Korean approach is the use of an emission trading system for NO_x and SO_x emissions. The air quality target of the capital was set based on the result of air quality prediction modelling and the regional environment capacity was then estimated considering terrain and weather conditions. Based on this analysis, the

Figure F.1. **Targets for ambient NO$_2$ and PM$_{10}$ concentrations**

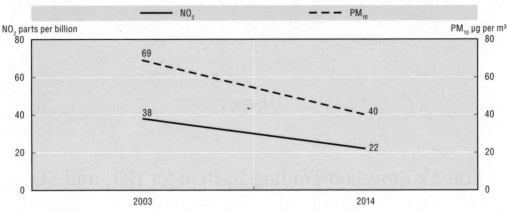

Source: OECD (2009).

StatLink http://dx.doi.org/10.1787/888932317730

Ministry of the Environment issues regional permission capacity. The city and provincial governments issue the pollutant emission limit of each industry according to the determined emission limit of the cities or provinces. In addition, they are in charge of total emissions management, reduction emissions investigation and reduction assessment.

The cap-and-trade programme targets 136 factories in Seoul, Incheon, and the Gyeonggi area, over 24 counties. Every year, maximum emission limits are issued under the cap-and-trade programme and business owners manage emissions of air pollutants within the permitted limits. For the first year, the target will be determined at the level of five-year average pollutant emission. For the final year, the goal is to reduce the pollution concentration to the level when optimal control device (the facility with superior reduction efficiency that is technologically and economically feasible) is established.

The targets of the cap-and-trade programme in Korea are for NO$_x$, SO$_x$, PM, and date of effect is January 2008 for Stage 1 and July 2009 for Stage 2. Table F.1 outlines the progression of the implementation of the cap-and-trade programme.

Table F.1. **Implementation progression of cap-and-trade programme**

Date of effect	Emission level (tonnes per year)		
	NO$_x$	SO$_x$	PM
January 2008 (Stage 1)	Over 30	Over 20	Postponed
July 2009 (Stage 2)	Over 4	Over 4	

Source: OECD (2009).

StatLink http://dx.doi.org/10.1787/888932318414

Stage 1 targets emission facilities of business category 1 (emissions greater than 80 tonnes) according to the atmosphere environmental conservation law, gas or light oil boiler with the capacity of evaporation amount two tonnes per hour or more, and indirect heating combustion facility with capacity of 1 238 000 kcal per hour or more (annual emissions standard). The targets of Stage 2 are facilities of the business categories 1-3 (emissions greater than 10 tonnes), gas or light oil boiler with the capacity of evaporation amount two tonnes per hour or more, and indirect heating combustion facility with capacity of 1 238 000 kcal per hour or more (annual emissions standard). With the implementation of Stage 2, management of 84% of NO_x emissions, 78% of SO_x emissions and 57% of PM is possible within the Metropolitan air quality management district.

To undertake a trade, factories request an allowance trade and seek approval from the government. The Metropolitan Air Quality Management Office then reviews the request, approves the trade request and recalculates the allowance account. Finally, the emission trade is completed.

This programme builds upon previous measures by the government. For example, the Ministry of Environment instituted total quantity management in parallel with emission concentration regulation (such as through a stack telemonitoring system).

Environmental trends in air pollution

Figures F.2 and F.3 show NO_x emission trends and NO_2 concentration trends in the Metropolitan area. NO_x emissions increased from 1999 to 2003, but started dropping in 2004. Decreases in NO_x emissions seem to be caused by the reduction of the energy consumption by large companies and transportation vehicle combustion. NO_2 concentrations fluctuated every year. It is interesting to note that the NO_2 concentration in Seoul is higher than other areas (although emissions are less) and exceeded the target concentration level specified by the basic plan and atmospheric standard. It should be noted that Seoul is also affected by long-range transboundary air pollutants.

SO_x emission and concentration trends in the Metropolitan areas are shown in Figures F.4 and F.5, respectively. SO_x emissions decreased from 1999 to 2004, but they have been growing since 2005. Increasing SO_x emissions seem to be caused by increases in the sulphur content of some oil. Of note, the SO_2 concentration in Incheon is higher than other areas due to high sulphur oil use by ships.

Figure F.2. **NO_x emission trends in Korea**

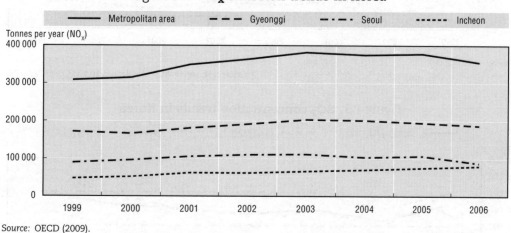

Source: OECD (2009).

StatLink ⬛⬛⬛ http://dx.doi.org/10.1787/888932317749

Figure F.3. **NO$_2$ concentration trends in Korea**

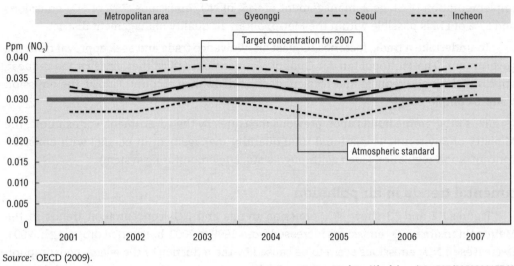

Source: OECD (2009).

StatLink http://dx.doi.org/10.1787/888932317768

Figure F.4. **SO$_x$ emission trends in Korea**

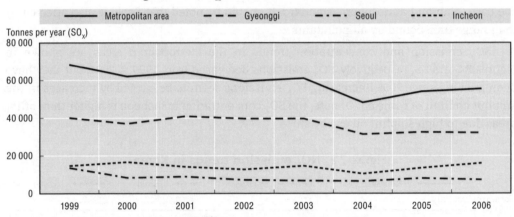

Source: OECD (2009).

StatLink http://dx.doi.org/10.1787/888932317787

Figure F.5. **SO$_2$ concentration trends in Korea**

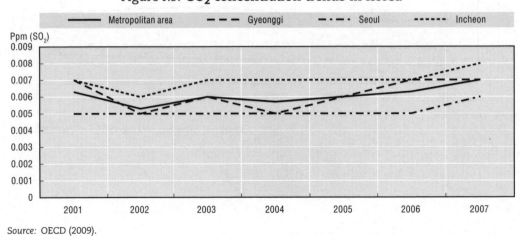

Source: OECD (2009).

StatLink http://dx.doi.org/10.1787/888932317806

Innovation trends in air pollution

De-NO$_x$ technologies include Flue Gas Recirculation Technology (FGR), Low-NO$_x$ Burners (LNB), Selective Catalytic Reduction (SCR), and Selective Non-Catalytic Reduction (SNCR). Korea plans to provide LNB, which is a technology to decrease NO$_x$ emissions, to improve fuel efficiency, reducing thermal NO$_x$ and fuel NO$_x$ concurrently. Generally, the NO$_x$ reduction of LNB is between 30 and 50%. Table F.2 shows the effectiveness of LNB, for example, in reducing air pollutants in 2007.

Table F.2. **Pollution impact of low-NO$_x$ burners**

Region (for year 2007)	Number of low-NO$_x$ burners supplied	Calculated NO$_x$ reduction (tonnes per year)
Metropolitan Area	205	146
Incheon	72	58
Gyeonggi	133	88

Source: OECD (2009).

StatLink http://dx.doi.org/10.1787/888932318433

The reduction efficiencies of low-NO$_x$ burners are diverse, ranging from 30% to 89%, depending on the type of fuel and the vintage of the burner. NO$_x$ reduction efficiencies of LNB in some examples of small-scale boilers are shown in Table F.3.

Table F.3. **NO$_x$ reduction efficiencies by low-NO$_x$ burners**

Burner (for year 2007)		NO$_x$ emission (kg per year)	Fuel quantity (Gcal per year)	Fuel cost (1 000 won per year)	NO$_x$ reduction efficiency (%)
Fuel	Type				
Heavy oil	Existing	1 091.3	1 626.4	80 507	–
	New(LNB)	768.9	1 585.3	78 472	30
LNG	New(LNB)	123.5	1 587.1	76 340	89
Light oil	Normal	435.9	1 626.4	168 170	–
	New(LNB)	212.4	1 585.3	163 920	51
LNG	New(LNB)	122.9	1 579.9	75 993	72
LNG	Normal	309.0	1 626.4	78 230	–
	New(LNB)	123.3	1 585.3	76 253	60

Source: OECD (2009).

StatLink http://dx.doi.org/10.1787/888932318452

With respect to technology development, Japan was the leader in obtaining combined SCR de-NO$_x$ technology patents in the past, specifically the 1970s, when its patent share reached 45.2%. Recently, it has been replaced by others such as United States and Korea. In the early 1990s, Japan maintained the leading position, averaging 38.6 patents annually, but has decreased since 1996. On the other hand, patents of the United States and Korea have increased steadily. In particular, Korea has been active in combined SCR de-NO$_x$ technology since 1999 and Europe has increased patenting tenfold since 1985.

It is interesting to also look at the temporal aspects of the development of combined SCR de-NO$_x$ technology applicable to power plants. For total patents of combined SCR de-NO$_x$ technology, Japan has secured the largest amount with 45.2%, followed by the United States with 27.8%, Europe with 15.0% and Korea with 11.9%. For the most recent five-year period, by contrast, the US has taken the largest share with 32.7% followed by Korea with 23.1% and Europe with 16.7%, while Japan's share has dropped to 27.6%.

Among all types of NO_x emission reduction strategies, Korea has been most active in SCR technologies as seen in Table F.4. In the case of the combined SCR de-NO_x technology applicable to power plants in Korea, SCR technologies occupied the largest share with 50.0%, followed by SNCR technologies with 20.1% (SCR/SNCR hybrid technologies represented only 1.5%). Moreover, N_2O removal technologies occupied 5.9%, mercury removal technologies 0.1% in the SCR and SNCR technical field, corrugate-type catalyst technologies accounted for 12.3%, and nano-type catalyst for 9.3% in corrugated-type de-NO_x catalyst power and forming technologies.

Patents in SCR technology have increased significantly since 1998 and have been more active than for other technologies, although N_2O removal, mercury removal and nano-type TI catalyst technologies became active in 2001. The first corrugate-type catalyst technology patent was granted in 1978. Since 1986, patents of related technologies have been growing gradually. More recently, Korea has moved forward significantly with SCR, SNCR and nano-type TI catalyst technologies. Corrugate type catalyst technologies showed gradually improvement over the 1985-89 period, as outlined in Figure F.7.

Another method of looking at innovation is to use the International Patent Classification codes to identify relevant patents deposited at the Korean Intellectual Property Office, classified by priority year and inventor country. Figures F.6 and F.7 identify

Table F.4. **Patents by technical field in Korea**

	SCR	SNCR	SCR/SNCR hybrid	Corrugated-type catalyst	Nano-type Ti catalyst	Mercury removal	N_2O removal
1975-79				1			
1980-84							
1985-89	5			4			
1990-94	3	4	1	9			1
1995-99	27	19	1	6	2		
2000-04	67	18	1	5	17	2	11

Source: KIPO (2007).

StatLink ⟨⟨⟩⟩ http://dx.doi.org/10.1787/888932318471

Figure F.6. **SO_x abatement patents in Korea**

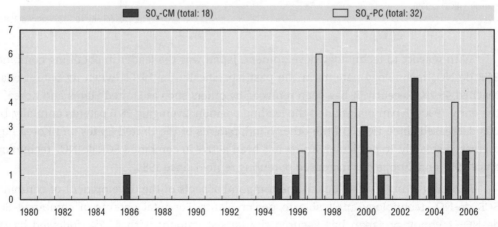

Source: OECD (2009).

StatLink ⟨⟨⟩⟩ http://dx.doi.org/10.1787/888932317825

Figure F.7. **NO$_x$ abatement patents in Korea**

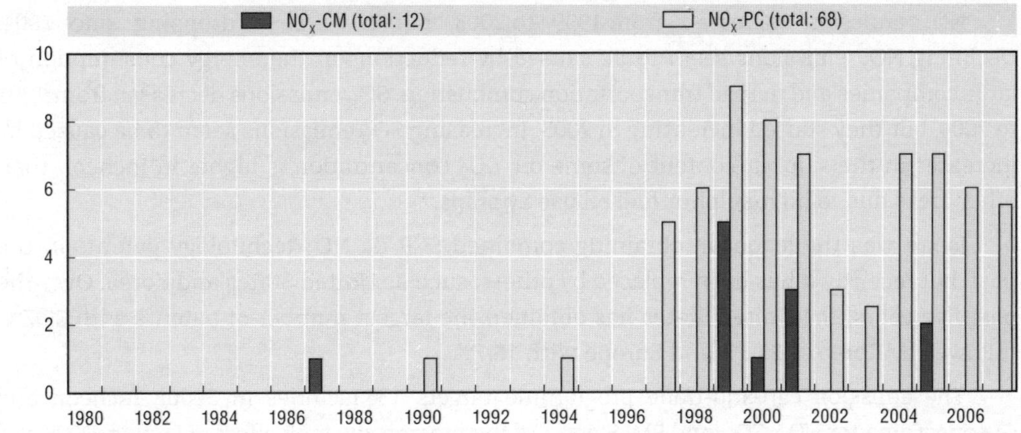

Source: OECD (2009).

StatLink ⟐⟐ http://dx.doi.org/10.1787/888932317844

patents for SO$_x$ and NO$_x$, respectively, emissions related to combustion modification technologies (CM) and post-combustion technologies (PC).

Finally, one can investigate changes in the environmental and the environmental technology R&D budgets. As outlined in Figure F.8, the R&D budget of environmental technologies has posted gradual increases since 2000.

The five largest power plants in Korea account for about 25% of domestic fossil consumption every year, and 15% of domestic air pollution emissions. As such, they make large investments in air pollution control. In the case of domestic desulphurisation facilities, investments of the five largest power plants amounted to KRW 2.3 trillion in 58 equipment purchases since the late 1990s. The amount fell between 2004 and 2005 but has increased since then. In addition, investment in air pollution control instrument construction increased annually between 2003 and 2006.

Figure F.8. **Budget for environmental R&D**

Hundreds of million won

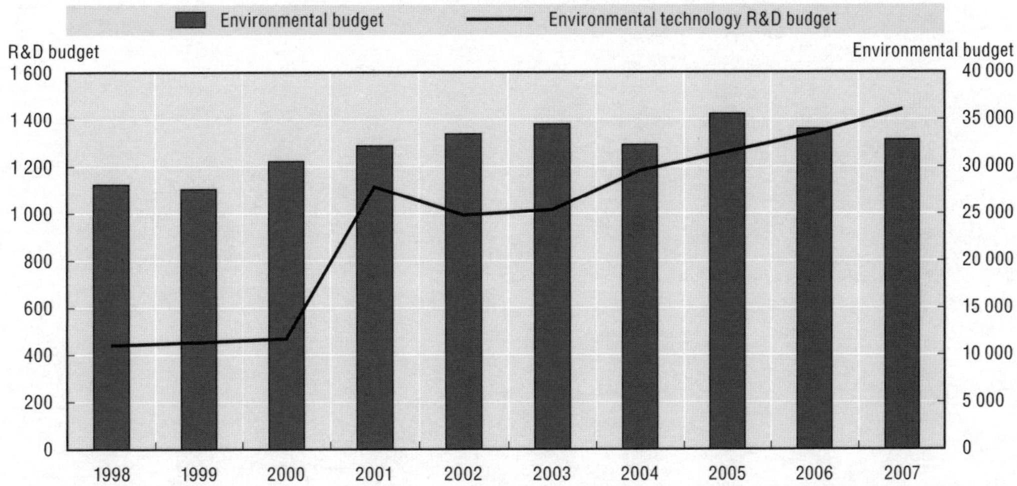

Source: OECD (2009).

StatLink ⟐⟐ http://dx.doi.org/10.1787/888932317863

Conclusions

NO_x emission increased from 1999 to 2003, but have been dropping since 2004. Declining NO_x emissions seem to be caused by reductions in the energy consumption of large companies and in road transportation combustion. SO_x emissions decreased from 1999 to 2004, but they started increasing in 2005. Increasing SO_x emissions seem to be caused by increases in the sulphur content of some oil; SO_2 concentration is higher in Incheon than other areas due to high-sulphur fuel oil use by ships.

Japan was the leader in obtaining combined SCR de-NO_x technology patents in the past, but recently it has been replaced by others, such as United States and Korea. Over the past five years, the United States has obtained the largest number of patents with 32.7%, followed by Korea at 23.1% and Europe with 16.7%.

The emission cap-and-trade programme targets 136 factories in Seoul, Incheon and Gyeonggi area for NO_x, SO_x, and PM. Stage 1 of the programme took effect in January 2008 and Stage 2 was enacted in July 2009. However, it is premature to assess NO_x and SO_x emission reductions, air quality improvements and the patent trend in Korea since the introduction of cap-and-trade programme. To be able to undertake a full assessment of the impact of the system, data over five to six years is necessary. Future work in this area will be needed.

References

Korean Intellectual Property Office (KIPO) (2007), *Application Trends of Combined SCR De-NO_x Technology in Korea*, KIPO, Korea.

OECD (2009), *A Case Study of the Innovation Impacts of the Korean Emission Trading System for NO_x and SO_x Emissions*, OECD, Paris.

ANNEX G

UK Firms' Innovation Responses to Public Incentives: An Interview-based Approach

This case study investigates UK firms and the influence of various policy and market forces on their innovation response. There was a strong correlation between firm-level targets for energy use or greenhouse gas emission targets and R&D (both general and climate change related). Investor and customer pressure also appear to drive process innovation. The effect of the EU ETS was positive for overall innovation but not for climate change related innovation, highlighting the potential issues of predictability of the trading system or the issue of measuring innovation.

Rationale for the study

Significant improvements in energy efficiency are a key component of any climate policy that seeks to achieve substantial carbon emission reductions in the business sector. Yet, more than three decades of research have demonstrated that it is difficult to reconcile neo-classical theories of energy use with the data. For example, both consumers and firms apply extremely high discount rates when evaluating investments into energy efficiency which lead to unfavourable payoff profiles, as such investments typically offer a flow of energy savings over their lifetime. This seemingly irrational phenomenon is often referred to as the "energy efficiency paradox". One theory is that there are certain types of market failure (environmental and innovation-related) that prevent efficient investments in the context of energy efficiency improvements. By contrast, it is possible to look beyond the paradigm of the neo-classical firm altogether and focus on frictions within the firm. This approach highlights the role of organisation structure and management practices in creating barriers to energy efficiency investments. Such investments remain inefficient if, for instance, management engages in short-run optimising behaviour, or if there is a lack of managerial resources and/or attention for cost-cutting projects outside the scope of the firm's main business. Yet, empirical evidence on the magnitude of these effects, and on their precise workings, is scarce. This study will attempt to look at how these factors potentially affect firms' decision regarding the environment.

In order to assess the importance of and interrelationships between public incentives/ regulations for energy use and business management and practice, managers of UK manufacturing firms were interviewed about a range of management practices relevant for climate policy, energy use, innovation and competiveness. In a second step, this information is linked to, and jointly analysed with, firm-level data on economic performance and energy use.

Design of the study

This study deviates from traditional approaches to investigating energy efficiency investments, such as interpreting observed decisions as revealed preferences of economic agents. A straightforward way of eliciting information about people's motivation and behaviour is by asking them. Unfortunately, data obtained in questionnaires are vulnerable to various kinds of survey bias. One way to mitigate survey bias is to conduct loosely structured interviews with informants, rather than collecting information via questionnaires. Thus, managers of British manufacturing facilities were interviewed for this study (Martin et al., 2009).

Structured telephone interviews with managers at randomly selected UK production facilities belonging to the manufacturing sector were undertaken. The defining characteristics of this research design are as follows. First, the interview process follows a double-blind strategy, in that interviewees do not know that they are being assessed on ordinal scales, and interviewers do not know the performance characteristics of the firm they are interviewing. Further, the interviewer engages interviewees in a dialogue with open questions which are meant not to be answered by "yes" or "no". On the basis of this dialogue, the interviewer then assesses and ranks the company along various dimensions on an ordinal scale from one to five. This process helps reduce several sources of potential bias – by using open-ended questions, the question order is less important and respondents are less inclined to answer what is "socially acceptable". The results of the interviews are also linked to independent data on economic performance as a validation exercise and some interviews are double-scored for validation purposes.

The survey seeks to gather information on three main factors concerning the effectiveness of climate change policies:

- The drivers behind a firm's decision to reduce GHG emissions. These include management's awareness of climate change issues (including whether it is a potential business opportunity) and whether they sell related products. Participation in the EU ETS, the UK's CCL/CCA and the effect of other government policies are queried as is the difficulty in complying. Customer and investor pressure related to climate change is investigated.

- The specific measures firms adopt both voluntarily and in response to mandatory climate change policies. These in include monitoring GHG emissions and the setting of targets, the adoption (or not) of technologies and the pay-back criteria used. Firms are also queried about their R&D policies, and the organisation of the firm.

- The relative effectiveness of various measures.

Overall, 190 firms from different subsectors of the manufacturing sector (such as paper mills, ship repair, semiconductors, etc.) were interviewed. The size of the firms in terms of employees in the United Kingdom ranges between 20 and more than 45 000, while global and plant size also show a strong disparity – 70% are multiunit firms, while 80% of firms are ultimately owned by foreign multinationals of different origins, such as South Africa, Korea, France or the United States.

Net income and turnover, as reported in their annual accounts, show as much variation. Firms also differ greatly in their age, with some very young firms (one year old) and one more than two centuries. The degree of competition faced by firms both in the United Kingdom and internationally ranges from inexistence to very high levels. Most firms export their products and import a share of their inputs, although again the intensity of this varies widely. Union membership varies between none and all employees, and the fraction of managers in the firm is usually below 15%. Firms in the sample therefore represent a wide variety of activities, size, profitability, age, international activity and ownership.

The fraction of energy costs relative to total costs was reported by half the interviewees and ranged from 0 to 80%, while some reported the energy as a fraction of turnover, representing between 0 and 32%. Total carbon emissions exhibit large disparities among the 27% of firms that reported them, ranging from less than a tonne to over 400 000 tonnes. Sixty-eight per cent of sites interviewed have implemented an ISO 14000 environmental management system.

To condense the information obtained through the interviews, the raw data are aggregated into summary indices of interview responses on different topics. Table G.1 provides a graphical representation of the construction of each summary index and of two overall indices of climate friendliness. All summary indices are constructed as unweighted averages of the underlying scores, which will differ across sectors. In the regressions that follow, three-digit SIC sector dummies that control for such systematic differences are included.

Linking climate policy to environmental outcomes and economic data

Table G.2 provides the regression results of the various interview data on identifiable environmental and economic statistics. The performance variables in these regressions are obtained by matching the management interview to business microdata maintained by the UK Office of National Statistics. Each panel and each column represent a separate regression. The dependent variable in the first four columns is (the log of) energy intensity, defined in columns 1 and 2 as energy expenditure divided by gross output or, in columns 3 and 4, as energy expenditure over non-capital expenditure (wage costs and materials expenditure). Capital is added as an additional control variable in columns 2 and 4. Column 5 looks at total factor productivity (TFP). An overall index of climate friendliness derived from the survey (Overall Score) is strongly negatively correlated with energy intensity, controlling for capital, which is expected. Interestingly, it is also positively correlated with productivity, which is consistent with the notion that firms with better management are both more productive and less energy-intensive.

From the wide range of variables available in the survey, two factors stand out in particular. First, the existence and stringency of energy quantity targets is negatively associated with energy intensity. That is, firms with targets (or more stringent targets) are clearly less energy-intensive than their peers. Second, there is a strong negative correlation between energy intensity and the relative stringency of investment criteria firms apply. That is, firms that are more demanding concerning hurdle rates or pay-back time when it comes to investments that might save energy are indeed more energy-intensive. Despite needing to be cautious in attaching a causal interpretation, the results are consistent with the well-known finding in the "energy efficiency gap" literature that some firms are not applying "rational" investment criteria.

Table G.1. **Drivers of innovation and construction of indices**

Question	Sign	Index	Overall Indices (1)	(2)
Awareness of climate change score	+		+	
Climate change related products score	+	Awareness	+	+
Positive impact of climate change	+		+	
Participation in ETS (0/1)	+			
Stringency of ETS target score	+		+	
ETS target (in per cent)	+	ETS	+	+
Length of participation	+		+	
Rationality of behaviour on ETS market score	+		+	
Participation in CCA(0/1)	+		+	
Stringency of CCA target score	+	CCA	+	
CCA target (in per cent)	+		+	+
Length of participation	+		+	
Competitive pressure due to climate change score	+	Competitive pressure	+	
Competitive relocation due to climate change score	+		+	+
Customer pressure score	+	Other drivers	+	
Investor pressure score	+		+	+
Energy targets presence (0/1)	+		+	
Energy monitoring score	+		+	
Energy consumption targets score	+	Energy quantity targets	+	
Energy consumption target (in per cent)	+		+	+
Length of target existence	+		+	
Target enforcement score	+		+	
GHG targets presence (0/1)	+	Targets	+	
GHG monitoring score	+		+	
GHG emissions targets score	+	GHG targets	+	
GHG emissions target (in per cent)	+		+	+
Length of target existence	+		+	
Target enforcement score	+		+	
Measures on site score	+		+	
Hurdle rate for energy efficiency investments	−		−	
Payback time for energy efficiency	+	Onsite measures	+	+
Barriers to investments in energy efficiency score	−		−	
Carbon Trust energy audit participation (0/1)	+	Carbon Trust audit	+	
Carbon Trust energy audit (how long ago)	+		+	+
Enhanced Capital Allowance scheme participation (0/1)	+	Enhanced Capital Allowance	+	
Enhanced Capital Allowance scheme (how long ago)	+		+	+
Research and Development – broad innovation score	+		+	
Process innovation score	+	Innovation	+	
Product innovation score	+		+	+

Source: Martin et al. (2009).

First, as indicated by the high hurdle rates, the problem could be external to the firm; *e.g.* banks and financial institutions might demand very stringent payback criteria for such investments. This is in line with the finding for the Enhanced Capital Allowance Scheme (ECA). The ECA is a government subsidy for investments in energy-saving equipment. For ECA-users, there is a strong negative correlation with energy intensity and a strongly positive correlation with productivity. This finding may indicate that capital market imperfections prevent firms from undertaking investments, which are mitigated by the ECA scheme. Second, the problem might in fact be internal to the firm. For example, if energy-intensive firms are not even taking simple measures such as target setting, they

Table G.2. **Survey results and energy intensity**

Index/score	Energy over output		Energy over variable costs		TFP
	(1)	(2)	(3)	(4)	(5)
Awareness	−0.07	−0.11	−0.078	−0.104	0.068
	(0.088)	(0.136)	(0.091)	(0.141)	(0.056)
ETS	0.18	0.151	0.14	0.038	0.051
	(0.149)	(0.175)	(0.137)	(0.155)	(0.105)
CCA	0.193	0.189	0.219*	0.185	0.075
	(0.134)	(0.133)	(0.131)	(0.131)	(0.064)
Competitive pressure	−0.067	−0.17	−0.078	−0.164*	−0.002
	(0.055)	(0.104)	(0.056)	(0.098)	(0.062)
Other drivers	−0.086	−0.109	−0.076	−0.08	0.048
	(0.089)	(0.105)	(0.097)	(0.118)	(0.049)
Energy quantity targets	−0.138	−0.272***	−0.097	−0.202**	0.216***
	(0.090)	(0.095)	(0.090)	(0.097)	(0.066)
GHG targets	0.085	−0.026	0.118	0.048	0.190**
	(0.096)	(0.121)	(0.099)	(0.115)	(0.073)
Targets	−0.081	−0.255**	−0.032	−0.159	0.277***
	(0.098)	(0.110)	(0.107)	(0.119)	(0.085)
Onsite measures	−0.062	−0.072	−0.029	−0.019	0.182***
	(0.072)	(0.089)	(0.073)	(0.092)	(0.064)
Carbon trust audit	−0.068	−0.104	−0.130**	−0.145**	0.022
	(0.061)	(0.063)	(0.062)	(0.067)	(0.046)
Enhanced capital allowance	−0.288**	−0.296**	−0.208*	−0.228*	0.131**
	(0.113)	(0.128)	(0.109)	(0.126)	(0.055)
Investment criteria stringency	0.331***	0.432***	0.344***	0.526***	−0.132
	(0.106)	(0.111)	(0.108)	(0.119)	(0.109)
Innovation	0.044	−0.028	−0.064	−0.093	−0.001
	(0.107)	(0.123)	(0.106)	(0.123)	(0.064)
Overall Score 1	−0.18	−0.418**	−0.191	−0.360*	0.331**
	(0.150)	(0.184)	(0.169)	(0.203)	(0.130)
Overall Score 2	−0.189	−0.429**	−0.215	−0.391*	0.341**
	(0.146)	(0.186)	(0.163)	(0.201)	(0.140)
Controlling for capital	No	Yes	No	Yes	Yes
Three-digit sector dummies	Yes	Yes	Yes	Yes	Yes
Year dummies	Yes	Yes	Yes	Yes	Yes
Observations	966	756	966	756	756

*   Significant at 10%.
** Significant at 5%.
*** Significant at 1%.
Source: Martin *et al.* (2009).

StatLink ᴍᴸᴸ http://dx.doi.org/10.1787/888932318490

may be less likely to take advantage of the ECA. In contrast, firms that set energy or carbon targets and apply for the ECA are on average more productive and less energy-intensive. Concerning government policy, another intriguing result is the negative correlation between participation in a Carbon Trust energy audit and energy intensity. Further analysis is required to determine if this is endogenous or indeed causal.

Using alternative outcome measures from matching the interview data to publicly available balance sheet data (from the *ORBIS Database*) the analysis of productivity can be repeated. Energy data is not contained in this database; however, matching this data with the interview data is more precise. The results are reported in Table G.3. Labour productivity is considered in the first column where each regression of the different scores on the logarithm

Table G.3. **Survey results and productivity**

	Labour productivity	TFP	
	(1)	(2)	(3)
Awareness	0.004	0.062**	0.070***
	(0.050)	(0.024)	(0.024)
EU ETS	0.190**	0.129*	0.109
	(0.096)	(0.068)	(0.069)
CCA	0.243***	0.023	0.022
	(0.063)	(0.039)	(0.037)
Competitive pressure	−0.039	−0.002	0.003
	(0.040)	(0.025)	(0.024)
Other drivers	0.024	0.050*	0.049*
	(0.047)	(0.028)	(0.027)
Energy targets	0.241***	0.087***	0.084***
	(0.054)	(0.029)	(0.029)
GHG targets	0.225***	0.058	0.065
	(0.064)	(0.039)	(0.040)
Overall targets	0.287***	0.102***	0.104***
	(0.065)	(0.038)	(0.037)
Onsite measures	0.104**	0.080**	0.079**
	(0.049)	(0.038)	(0.038)
Carbon Trust audit	0.129**	0.087***	0.076***
	(0.050)	(0.024)	(0.027)
Enhanced Capital Allowance	0.083	0.059**	0.079***
	(0.065)	(0.027)	(0.028)
Innovation	0.078	0.008	0.027
	(0.057)	(0.040)	(0.040)
Climate change friendliness (1)	0.340***	0.172***	0.173***
	(0.107)	(0.056)	(0.054)
Climate change friendliness (2)	0.326***	0.174***	0.172***
	(0.118)	(0.059)	(0.058)
Sector and time dummies	Yes	Yes	Yes
Interviewer dummies	No	No	Yes
Observations	1 387	1 106	1 106

Notes: The dependent variable in all regressions is the logarithm of turnover. In the first column, the logarithm of employment is included in the explanatory variables such as to capture labour productivity, while the second and third columns approximate total factor productivity by including also the logarithm of capital and materials. Each panel reports the coefficient and standard errors clustered at the firm level (i.e. robust to heteroskedasticity and autocorrelation of unknown form) relative to each explanatory overall index included in separate regressions.
* Significant at 10%.
** Significant at 5%.
*** Significant at 1%.
Source: Martin et al. (2009).

StatLink ᴬᴵᴸ http://dx.doi.org/10.1787/888932318509

of turnover includes the logarithm of employment as a control. Columns 2 and 3 consider total factor productivity (TFP) by including (logarithm of) employment, materials and capital in each regression on turnover. Column 3 adds as a control a dummy for the identity of the interviewer.

The derived climate friendliness index is strongly positively correlated with productivity. The coefficient nearly halves once capital and materials are included. Among other things, climate friendliness might affect productivity by increasing investment in capital through cleaner technologies. In this case, the coefficient on the overall index might be an underestimation when capital is included in columns 2 and 3.

Productivity is strongly positively correlated with targets and measures. The energy targets index includes the monitoring of energy use, stringency of targets and the period of time they have been in place. The positive correlation between TFP and the onsite measures index is comparable to the first set of results. Again, productivity is strongly and positively correlated with the Enhanced Capital Allowance and also the Carbon Trust energy audit indices.

The climate change awareness index – though insignificant in the first set of results – is now positively and significantly correlated with TFP. Hence, more productive firms are also more likely to have climate change related products, to expect positive impacts of climate change and to exhibit more awareness of climate change issues among its management. Further, at the 10% level of confidence, investor and customer pressure, summarised in the "other drivers" index, are positively correlated with TFP.

Linking climate policy to innovation

Of greatest importance is how management practices and government policies affect the innovative activity of the firm. To confirm the robustness of the R&D survey measure, it was regressed on the number of patents held by a firm (based on European Patent Office data) and found to be consistent.

Table G.4 displays results from linear regression models where the dependent variable is the score for climate change related (CCR) process innovation (in columns 1 and 2), the score for CCR product innovation, and the score for the importance of common R&D in the company. Other scores and indices from the survey data are included as explanatory variables, where each panel corresponds to a separate regression.

Climate change awareness: In panels 1 and 2, both types of CCR R&D are strongly correlated with both the degree of climate change awareness and the importance of CCR products for the firm. This corroborates the internal consistency of the survey responses. The insignificant finding for general R&D demonstrates that the sample is well stratified, in the sense that not all R&D-intensive firms happen to be highly aware of climate change or producers of CCR products.

Climate Change Agreements and Climate Change Levy: Panel 3 displays insignificant coefficient estimates for participation in a Climate Change Agreement. The other study on the UK's CCL (case study H) explains that, since all firms are subject to the Climate Change Levy, one can identify only the effect of the Climate Change Agreements which gave some firms a large discount on their tax liability if they promised to reduce their energy consumption. This insignificant result could lead the reader to believe that the combination of tax discount and quantity target embodied in the CCA provided very similar incentives for R&D as the Climate Change Levy – at least in this particular sample. However, the analysis of patent grants presented in the other study goes strongly against this conclusion. Firms in a CCA obtain significantly fewer patent grants than firms in the CCL. Moreover, the use of panel data methods is necessary to control for firm-specific unobservable factors that affect both the firm's innovative activity and its decision to participate in a CCA. Therefore, the results in the other case study regarding the impact of CCA membership on innovation should be used.

Competitive pressures: There is not a significant effect of the competitive pressures index on innovation. This is probably due to the fact that few firms expected strong effects of climate policy on competition and relocation in the first place.

Table G.4. **Survey results and innovation**

		R&D type					
		CCR process innovation		CCR product innovation		General	
		(1)	(2)	(3)	(4)	(5)	(6)
(1)	**CCR products**	0.422***	0.395**	0.825***	0.762***	0.1	0.11
	(Score)	(0.146)	(0.154)	(0.160)	(0.183)	(0.179)	(0.198)
(2)	**CCR awareness**	0.343**	0.301**	0.497***	0.500***	0.22	0.25
	(Summary Index)	(0.134)	(0.139)	(0.140)	(0.137)	(0.142)	(0.150)
(3)	**CCA stringency**	0.17	0.12	−0.12	−0.15	0.0	0.03
	(Summary Index)	(0.121)	(0.138)	(0.138)	(0.154)	(0.124)	(0.151)
(4)	**Competitive pressures**	0.08	0.10	−0.09	−0.13	0.14	0.15
	(Summary Index)	(0.070)	(0.070)	(0.152)	(0.158)	(0.128)	(0.139)
(5)	**Enhanced Capital Allowance**	0.09	0.14	0.14	0.12	0.165*	0.17
	(Summary Index)	(0.117)	(0.122)	(0.158)	(0.175)	(0.092)	(0.104)
(6)	**Energy quantity targets**	0.387***	0.395***	0.04	0.11	0.291**	0.300**
	(Summary Index)	(0.102)	(0.113)	(0.137)	(0.165)	(0.119)	(0.135)
(7)	**GHG targets**	0.495***	0.439***	0.395**	0.368*	0.443***	0.432***
	(Summary Index)	(0.143)	(0.155)	(0.178)	(0.197)	(0.139)	(0.155)
(8)	**Targets**	0.482***	0.451***	0.14	0.17	0.392***	0.383**
	(Summary Index)	(0.135)	(0.147)	(0.172)	(0.197)	(0.149)	(0.165)
(9)	**EU ETS**	−0.04	−0.1	−0.05	−0.02	0.287*	0.423*
	(Summary Index)	(0.18)	(0.218)	(0.157)	(0.191)	(0.170)	(0.237)
(10)	**Onsite measures**	0.269**	0.250**	0.11	0.14	−0.01	−0.07
	(Summary Index)	(0.105)	(0.103)	(0.12)	(0.128)	(0.127)	(0.128)
(11)	**Other drivers**	0.429***	0.409***	0.342**	0.300**	0.371***	0.385***
	(Summary Index)	(0.101)	(0.110)	(0.132)	(0.151)	(0.118)	(0.133)
(12)	**Carbon Trust audit**	0.13	0.06	0.08	0.05	−0.08	−0.05
	(Summary Index)	(0.101)	(0.104)	(0.112)	(0.117)	(0.107)	(0.106)
(13)	**Customer pressure**	0.427***	0.357**	0.343*	0.28	0.392**	0.471**
	(Score)	(0.159)	(0.175)	(0.185)	(0.203)	(0.188)	(0.205)
(14)	**Investor pressure**	0.464***	0.498***	0.408*	0.35	0.455**	0.434**
	(Score)	(0.172)	(0.176)	(0.212)	(0.251)	(0.18)	(0.206)
(15)	**Investment criteria stringency**	−0.11	−0.16	0.22	0.34	−0.03	0.23
	(Score)	(0.435)	(0.467)	(0.397)	(0.470)	(0.409)	(0.426)
	Sector controls	Yes	Yes	Yes	Yes	Yes	Yes
	Controls for capital stock	No	Yes	No	Yes	No	Yes
	Observations	181	163	176	157	183	164

* Significant at 10%.
** Significant at 5%.
*** Significant at 1%.
Source: Martin *et al.* (2009).

StatLink http://dx.doi.org/10.1787/888932318528

Enhanced Capital Allowance: In contrast to previous findings, there is no robust evidence that beneficiaries of the Enhanced Capital Allowance innovated more. A plausible explanation for this is that the allowance was granted for the adoption of existing technologies and not for R&D expenditures with uncertain outcomes. It is possible that the allowance freed up financial resources that firms subsequently deployed to R&D projects, yet this indirect effect is not estimated precisely enough to be conclusive.

Targets: Panels 6, 7 and 8 display a strong positive correlation between CCR process innovation with both energy quantity targets and GHG targets. This is intriguing and calls for a closer examination of the underlying mechanisms. For example, it is possible that senior management embarks on a CCR R&D project and then sets tight energy quantity targets to strengthen the incentives for a successful outcome of the R&D project. Conversely, it could also be that stringent targets are implemented first, and that their presence induces innovation in CCR processes. In view of the earlier finding that stringent targets are also associated with higher energy efficiency, it is hypothesised that only those firms that have already picked the "low-hanging fruit" in terms of energy efficiency improvements need to conduct proper R&D to further reduce the energy used in their production processes.

It is striking that CCR product innovation is positively correlated with GHG quantity targets but not so with energy quantity targets. The most immediate explanation for this is that CCR product innovation reduces energy consumption of the firms' customers, but does not necessarily help the firm itself to meet its energy quantity targets. Nevertheless, for a firm that tries to sell a CCR product, it may be important to be perceived by their customers as "climate-friendly", and hence the presence of GHG targets is a vital part of their marketing strategy. Notice that, according to this idea, the directions of causation for process and product innovation are diametrically opposed in that stringent energy and GHG targets cause process R&D, but CCR product innovation causes GHG targets.

EU ETS: Panel 9 shows that EU ETS participation had no significant effect on CCR process or product innovation. The lack of an innovation impact of this EU-wide policy can in part be explained by the low average allowance prices that have reigned on the carbon markets so far. Another issue is the high volatility of allowance prices during Phase 1 of the trading scheme, potentially being a real options problem. Uncertainty about future prices might induce firms to postpone, and even reduce, current R&D spending because they prefer to wait and see how the allowance price evolves. Similarly, firms may have been waiting for legal certainty about future tightening of ETS targets beyond the end of Phase 2 in 2012 before spending resources on CCR R&D.

ETS membership, however, is positively associated with general R&D. In the spirit of the "strong" Porter hypothesis, one could argue that ETS firms seek to advance their overall productivity in order to better compete in the future. Still, it seems odd that this effort does not affect CCR R&D at all. It could also be that generous allowance allocations at the beginning of the ETS along with grandfathering of allowances left ETS firms with a windfall of financial resources, part of which they diverted to their R&D departments. While this effect is significant at the 10% level only, it is robust to the inclusion of capital which controls for firm size.

Onsite measures: There is a significant, positive association between onsite measures and CCR process innovation, but not with product or overall innovation. This is intuitive because the survey questions about onsite measures refer to the adoption of new processes and technologies suitable for immediate abatement, and not to future abatement that could be brought about by full-fledged R&D projects.

Other drivers: Panel 11 displays a strong positive correlation between other drivers and all types of R&D. Since this index is an average of the scores for investor and customer pressure, panels 13 and 14 report results from separate regressions. It seems that both factors have an effect of equivalent size. Moreover, the relationship is stronger for CCR process R&D than for CCR product R&D. The coefficients for CCR product R&D in columns 3 and 4 of panel 11 are also significant, and the separate coefficients in panels 13 and 14 are not (or less so). This suggests that both customer and investor pressure must coincide to induce a firm to undertake R&D in CCR products.

Carbon Trust audit: Panel 12 shows that participation in a Carbon Trust audit is not associated with any significant change in R&D efforts, in line with the audits' purpose to identify opportunities for energy efficiency improvements near zero cost.

Investment criteria stringency: The last panel of Table G.4 shows that the stringency of investment criteria has no effect on R&D across the board, different than the previous analysis on hurdle rates. Hence, it seems that these criteria are applied to guide decisions on the adoption of existing technologies, but not on the invention and commercialisation of new technologies. This makes sense, since R&D spending is a long-term and often strategic investment with uncertain returns, so that simple rules-of-thumb hardly seem appropriate.

Summarising, some climate policies are effective at improving energy efficiency. In the survey data, there is suggestive evidence of this for policies that promote the transfer of known practices and the adoption of existing technologies, such as the Carbon Trust Audit or the Enhanced Capital Allowance Scheme. In the other, related case study of the Climate Change Levy, the levy caused larger reductions in energy use and increases in energy efficiency than the CCA. Since neither of these policies was in place before 2001, it appears that the Climate Change Levy fostered both energy efficiency and innovation in energy efficiency. Concerning other climate policies, the survey data suggest that none of them was successful in promoting innovation. The most plausible explanation for this is that either these policies were geared at short-term improvements in energy efficiency (*e.g.* the Carbon Trust audits and the Enhanced Capital Allowance Scheme) or that their design did not give the strong price signals and stable planning horizon necessary for R&D spending with highly uncertain returns (in the EU ETS case).

An econometric approach is best suited if the goal is to derive the causal effects of these policies, as is done in the related case study. The distinctive advantage of this research design, however, is that one can identify new transmission channels for government policy based on the detailed data on management practices and other firm characteristics gathered in the interview process. The most salient effect is the presence of energy quantity targets that, when combined with adequate monitoring and enforcement, are strongly associated with higher energy efficiency and with R&D into even better processes and into general-purpose R&D. On the one hand, this finding gives some confidence in the assessment that quantity targets under the CCA and the EU ETS have been too lax to foster innovation. On the other hand, this finding also suggests that, more stringent target setting aside, policy measures that facilitate the monitoring process and that streamline enforcement might be necessary to foster innovation effects. Moreover, it appears that suppliers of carbon-saving intermediate goods or final products adopt GHG emission targets as a part of their marketing strategy, while pursuing R&D in the development of such products. The success of such a marketing strategy is likely to depend on the availability of an institution that monitors and certifies the carbon footprint of the firm. Policy can thus not only help with the creation of markets for carbon-saving products but also with an independent agency that certifies the carbon savings derived from them. Finally, the positive relationship between CCR innovation and consumer and investor pressure suggests that the presence of such a carbon certification agency could leverage consumer and investor pressures on the firm. The higher the degree of accuracy in the agency's rating of the firm's "climate friendliness", the clearer defined are the firm's incentives to undertake R&D aimed at improving its rating.

In a nutshell, the three policy recommendations coming out of this research are that: i) price incentives for carbon saving should not be watered down by discounts; ii) quantity targets must be stringent, easy to monitor and not just for the short term; and iii) the innovation impact of both emission targets and green preferences by the public can be leveraged by implementing independent assessments of firms' carbon footprint.

Conclusion

This case study has sought to improve the understanding of the interdependencies between climate change policies, management practices, and innovation. In order to assess firm-level responses to climate change policies, management practices related to climate change were empirically analysed, using a survey tool recently developed in the productivity and management literature.

In looking at the drivers of energy intensity and productivity, regression analysis has shown that an index of overall climate change friendliness is positively correlated with energy efficiency and productivity in a robust and very significant way. Upon analysing the different components in more detail, two main elements seem to be driving this result. First, a firm's use of targets and its monitoring of energy consumption have strong positive correlations with both the firm's energy efficiency and its total factor productivity. Second, there is also a strongly significant and negative correlation between energy intensity and the relative stringency of investment criteria firms apply, meaning that firms that are more conservative in their investment criteria (and therefore unlikely to invest in energy-saving technologies) are likely to be more energy intensive.

Of most importance is the relationship between a firm's characteristics and the impact on innovation. There is a strong correlation between firms' use of targets for energy use or GHG emission targets and R&D (both general and climate change related). Although the direction of causality cannot be specifically tested here, the results are indicative of the causality going from stringent targets leading to process R&D, but product R&D causing targets. Investor and customer pressure also drive process innovation and – when combined – are positively correlated with product innovation. Firms' use of the Enhanced Capital Allowance, a corporate tax benefit for adopting capital equipment, appeared to have little effect on innovation. Finally, the effect of the EU ETS (which can be thought of as a tax-like measure) has positive effects on overall innovation but not for innovation specifically related to climate change. This somewhat counterintuitive result may be due to the overall impact of the variability of the permit price in affecting business decisions about investments in innovation.

For more information on UK firms' responses to various policy measures and market forces, the full version of the case study (OECD, 2009) is available at *www.olis.oecd.org/olis/ 2008doc.nsf/linkto/com-env-epoc-ctpa-cfa(2008)34-final*.

References

Martin, R. *et al.* (2009), "Climate Change Policies and Management Practices: Evidence from Interviews with Managers", Draft, Centre for Economic Performance, London School of Economics, UK.

OECD (2009), *Survey of Firms' Responses to Public Incentives for Energy Innovation, including the UK Climate Change Levy and Climate Change Agreements*, OECD, Paris, available at *www.olis.oecd.org/olis/ 2008doc.nsf/linkto/com-env-epoc-ctpa-cfa(2008)34-final*.

ANNEX H

The UK's Climate Change Levy and Climate Change Agreements: An Econometric Approach

This case study examines the role of the UK's Climate Change Levy (and associated negotiated Climate Change Agreements with industry) on innovation. Firms with CCAs, who were granted an 80% reduction in the rate of the CCL, tended to be more energy intensive and use more electricity (which was taxed the highest within the levy scheme) than similar firms paying the full rate. Firms paying the full rate did not appear to experience adverse financial or economic effects. Moreover, CCA firms were significantly less likely to innovate than firms paying the full rate, including in areas related to climate change.

Rationale for the instrument

Addressing climate change means reducing carbon levels (and those of other greenhouse gases as well) in the atmosphere. Combustion of fossil fuels – whether in industry, transportation, or for electricity generation – is the main culprit in anthropomorphic greenhouse gas emissions. Taxes on fossil fuels, such as the Climate Change Levy (CCL), provide incentives for energy efficiency as well as for the development of less carbon-intensive power sources.

Design of the instrument

The CCL was first announced in March 1999 and came into effect in April 2001. The CCL is a per unit tax payable at the time of supply to industrial and commercial users of energy. Taxed products include coal, natural gas, electricity, and non-transport liquefied petroleum gas (LPG). Table H.1 displays, for each fuel type subject to the CCL, the tax rates per kilowatt hour (kWh), the average energy price paid by manufacturing plants in 2001 and the implicit carbon tax. It is evident that energy tax rates vary substantially across fuel types, ranging from 6.1% on coal to 16.5% on natural gas. The tax thus establishes a meaningful price incentive for energy conservation overall.

Since the CCL is a tax on energy and not a carbon tax, the varying carbon contents among fuels means that the implicit carbon tax rate also varies, *e.g.* gas and electricity is taxed at almost twice the rate as carbon contained in coal. This can be attributed to political

Table H.1. **Rates of the Climate Change Levy**

Fuel type	Tax rate[1]	Fuel price	Implicit carbon tax
	GBP per kWh		GBP per tonne carbon
Electricity	0.0043	0.0425	31
Coal	0.0015	0.0246	16
Gas	0.0015	0.0091	30
LPG	0.0007	0.0085	22

1. In FY 2010-11, the tax rates have increased to GBP 0.0047 per kWh electricity, GBP 0.00164 per kWh natural gas, GBP 0.00105 per kilogram LPG and GBP 0.01281 per kilogram coal and other solid fuels.
Source: Martin et al. (2009).

StatLink ⟨⟨⟨⟨ http://dx.doi.org/10.1787/888932318547

pressures arising from historical ties between the government and the coal industry. Some fuel types are tax-exempt based on their low carbon content, notably electricity generated from renewable sources and from combined heat and power generation.

The revenue generated by the CCL is, for the most part, recycled back to industry in the form of a 0.3 percentage point reduction in employers' National Insurance Contributions. A small part of the revenue is used to fund the Carbon Trust, an institution set up by the government to foster research and development into energy efficiency schemes and renewable energy resources.

In order to address concerns about possible adverse effects of the CCL on competitiveness and economic performance of energy-intensive industries, the government set up a scheme of negotiated agreements, the Climate Change Agreements (CCA). Participation in a CCA entitles facilities in certain energy-intensive sectors to an 80% discount[1] on their tax liability provided that they adopt a binding target on their energy use or carbon emissions. The participation process involved two stages. First, the trade association of an energy-intensive sector negotiates a so-called umbrella agreement with the government to determine a sector-wide target for energy use or carbon emissions in 2010, as well as interim targets for each two-year milestone period. Targets are defined either in absolute terms or relative to (often physical units of) output. At the second stage, firms in eligible sectors apply for a reduced-rate certificate that entitles them to the discount on the levy paid at a qualifying site. If the application is approved, these firms enter a so-called underlying agreement which defines the target unit, i.e. the facility or group of facilities benefiting from the tax discount, and stipulates a specific reduction to be achieved by the target unit.

At the end of each milestone period, the associations report whether the sector-wide target has been met. Only if a sector-wide target has been missed does the government verify compliance at the unit level. A facility that is found in non-compliance is not "re-certified" for the reduced rate in the following milestone period. If the facility misses the 2010 target, it faces the threat of repaying all rebates on the levy it has accumulated in previous periods. CCA participants who were in danger of missing their target could buy emission allowances on the UK Emissions Trading Scheme (UK ETS), a market for carbon permits that was launched in 2002 and ended in December 2006.[2] Conversely, excess carbon or energy reductions could be sold in the UK ETS or "ring-fenced" (banked) for use towards future targets.

While the CCL and the CCA share the common objective of enhancing the efficiency of energy use in the business sector, it is important to note that there are fundamental differences between these policy instruments:

- The CCL increased energy prices faced by the typical business in 2001 by approximately 15%. If energy demand is price sensitive, the increased relative price of energy should lead to improvements in energy efficiency. Unless there is a strong rebound effect, or an exogenous increase in economic activity, this should reduce energy use in the CCL sector. In theory, however, the levy's impact on carbon emissions is ambiguous because even an absolute reduction in energy use could come with a shift towards more carbon-intensive fuels.

- The CCA, by contrast, combines a much more diluted price signal with quantity regulation, mostly in the form of efficiency targets. The CCAs' impact on energy use thus depends critically on whether the target places a binding constraint on a plant's production choices. If not, then the plant has less of an incentive to conserve on energy than it would under have under the full tax rate. Furthermore, since most targets are specified in terms of energy units rather than carbon emissions, there is no guarantee that even a stringent target leads to reductions in GHG emissions.

There is the strong possibility that the negotiated targets were rather lax. The proportion of target units that were re-certified was consistently high, rising from 88% in the first period to 98% and 99% in the second and third target periods, respectively. As a rule, CCA participants reached their targets or purchased allowances on the UK carbon trading market to ensure compliance at low cost. In fact, a lower bound on compliance cost is zero. This is true for a considerable amount of target units that missed their target but were re-certified due to the sector as a whole being in compliance.

A large degree of flexibility was built into the target negotiations both prior and subsequent to the compliance review. Target units could call upon several risk management tools that made it easier to meet their targets. For example, adjustments to targets could be made to reflect a more energy-intensive product mix, declining output or relevant constraints arising from other types of regulation. In some sectors, performance was measured against a tolerance band in lieu of a fixed target. Moreover, sectors were permitted to choose their baseline year. More than two-thirds of all sectors chose baseline years of 1999 or earlier. Hence, carbon savings that had occurred before the policy package was implemented could be counted towards the target achievement. Finally, in some instances, growing companies that belonged to a sector with an absolute target successfully bargained for a relative target (and *vice versa*) as this made it easier to comply. In sum, there is ample evidence that the negotiated targets are unlikely to have placed binding constraints on energy use by CCA companies.

In order to explore these issues, a novel data set has been created by matching two confidential business data sets and augmenting it with publicly available data on participation in the CCA. In particular:

- The *Annual Respondents Database* (ARD) from the Office of National Statistics (ONS) has data on output and factor inputs, including energy expenditure, for about 10 000 manufacturing plants between 1999 and 2004.

- The Quarterly Fuels Inquiry (QFI), provided by the ONS, holds energy consumption data (kWh, tonnes, etc.) for about 1 000 firms for 1997-2004.

- Data on CCA participation for about 5 000 agreements is available online from the Department of the Environment, Food and Rural Affairs and HM Revenue and Customs.

- Data on pollution emissions (thresholds and actual discharges to air and water for over 50 pollutants) by UK facilities is reported to the European Pollution and Emissions Register (EPER).

Table H.2 shows the descriptive statistics for all samples in 2000, broken down by CCA participation status, as well as the results of a t-test of equality of the group means. CCA plants are, on average, older, larger and more energy intensive and, for most of these plant characteristics, equality is rejected at the 1% significance level. In view of the strong correlation of CCA participation with observable plant characteristics, the possibility that unobservable plant characteristics also influence selection cannot be ruled out. In the analysis below, an identification strategy that takes due account of the sample selection issue, so as to avoid inconsistent estimation, is used.

The analysis below focuses on the manufacturing sector and the first two target periods, running from April 2001 until December 2004. On the one hand, this is dictated by the time coverage in the data set. On the other hand, it avoids possible complications due

Table H.2. **Descriptive statistics by CCA participation status**

For year 2000	ARD			QFI			QFI and ARD		
	CCA = 0	CCA = 1	diff	CCA = 0	CCA = 1	diff	CCA = 0	CCA = 1	diff
Variables	(1)	(2)	(3)	(4)	(5)	(6)	(7)	(8)	(9)
Age	13.55	17.53	***	21.86	22.87	–	21.54	22.84	*
Employment	151.49	536.44	***	372.05	548.98	***	373.14	548.98	***
Gross output	19.08	86.08	***	49.07	91.56	***	49.29	91.56	***
Energy expenditure	0.22	1.95	***	0.59	3.79	***	0.59	3.79	***
Variable costs	15.99	75.14	***	42.19	78.46	***	42.39	78.46	***
Capital stock	9.64	58.17	***	23.12	65.44	***	28.89	72.78	***
EE /variable costs	1.92	3.01	***	1.99	3.60	***	1.99	3.60	***
Electricity	8 701.55	38 191.39	***	8 888.03	34 210.84	***	8 701.55	38 191.39	***
Electricity expenditure	306.93	1 162.83	***	292.64	1 050.91	***	306.93	1 162.83	***
Gas	14 144.07	75 098.82	***	14 859.74	68 213.13	***	14 144.07	75 098.82	***
Share of gas/gas and electricity consumption	0.19	0.24	***	0.18	0.25	***	0.19	0.24	***
Solid fuels	0.01	0.34	–	0.39	1.44	–	0.21	1.66	–
Solid fuels expenditure	1.91	44.30	*	55.98	191.24	–	36.42	219.43	–
Liquid fuels	0.01	0.36	–	0.21	2.02	*	0.28	1.78	–
Liquid fuels expenditure	0.71	20.45	**	10.74	132.41	**	13.52	101.28	*
Total kWh	27 261.95	146 775.90	***	29 834.32	135 378.51	***	27 261.95	146 775.90	***
Total kWh expenditure	23.23	390.91	***	446.06	1 784.71	***	443.30	1 936.10	***
Total kWh/gross output	0.01	0.03	***	0.01	0.03	***	0.01	0.03	***
CO_2	10 673.51	54 239.67	***	11 454.80	50 219.85	***	10 673.51	54 239.67	***
CO_2/total kWh	0.45	0.44	–	0.45	0.43	*	0.45	0.44	–
CO_2/gross output	326.82	750.21	***	326.82	750.21	***	326.82	750.21	***
Number of plants	8 282	1 050		701	251		434	212	

Notes: Gross output and all the expenditure variables are in thousands of pounds. Total kWh, gas and electricity are in thousands of kWh. Solid and liquid fuels are in thousands of tonnes. The CO_2 variable measures total CO_2 emissions in thousands of tonnes based on fuel use. Columns 3, 6, and 9 report significance levels from a t-test of differences in group means with unequal variance.
* Significant at 10%.
** Significant at 5%.
*** Significant at 1%.
Source: Martin et al. (2009).

StatLink http://dx.doi.org/10.1787/888932318566

to: *i)* an overlap with the EU ETS which affected about 500 CCA plants from 2005 onwards; *ii)* adjustments of CCAs targets for the third milestone period; and *iii)* new entry of sectors in 2006 after eligibility had been changed.

Environmental impacts of the CCL and CCA

Table H.3 summarises final regression results for various environmental/energy outcome variables (in rows) under different assumptions about the error term (in columns). The explanatory variable of interest is CCA participation after 2000. To address potential endogeneity of CCA participation with respect to the various outcome variables, all regressions are conducted in terms of differences. In addition, an instrumental variables approach (IV) is used. The instrument is an indicator variable equal to one if a business is reporting emissions other than combustion-related emissions in the EPER. The legal basis for the EPER is European Pollution Prevention and Control legislation which the UK government used as the eligibility criterion for CCA participation. Thus, unlike CCA participation, which depends on endogenous decisions of firms in the sample, EPER is based on factors exogenous to post-2000, firm-specific shocks.

The first column contains results from a pooled ordinary least squares (OLS) estimation without plant fixed effects. In column 2, the CCA participation variable is replaced with the instrumental variable EPER to estimate a reduced-form equation. Column 3 reports results from a pooled two-stage, least squares specification. Columns 4 to 6 repeat this sequence while including plant specific fixed effects (FE). Consequently, column 6 reports the most general estimate of the average treatment effect on the treated, using an instrumental variable (IV) specification.

Energy: The first two panels report the results for energy intensity measured as energy expenditures over gross output and as the share of energy expenditures in variable costs (the sum of expenditures on materials, energy and wages), respectively. The results are very similar for both variables. Plants in a CCA increased their energy intensity by more than 20% relative to plants that paid the full levy after 2000. This effect is both economically and statistically significant. The point estimates change very little when moving to the regressions with fixed effects in columns 4 to 6. This suggests that normalising energy use by some measure of plant size goes a long way to control for unobserved heterogeneity between plants. Further, the importance of controlling for selection is evident from the sizable differences between the OLS and IV estimates.

Panel 3 reports the results for energy expenditure. A statistically significant and positive effect is found only with fixed effects, potentially indicating declining trends in energy use within industries that are correlated with both CCA participation and EPER coverage. For instance, parts of the steel industry experienced a seminal downturn that coincided with the introduction of the CCL package, yet did not affect all quality tiers in steel production equally. Naturally, this issue disappears when dividing by a size control (as in panels 1 and 2) or when controlling for plant specific fixed effects (as in columns 4 to 6). The point estimate in the IV regression (column 6) implies that participation in a CCA led plants to increase their expenditures on energy by more than 15% relative to plants that were subject to the full tax.

Fuel Substitution: The above results leave open the question whether CCL plants lowered their energy expenditures in a way that would be considered a success for climate change policy. *A priori,* this is not clear because this measure of energy use lumps together

Table H.3. **CCA participation and environmental performance**

Dependent variable	Explanatory variable	OLS (1)	Reduced form (OLS) (2)	IV (3)	Fixed effects (4)	Reduced form (FE) (5)	Fixed effects IV (6)	Observations/ plants (7)
Energy expenditure over gross output $\Delta\ln(EE/GO)$	CCA/EPER	0.026** (0.013)	0.086*** (0.028)	0.220*** (0.072)	0.025 (0.019)	0.111*** (0.040)	0.231*** (0.084)	14 336 4 209
Energy expenditure over variable costs $\Delta\ln(EE/VCost)$	CCA/EPER	0.026** (0.012)	0.104*** (0.026)	0.266*** (0.069)	0.015 (0.018)	0.137*** (0.037)	0.285*** (0.080)	14 336 4 209
Energy expenditure $\Delta\ln(EE)$	CCA/EPER	0.019 (0.012)	0.033 (0.024)	0.085 (0.061)	0.036** (0.017)	0.075** (0.029)	0.156** (0.061)	14 336 4 209
Total kWh $\Delta\ln(kWh)$	CCA/EPER	0.068** (0.027)	−0.000 (0.049)	−0.001 (0.115)	0.079** (0.035)	−0.004 (0.068)	−0.007 (0.135)	4 452 928
Electricity $\Delta\ln(El)$	CCA/EPER	0.026 (0.021)	0.085* (0.046)	0.206* (0.118)	0.028 (0.024)	0.128** (0.058)	0.258** (0.127)	4 452 926
Gas $\Delta\ln(Gas)$	CCA/EPER	0.016 (0.037)	0.014 (0.052)	0.036 (0.127)	0.012 (0.047)	−0.035 (0.080)	−0.066 (0.151)	3 602 764
Share of gas over gas and electricity consumption $\Delta(Gas/(Gas+El))$	CCA/EPER	0.018** (0.008)	−0.044 (0.031)	−0.107 (0.078)	0.022** (0.009)	−0.048 (0.039)	−0.097 (0.084)	4 435 926
Share of gas over kWh $\Delta(Gas/kWh)$	CCA/EPER	0.013 (0.011)	−0.007 (0.023)	−0.018 (0.055)	0.018 (0.015)	−0.010 (0.032)	−0.021 (0.065)	4 449 928
Solid fuels $\Delta\ln(SO)$	CCA/EPER	−0.155 (0.101)	−0.226 (0.224)	−0.649 (0.597)	−0.091 (0.115)	−0.290 (0.266)	−0.542 (0.486)	1 467 344
Solid fuels over kWh $\Delta(SO/kWh)$	CCA/EPER	0.003 (0.004)	−0.016 (0.011)	−0.039 (0.025)	0.005 (0.006)	−0.022 (0.015)	−0.044 (0.030)	4 452 928
CO$_2$ $\Delta\ln(CO_2)$	CCA/EPER	0.050** (0.021)	0.018 (0.040)	0.044 (0.094)	0.053** (0.026)	0.024 (0.051)	0.048 (0.101)	4 452 928
Employment $\Delta\ln(L)$	CCA/EPER	−0.014 (0.011)	−0.039* (0.021)	−0.101* (0.054)	0.021 (0.014)	−0.019 (0.036)	−0.041 (0.075)	14 336 4 209
Real gross output $\Delta\ln(Real\ GO)$	CCA/EPER	−0.008 (0.011)	−0.053** (0.022)	−0.136** (0.057)	0.011 (0.014)	−0.036 (0.035)	−0.076 (0.072)	14 336 4 209
Total factor productivity	CCA/EPER	−0.002 (0.006)	0.000 (0.015)	0.001 (0.038)	−0.007 (0.009)	0.009 (0.026)	0.018 (0.054)	14 288 4 194
$\Delta\ln(Gross\ output)$	$\Delta\ln(M)$	0.477*** (0.013)	0.477*** (0.013)	0.477*** (0.013)	0.468*** (0.017)	0.469*** (0.017)	0.468*** (0.017)	
	$\Delta\ln(EE)$	0.034*** (0.006)	0.034*** (0.006)	0.034*** (0.006)	0.036*** (0.008)	0.036*** (0.008)	0.036*** (0.007)	
	$\Delta\ln(L)$	0.257*** (0.013)	0.257*** (0.013)	0.257*** (0.013)	0.237*** (0.018)	0.237*** (0.018)	0.237*** (0.018)	
	$\Delta\ln(K)$	0.049*** (0.016)	0.049*** (0.016)	0.049*** (0.016)	0.069*** (0.020)	0.068*** (0.020)	0.068*** (0.020)	

Notes: Dependent variables are first-differenced from 1997 until 2000 and differenced at various intervals thereafter. Column 1 displays OLS coefficient, column 2 displays OLS coefficient on the instrumental variable in the reduced form, and column 3 displays the 2-stage least squares estimate. Columns 4 to 6 have the same setup while including a firm fixed effect. Column 7 reports the number of observations. All regressions include age, age squared, year dummies and region-by-year dummies. Columns 1 to 3 in addition include region and three-digit industry dummies.
* Significant at 10%.
** Significant at 5%.
*** Significant at 1%.
Source: Martin et al. (2009).

StatLink http://dx.doi.org/10.1787/888932318585

changes in the price and quantity of energy, as well as the effects of substitution between different fuel types. For example, instead of consuming less of all fuel types, CCL plants might substitute towards cheaper fuel sources, which might also be more polluting, e.g. coal. To investigate this, the next seven panels in Table H.3 report results from regressions using quantity changes in energy consumption by fuel type which are available in the QFI sample. Although this sample is smaller than the ARD sample, there exists economically and statistically significant evidence that CCA membership led plants to relatively increase their electricity use by about 26%. This is in line with the design of the CCL, which imposes the highest tax rate for electricity. For both gas and solid fuels (*i.e.* coal), negative point estimates on the CCA coefficient are obtained. There are also negative point estimates when looking at the share of these fuels in total kWh consumed. While these coefficients are not different from zero in a statistical sense, they hint at the possibility that some CCL plants switched from electricity to the lower-taxed fuels gas and coal. This would also explain why the overall effect on total kWh is not significant in the IV regressions. If plants switch from electricity to gas or coal they are likely to require more kWh of primary energy to achieve a given energy service. This could account for at least a partial offset of a tax-induced reduction in the demand for those services.

Carbon emissions: A significant increase in electricity consumption by CCA plants should translate into an increase in carbon emissions, given that a significant decline in the consumption of other fuel types was not found. Next, the focus is on whether this effect occurs when the outcome measure is the total sum of CO_2 emissions across fuel types. The eleventh panel of Table H.3 reports that CCA membership is associated with a 5% increase in total CO_2 emissions. The point estimate is very robust across specifications, yet loses statistical significance in the IV regressions. It seems likely that this is due to the noise in the estimated response by fuels other than electricity. In the absence of a larger sample that would enable estimating this effect with more precision, there are two possible interpretations. On the one hand, coefficients that are statistically insignificant at conventional levels could be disregarded altogether and the conclusion drawn that the unchecked increase of electricity consumption translates into an increase in CO_2 emissions of equal magnitude. On the other hand, a more cautious interpretation of the results would put the impact of CCA participation on carbon dioxide emissions at 5%, which accounts for the possibility that some CCL plants switched into dirtier fuels such as coal. Thus, the full-rate CCL – though not designed as a pure carbon tax – led plants to reduce growth in CO_2 emissions by between 5 and 26% more than the CCA targets did in combination with the discount on the levy.

Economic performance: Finally, whether the impacts on energy consumption and energy efficiency correspond to movements along the production isoquant, or stem from significant shifts in the scale at which plants operate, are investigated. The three panels at the bottom of Table H.3 look at various plant performance variables such as output, employment and total factor productivity (TFP) are presented. When estimating the difference regression without fixed effects, significantly negative coefficients for both employment and output are obtained. However, these effects disappear when controlling for plant-specific trends in columns 4 to 6. There are two issues of note. First, a key policy concern with unilateral implementation of energy taxes is that they might jeopardise the competitiveness of domestic industry. If this was the case, one should observe positive employment or productivity effects of CCA participation, because plants that pay the CCL scale down production and employment. Finding the opposite effect ought to dissipate such concerns. Second, the fact that the negative coefficients effects lose significance once plant fixed effects are included suggests that they are driven by pre-existing trends, unrelated to the policy intervention. Similar to what occurred in the steel industry, this could be due to plants in industries covered by both CCA and PPC regulations which were on a declining trend even before the arrival of the CCL policy package. The last panel suggests that CCA participation had no discernible effect on total factor productivity. In sum, there is no evidence that the CCL had any adverse effects on economic outcome variables.

Innovation impacts of the CCL and CCA

Assessing the innovation impacts of the CCL and CCA can best be done with patents. In doing so, the combination of a list of energy efficiency related patent classifications and keyword searches within the abstracts of patents have been utilised.

The principal reason for exploring abstract searches in addition to the energy efficiency patent classes is that climate change mitigation is not only about energy efficiency. Since many economic and non-economic activities contribute directly or indirectly to GHG emissions, it is plausible that innovation in a wide range of areas – not necessarily those classified as "energy saving" – may have an impact on GHG emissions. Consequently, climate change policies might induce innovation in a wide range of areas with a potential for further GHG emission reductions.

Using both methods, more than 45 000 climate change related (CCR) patents in the *EPO Database* were identified. The majority of those (about 77%) were identified via the patent classification system. A sizeable number (23%) were identified through abstract searches. Interestingly and quite surprisingly, the overlap between both types was less than 1%. Figure H.1 examines the evolution of total patents and CCR patents over time. It shows indices of new patent applications each year with base year 1980. Both the number of patents overall and CCR patents have been increasing dramatically since then. The increase in CCR patents was more pronounced and also more volatile.

Figure H.1. **Index of patents in the United Kingdom**

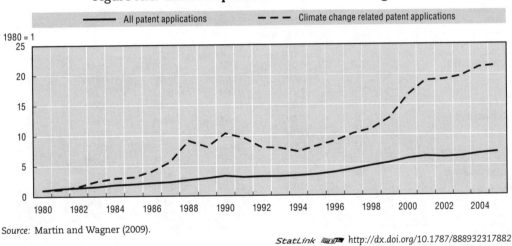

Source: Martin and Wagner (2009).

StatLink 🔗 http://dx.doi.org/10.1787/888932317882

The econometric approach attempts to estimate the impact of CCA participation on patenting activity of firms. Due to the discrete nature of patent counts, appropriate econometric models are critical. Two different models are commonly used in econometric analyses of discrete data. The first model performs a conditional logit regression on the binary event of a firm i applying for at least one patent in year t. Thus:

$$\Pr\left(I\left\{Patents_{it} > 0\right\}\right) = f\left(\beta_D D_{it} + x_{it}'\beta_x + \alpha_i\right) \tag{4}$$

where D_{it} is the treatment indicator, x_{it} is a vector of control variables which includes year dummies, α_{it} is a firm fixed effect and f is derived from the extreme value distribution.

The issue with binary outcomes is that they provide only an incomplete picture of the intensity of innovative activity. Therefore, a Poisson count data model is also employed that posits that the innovation process follows a stochastic process such that the expected number of patents of firm i in year t is given by:

$$E(Patents_{it}) = e^{\alpha_i} e^{\beta_D D_{it} + x_{it}' \beta_x} \tag{5}$$

The CCA and EPER variables described above are used as treatment indicator, D_{it}. As discussed there, CCA participation was contingent on coverage under the Pollution Prevention and Control legislation. That is, only firms that were releasing polluting substances into air, soil or water could apply for a CCA. To measure PPC coverage in practice, firm level data was combined with information from the EPER.

Table H.4 reports the main regression results. Columns 1 and 2 contain coefficient estimates from simple logit and Poisson models, respectively, without controls for firm-specific unobserved heterogeneity in the propensity to apply for patent protection of new inventions. To control for this, columns 3 and 4 display results from a conditional logit model and from a Poisson conditional maximum likelihood estimation (CMLE), respectively.

Table H.4. **CCA participation and innovation performance**

Patent type	Model	Logit	Poisson	Clogit	FE poisson	Observations/ firms
	Policy variable	I(Patent)	Patent count	I(Patent)	Patent count	
		(1)	(2)	(3)	(4)	
All patents	CCA	0.069*** (0.017)	1.382*** (0.295)	−0.109*** (0.035)	−0.510** (0.243)	134 320 8 395
	EPER	0.055*** (0.021)	1.326*** (0.376)	−0.161*** (0.048)	−0.585*** (0.186)	
CCR patents (all)	CCA	0.024 (0.024)	0.506** (0.228)	−0.135 (0.087)	−0.531 (0.388)	8 832 552
	EPER	0.033 (0.029)	0.474 (0.317)	−0.140* (0.082)	−0.432 (0.359)	
CCR patents (Popp)	CCA	0.021 (0.024)	0.491* (0.269)	−0.138 (0.088)	−0.513 (0.371)	8 576 536
	EPER	0.026 (0.029)	0.436 (0.304)	−0.172** (0.076)	−0.528** (0.221)	
Non-Popp patents	CCA	0.070*** (0.017)	1.375*** (0.236)	−0.106*** (0.035)	−0.510** (0.220)	134 224 8 389
	EPER	0.056*** (0.022)	1.328*** (0.375)	−0.167*** (0.048)	−0.586** (0.277)	

* Significant at 10%.
** Significant at 5%.
*** Significant at 1%.
Source: Martin and Wagner (2009).

StatLink ⟶ http://dx.doi.org/10.1787/888932318604

The first panel deals with all patents. Without controlling for firm-level heterogeneity, treated firms patent significantly more. This result appears regardless of whether CCA or EPER status is used as the treatment variable. Column 1 reports marginal effects instead of coefficients from the binary choice regression. These effects correspond to the marginal effect of the treatment on the propensity to patent. For example, the results imply that treated firms are 5.5 to 6.9 percentage points more likely to apply for a patent than other firms (depending on whether CCA or EPER is used as the treatment variable). Likewise, the Poisson regression indicates that CCA participation has a positive and significant effect on the expected number of patent applications.

However, this result is not robust. When controlling for unobserved heterogeneity, it is found that, to the contrary, the propensity of CCA firms to patent innovation is up to 16 percentage points lower than that of non-CCA firms after 2001 (column 3). The Poisson regression in column 4 confirms that the number of patents filed by CCA firms dropped relative to that of non-CCA firms following the introduction of the CCL package in 2001. As was the case with the results in columns 1 and 2, the differences between the results obtained with CCA and EPER are small and well within the margin of error. This demonstrates that there are important unobserved differences between treated and non-treated firms which need to be controlled for in order to gauge the effect of CCA participation.

In the subsequent panels, results from the same set of specifications using different dependent variables are presented. In panel two, all CCR patents identified using the combination of abstract searches and patent classifications described above are used. In the results in panel three, only patent classes suggested by Popp (2002, 2006) are used. Similar to the regressions with all patents, there is evidence of a relative decline in patenting by CCA firms after 2001.

Finally, in the last panel, non-CCR patents (all patents minus the patents identified using Popp's mapping of patent classes) are presented. Perhaps not surprisingly, the pattern emerging from this exercise is very similar as for all patents, as non-CCR patents dominate the sample.

Overall, firms in a CCA face less stringent regulation and therefore have lower incentives to respond to the regulation with innovation. This could, in principle, generate the negative coefficient on patenting found in columns 3 and 4. However, any such innovation responses would be concentrated in areas related to climate change, so that negative effects should only arise in panels 1 to 3 and not in panel 4. There are two possible explanations for the results in panel 4. First, the CCL had indeed an impact on innovation across the board. An explanation for why this could happen is as follows: Suppose there is a known technology that allows the firm to produce a given output using less energy but increasing its labour input. For a firm that shifts to this technology in response to the CCL, the eventual effect of the CCL is to increase the incentives for labour saving R&D rather than energy saving R&D. Second, it might be the case that the measure of climate change related patents is incomplete or subject to measurement error. For instance, there are concerns that the EU and US patent classification systems are too different, so that using a concordance table, as was done above, leads to misclassifications.

Additional analysis was done, looking at the time effects using year dummies. It appears that the impact emerges primarily from 2002 onwards, which is consistent with there being a short lag between the introduction of a policy (in 2001) and its impact on patenting.

Conclusions

The results of this case study support a strong case for the introduction of moderate energy taxes to encourage electricity conservation, to improve energy efficiency and to curb greenhouse gas emissions.

The use of negotiated agreements, which provide discounts in the energy tax to certain businesses and industries, may not bring about emission targets that are stringent. One rationale behind the CCA tax discount is that the unilateral implementation of a major climate change policy could jeopardise the economic performance of energy-intensive UK firms. Having investigated this empirically, neither a discernible loss of jobs nor a decline in output or productivity for the average plant paying the full rate was found. From this, the discount granted to plants in a CCA cannot be justified as a means to avoid alleged negative impacts on economic performance arising from the CCL. As such, further cuts in energy use of substantial magnitude could have been achieved without negative impacts on economic performance had the full CCL been implemented for all businesses.

Since climate change is a long-term problem, it is often emphasised that climate policy must stimulate technical change that will allow further reductions in GHG emissions in the future. Evidence from an empirical investigation of the impacts of CCA participation on firm-level counts of patents strongly suggests that a moderate energy tax on the business sector leads to increased innovative activity overall. In particular, it was found that a firm subject to the full CCL was up to 16 percentage points more likely to patent that a similar firm subject to a reduced tax rate through a CCA. The results also indicate that this difference in patenting is most likely driven by patents for energy efficiency equipment, but also for things not related to climate change. Based on these findings, it appears that more such innovation would have occurred had the CCL been implemented at full rate for all businesses, instead of with the 80% reduction during the period of study.

For more information on the UK's Climate Change Levy, the full version of the case study (OECD, 2009) is available at *www.olis.oecd.org/olis/2008doc.nsf/linkto/com-env-epoc-ctpa-cfa(2008)33-final*.

Notes

1. The CCA discount is scheduled to be reduced to 65% from the current 80% as of 1 April 2011.

2. THE UK ETS market is now defunct: there is no trading and allowances have no value.

References

Martin, R., L.B. de Preux and U.J. Wagner (2009), "The Impacts of the Climate Change Levy on Business: Evidence from Microdata", *Centre for Economic Performance Discussion Paper*, No. 917, London School of Economics, UK.

Martin, R. and U.J. Wagner (2009), "Climate Change Policy and Innovation: Evidence from Firm Level Patent Data", Mimeograph, Centre for Economic Performance, London School of Economics, UK.

OECD (2009), *Econometric Analysis of the Impacts of the UK Climate Change Levy and Climate Change Agreements on Firms' Fuel Use and Innovation Activity*, OECD, Paris, available at *www.olis.oecd.org/olis/2008doc.nsf/linkto/com-env-epoc-ctpa-cfa(2008)33-final*.

Popp, D. (2002), "Induced Innovation and Energy Prices", *American Economic Review*, Vol. 92(1), pp. 160-180.

Popp, D. (2006), "International Innovation and Diffusion of Air Pollution Control Technologies: The Effects of NO_x and SO_2 Regulation in the US, Japan, and Germany", *Journal of Environmental Economics and Management*, No. 51, pp. 46-71.

ANNEX I

Japan's Tax on SO_x Emissions

> This case study examines the Japanese tax on SO_x emissions that was implemented in the 1970s to finance compensation to victims of air pollution. The tax rate rose very quickly after its introduction, peaking in 1987, when the system was reformed. Due to the tax, firms undertook significant abatement and adopted existing abatement technologies. However, the design of the tax brought about unpredictability of the tax rate and a lack of credibility of the overall system, ushering in a period of declining patenting in related technologies, despite increasing tax rates.

Rationale for the instrument

From the end of World War II through the 1960s and 1970s, rapid industrialisation of Japan's economy brought significant economic gains concomitant with increasing pollution. One of the most significant airborne pollutants was SO_x. While the government had embraced many policy mechanisms during this period to address SO_x emissions, citizens were nevertheless developing respiratory problems, such as asthma, and linkages between illness and pollution were discussed.

At the same time, victims of air pollution were using the courts to seek damages from the major emitters, winning some initial victories. Because early victories acknowledged joint liability, it opened up the possibility that only large companies with the money to pay damages would be sued. In response, Keidanren, an umbrella organisation of industrial associations, sought to introduce an administrative compensation system that possessed both a lawsuit deterrent effect and the effect of insurance that spreads the burden broadly and thinly. The victims also sought an administrative compensation system that would compensate them more quickly and with less expense than a lawsuit. Thus, a tax on SO_x emissions intended to compensate victims was sought by all concerned parties and approved in 1973.

Design of the instrument

The *Compensation Law for Pollution-Related Health Damage* ("Compensation Law" or CL) was approved in 1973 and put into force in following year. The national government specifies health damage due to air pollution as a designated illness, and it specifies areas where the incidence of such illness was high as "designated areas". Persons who live in a designated area, or work or go to school there, and who suspected that they have a designated illness can apply to their local government to be certified as a patient with a designated illness, and undergo a medical examination. Other potential causes such as tobacco are not considered; epidemiological causality is applied. A certified victim receives, in proportion to the severity of the damage, healthcare expenses, damage compensation expenses (for living expenses), treatment expenses (for transportation to the hospital, etc.) and so on. In the most severe cases, damage compensation expenses equivalent to 80% of average wages are paid.

The source of 80% of the compensation funds is the CL levy paid, in proportion to the emissions volume, by companies across Japan with facilities that produce soot and smoke with SO_x. Companies that pay the levy are those with maximum gas emissions of 5 000 Nm^3 per hour or more from a given worksite in a designated area or 10 000 Nm^3 per hour or more in other areas. The remaining 20% comes from part of the automobile weight tax paid by automobile owners, which is levied according to the type and the weight of the vehicle. Because the tax is levied according to the weight, its effect in reducing emissions of pollutants and inducing technological development is only indirect.

Four illnesses were recognised as designated illnesses when the system was put into effect, and by 1978, 41 areas had been recognised as designated areas. In the first three years of the system, 50 000 patients were certified; by 1987, when the law was revised to halt new designations, over 100 000 patients had been certified, and the total annual compensation payments exceeded JPY 100 billion or about USD 750 million. Approximately 9 000 plants and other worksites were paying the CL levy out of a total of approximately 60 000 facilities in 1980.

To determine the rate, the government, in fiscal year $(t - 1)$, estimates the amount of compensation to be paid in fiscal year t. The average levy rate of the nine regional blocks of designated areas is set to be nine times that of other areas, and among the nine blocks of designated areas, the levy rate is set so that the higher the compensation amount per SO_x emissions volume in a given area, the higher the rate.

The first characteristic is that the CL levy is intended to secure a funding source for compensation, with the total amount of revenue being decided first and then the levy rate being decided. For this reason, SO_x emitters do not know the levy rate when the emissions occur. Second, current SO_x emitters incur excessive burden because the levy is applied only to SO_x currently being emitted and no levy is applied to NO_x, particulate matters or other air pollutants, past or present, which also have an impact on respiratory health.

This produces unfairness in light of the "polluter pays" principle, and simultaneously, this means that no matter how much SO_x is reduced, not only in other areas but also in the designated areas, compensation payments will not decline, and consequently the levy rates skyrocket. However, if part of the burden is borne by SO_x emitters in other areas that are not connected with health damage in the designated areas and the burden is spread broadly and thinly, that is what the system intended.

Since patients must apply for certification, the number of certified patients increases as the system becomes more well known, and this contributes to a subsequent increase in the compensation amount. Another major factor in the increase of compensation payments was the fact that damage compensation expenses and survivors' compensation expenses are linked to average wages.

As a result, despite the reduction in SO_x emissions volume, the amount of compensation payments continued to grow, and the levy rate skyrocketed, reaching 134 to 339 times the FY 1974 amount in FY 1987, as in Figure I.1. The GDP deflator growth during this time was 1.6 times. In the Osaka Block, which has the highest rates, a company burning heavy fuel oil C with 3% sulphur paid a levy equivalent to nearly seven times the price of the fuel. Even in the area with the lowest rates, a levy equivalent to approximately 40% of the fuel price was paid.

Figure I.1. **Tax rates for current SO_x emissions**

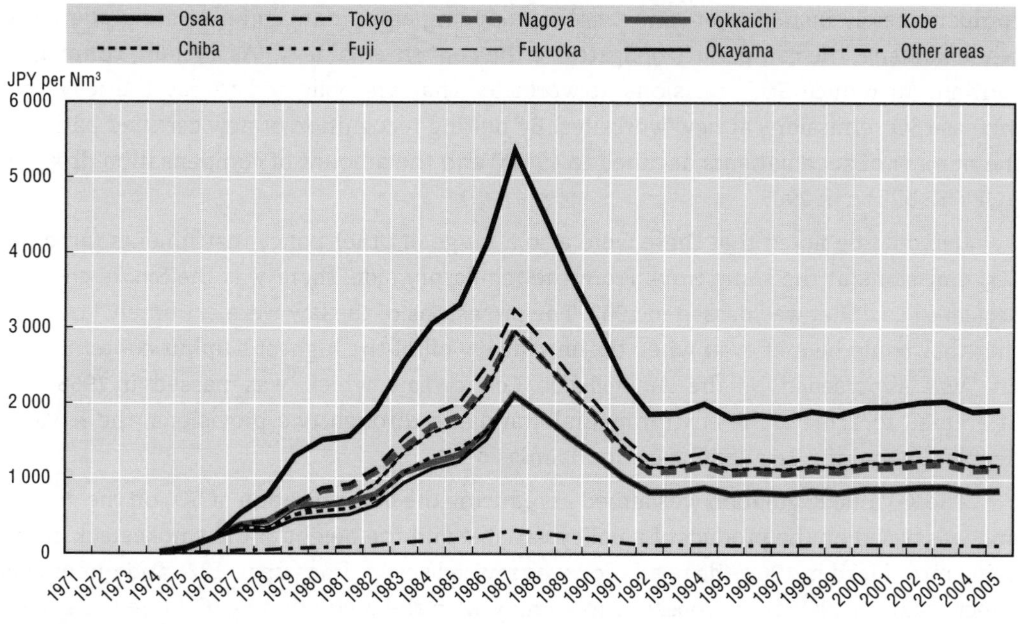

Source: OECD (2009).

StatLink http://dx.doi.org/10.1787/888932317901

Starting in 1987, significant reforms were introduced in response. The regional designations related to air pollution were removed and therefore no more new certified patients were recognised. This move was made in response to the fact that SO_x emitters could no longer bear the burden of compensation payments, which were increasing in spite of the significant decline in SO_x emissions.

Although area designation was halted and recognition of new certified patients was stopped, the Compensation System continued to exist in a different form. Compensation to pre-existing certified patients continued as before, and the funding burden continued to be apportioned between fixed sources and mobile sources. However, a new levy was raised on past SO_x emissions from fixed sources. Specifically, the levy rate on the current emissions and past emissions were now determined so that revenue on emissions during the five years immediately previous to the decision to halt area designation (i.e. 1982-86) would amount to 60%, with the other 40% being composed of revenue levied on annual emissions. Temporary measures smoothed this transition. Because the original reason for halting area designation was the reduction of air pollution that causes the designated illnesses, it is contradictory to have a levy on current emissions of SO_x, but the levy on current emissions remained in place and it was explained as an incentive for corporations to prevent pollution. As long as the system continues to exist, there will be a levy will be levied on emissions during the five-year period from 1982 to 1986, and so the cumulative levy rate for this period will continue to increase into the future.

However, the most important change is that the only parties obligated to pay the levy were firms which had plants or other worksites with facilities that create soot or smoke with SO_x as of 1 April 1987. This means that all such companies will continue to pay a levy on past emissions as long as they continue to exist as companies, even if they close their plants and other worksites. On the other hand, a new plant or other worksite opened after 1 April 1987 will not pay a levy, regardless of how much SO_x it emits. (However, if a company newly installs soot- and smoke-producing equipment in an existing plant and other worksite, the company is obligated to the conventional levy.) As a result, companies continue to reduce SO_x emissions at worksites that are obligated to pay the levy, and increase SO_x emissions at new worksites. By halting recognition of new certified patients, the number of such patients declined to 45 000 and the amount of compensation dropped to JPY 52 billion in 2005.

It should be noted that there were also a range of other policy instruments targeting SO_x emissions at the same time. From the regulatory side, there was the *Smoke and Soot Regulation Law* that was passed in 1962. The provisions of the law were extremely lax, such that they could be met even when burning heavy oil of the highest sulphur content. That law was superseded by the *Air Pollution Control Law* which was passed in 1968 and attempted to achieve the environmental standards through two provisions, the K-Value Regulation and the Regulations on Total Emissions.

The K-Value Regulation was aimed at reducing the concentration of SO_2 on the ground and was based on the product of a policy variable and the height of the smokestack. Since being introduced in 1968, the regulation was revised seven times up to 1976 and remains in effect today. In 1968, it was possible to comply with the regulation by building tall stacks, even when using high-sulphur fuel, and stacks over 100 metres appeared. However, the introduction of regulations in 1976 required additional actions.

Where achieving SO_2 standards was difficult using the K-Value Regulation, the law required prefectural and metropolitan governors to additionally introduce Regulations on Total Emissions. Whereas the K-Value Regulation applied to each facility, Regulation on Total Emissions is applied to each plant and other worksite and target large-scale plants and other worksites which accounted for over 80% of SO_x emissions prior to regulation. The regulations were based around a policy variable and the amount of heavy oil equivalent of fuel use (kilolitres per hour) in determining the maximum emissions per hour. For new facilities, the permitted amount of emissions is stricter for the amount of fuel use but can be larger than for older firms.

Subsidy-like measures have also been used: low-interest loans, special depreciation, shortening of legal durable life, and an exemption from fixed asset taxes in return for investment in pollution prevention. With the Taxation System for the Energy Reform of 1978 which is still in place today, a corporation that has installed targeted facilities may choose a corporate tax reduction through either a 7% tax exemption on the base acquisition price or special depreciation of 30% in the first fiscal year (this component was available from 1975). These measures were for energy saving, not for pollution control, and were applicable to the facilities.

Voluntary measures – pollution control agreements (PCA) – have also been widely utilised in Japan. Local governments started using PCAs as a last resort in the 1960s when they still lacked regulatory authority. PCAs proliferated rapidly starting in the late 1960s and have come to number more than 30 000 in effect today. Given the six million worksites in Japan, there is an average of one PCA per 200 worksites; it is estimated that coverage by PCAs is very high at large-scale worksites. Most PCAs that are concluded with such large-scale worksites incorporate restrictions in addition to the legal regulations, and such worksites must observe the emission levels stipulated in the PCAs. The Standardised K-Value Regulation and Regulations on Total Emissions are applied indiscriminately because it is difficult for them to take account of local natural conditions and the spatial distribution of the worksites. For this reason, local governments use PCAs to place more stringent restrictions on emissions from large-scale worksites which have great financial resources, while allowing the standardised regulatory levels to remain relatively lax.

Theoretically, businesses do not have to conclude PCAs which require them to take measures beyond legal requirements. However, local governments, which have authority to give different kinds of permissions to businesses, have strong bargaining power and businesses appear to accept individual PCAs in order to avoid strife. Recently, numerous companies have started publicising in their environmental reports the fact that they have concluded and are observing PCAs which are considerably stricter than the legal regulations. The balance between bargaining powers of local governments and of businesses vary from case to case and businesses sometimes refuse PCAs.

Environmental impacts of the instrument

By the end of the 1970s, the amount of SO_x had been adequately reduced, and this achievement is lauded as an example of successful environmental policy. Figure I.2 shows the estimated levels of SO_x emissions, derived from data from the Ministry of the Environment (MOE) and from the Compensation Law levy.

This figure shows that there was a sharp drop in SO_x in Japan from the early 1970s through the mid-1980s, followed subsequently by a slight increase or a levelling off, and then further decline. This appears to indicate that the CL levy exerted some effect on SO_x

Figure I.2. **Trends in SO_x emissions**

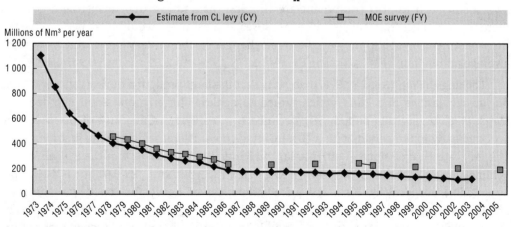

Source: OECD (2009).

StatLink ᴍᴬᴾ http://dx.doi.org/10.1787/888932317920

reduction, at least during a certain period prior to the halting of area designation in 1987. Following the halting of area designation, the fact that the levy estimate's SO_x continues to decline while the MOE survey's SO_x is not declining is most likely due to the increase in SO_x emissions volume from worksites which are not subject to the levy.

One can also consider the factors affecting emissions: GDP as production volume, energy usage per unit of GDP, oil and gas as percentage of total energy use, and units of SO_x per unit of oil and gas. These relative factors are displayed in Figure I.3 and it is clear that there is significant decoupling of SO_x emissions from GDP.

Figure I.3. **Factors of SO_x emissions**

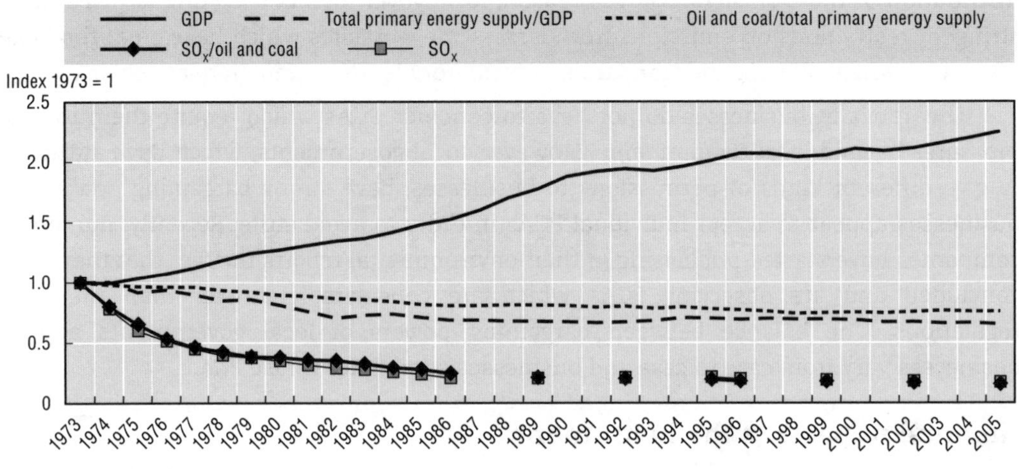

Source: OECD (2009).

StatLink http://dx.doi.org/10.1787/888932317939

The rest of this section will attempt to consider and differentiate the impact of the CL levy and other policy instruments (legal restrictions and PCAs) from the object and timing of each instrument and SO_x reduction cost. Subsidy-like measures will not be investigated because they do not provide incentives for SO_x reduction but have an impact only when used in conjunction with a pollutant emissions charge. Although investment costs depend on their depreciation time, it seems that more important costs for pollution control investment such as FGD equipment are running costs rather than investment costs. If there was an effect from subsidy-like measures, it most likely would have been to hasten the investment by about one year because changes in the CL levy rates were very rapid (average annual rate of 110% during FY 1974 to FY 1979 and 20% during 1979 to 1987).

Looking at FY 1973 to FY 1978, it appears that the large reduction during this period was due to regulations and PCAs which aimed to achieve the environmental standards by March 1978. During this period, SO_x declined by 60%, but the majority of the reduction occurred early in the period, with a 40 percentage point decrease during the first two years. The levy rate on emissions in other areas (which emitted 87% to 88% of the SO_x emissions volume) during 1974 and 1975 was JPY 8.59 and JPY 23.33 per Nm^3 (JPY 6.01 and JPY 16.33 per kg sulphur), respectively. Also, the price differences of heavy fuel oil C with different amounts of sulphur during this period was JPY 108 per kg sulphur even for that with the lowest sulphur content of 3% to 2.5%. So, it did not contribute to SO_x reduction since it was cheaper to pay than to abate. The situation was the same even in designated areas, where the levy rate was nine times larger, because the regulations were also stricter.

Meanwhile, the K-Value Regulation was steadily revised in 1969 and 1973 and Regulations on Total Emissions were also introduced. Regulations on Total Emissions were planned to reduce 50% of the overall SO_x in the designated areas, which comprised 20 to 30% of the SO_x emissions nationwide from 1974-76 through 1976-78. PCAs with strict emissions standards were concluded starting in the late 1960s, and the number increased sharply in the 1970s. As such, it appears that the development and spread of SO_x reduction technology, such as heavy oil desulphurisation and FGD which led to significant SO_x emission reductions during this period, was triggered mainly by regulations and PCAs. Meanwhile, energy saving in the broad sense was suddenly observed in FY1975, and it seems that the soaring energy prices caused by the first oil shock clearly had an impact on that. The increase in gas and nuclear power may also be such a response.

Looking at FY 1978 to FY 1986, the environmental standards were achieved in FY 1979 and therefore regulations were not strengthened. It is possible that the PCAs of companies which expanded production and energy use were strengthened so as to reduce SO_x emissions per energy usage volume. But, because environmental standards had been achieved, there was no need for rational regulatory authorities to further reduce the absolute amount of SO_x emissions in the areas. Yet, the absolute amount of SO_x emissions was declining, suggesting that the SO_x reduction during this period was caused primarily by the CL levy. During this period, SO_x declined by 69% in the designated areas and by 51% in other areas.

Again, looking at the price differences in heavy fuel oil C with differing sulphur contents and the levy rate in other areas, the levy rate was relatively too low to justify fuel switching until 1985; in 1986, the levy rate was set to justify the reduction of sulphur content of fuel to 1.5%. This compelled a reduction of around 50% in the case of a pollution source that continued to use heavy oil with a sulphur content of about 3% and that cleared the loosest K-Value Regulation by building a smokestack with an effective height of stack exceeding 200 meters. However, the above applies in the case where a company accurately predicted the levy rate for the following fiscal year which applied to 1986 emissions, and those that predicted a higher rate would have been compelled to make a larger reduction in emissions.

Meanwhile, the average cost of FGD to reduce gas with a SO_x concentration of 2 000 ppm (equivalent to heavy oil with sulphur content of 2.9%) to a concentration of 200 ppm (0.29%) was JPY 86 to 90 per kg sulphur. This is lower than JPY 105 per kg sulphur, the levy rate for other areas in FY 1983. Because other methods were less expensive than this specific FGD method, it is likely that there were cheaper options overall. Though the figures obtained here are average, not marginal costs, it means that if such reduction were carried out, the benefit would be larger than the cost.

Referring to the available data for heavy fuel oil C for electric power generation with regard to heavy oil with low sulphur content of less than 0.3%, there is the same price difference between 0.2% and 0.3% as for that with higher sulphur content. However, the price differences which were initially extremely large rapidly shrank, with the price differences between 0.2% and 0.1% dropping from JPY 5 215 to 430 per kg sulphur between 1980 (October) and 1986 (October). For this reason, use of heavy fuel oil C with 0.1% sulphur was justified in the Osaka area, which had the highest levy rate, in 1983 and in all designated areas in 1986.

At the end of the 1970s, a price reversal occurred, making liquefied natural gas (LNG), which contains no sulphur, cheaper than heavy fuel oil C with 0.1% sulphur for electric power and steel companies which have their own long-term contracts with LNG exporters. In this case, LNG seems to be used without any institutional pressure such as legal emissions standards, PCA, or the CL levy. That is, SO_x emissions reduction due to the use of LNG seems not to be an effect of policy instruments. However, if polluters could buy any amount of cleaner fuel, LNG, at a cheaper price, they would have completely replaced other kinds of fuels by LNG. But they did not do so. Therefore, there must have been certain costs to do so, such as weaker energy security, higher price because of larger demand, and so on. It is then possible that the CL levy whose rates got extremely high contributed to the additional expansion of LNG use, by making it far less costly compared with fuels containing sulphur. It is also possible that the LNG prices offered by gas companies for large consumers of gas who do not import it themselves became sufficiently low and contributed to the reduction of SO_x emissions.

However, as seen in the Figure I.3, the main factor of SO_x reduction in this period is the decrease of SO_x/(Coal + Oil) ratio, which implies fuel desulphurisation and FGD technologies. The decrease of (Coal + Oil)/Energy ratio, which implies the expansion of the use of gas, nuclear and other sources of energy, was a less important, third factor following that of energy saving. Much the same factors relate to the decisions about coal as well.

Although it is possible that SO_x reduction in this period was, to some extent, brought about by the relatively lower prices of low sulphur fuels, without any relation to any policy instrument, this is believed to have been limited and the CL levy is thought to have contributed to a large extent to the reduction of SO_x emissions. This is consistent with the fact that, when the levy rates started to fall after 1987, SO_x emissions stopped decreasing. Ordinary companies have multiple options for SO_x reduction and can find less expensive means than regulatory authorities, and researchers also support the CL levy's effectiveness in this period.

The diffusion of SO_x reduction technology during this period seems to have been primarily caused by the CL levy. However, analysing interviews with environmental equipment manufacturers and at patent data, it was not a period of flourishing development of SO_x reduction technology although certain developments were made. That is, FGD technology was adapted to coal use, energy efficiency of FGD was improved, and so on. In short, this was a period in which technology developed in the past spread due to the levy.

Following the halting of area designation, the decline in total SO_x emissions nationwide markedly decelerated. The lower rate of decline is believed to be due to the halting of area designation under the Compensation Law, but the actual situation is slightly complex. In the periods FY 1986-FY 1996 and FY 1996-FY 2005, SO_x emissions according to the MOE's survey declined at an annual rate of 0.4% and 1.7%, respectively, slower than in prior periods. However, the levy's estimate of SO_x declined at a faster rate both in designated and in other areas (see Table I.1). One reason for this is that the number of pollution sources that pay the levy systematically fell along with a decline in the emissions per pollution source.

Due to such factors, the gap is growing between the MOE's survey and the levy's estimate of SO_x. There are two possible explanations for this. One is that companies increased production at new plants and other worksites (thereby also increasing emissions) and cut production at plants and other worksites which were required to pay

Table I.1. **Annual average rate of change of SO$_x$ reduction**

MOE survey in fiscal year, estimate from CL levy in calendar year	MOE survey	Estimate from CL levy					
		Designated areas			Other area		
	National total (%)	Emissions (%)	Number of sites (%)	Emissions per site (%)	Emissions (%)	Number of sites (%)	Emissions per site (%)
1973-78	−16.9	−24.9	+12.9	−33.5	−17.3	+6.9	−22.6
1978-86	−7.8	−13.8	−1.7	−12.3	−8.6	+1.9	−10.3
1986-96	−0.4	−4.1	−0.4	−3.8	−1.5	−0.2	−1.3
1996-2005 (2003 estimate from CL levy)	−1.7	−6.8	−0.7	−6.2	−3.9	−0.5	−3.4

Source: OECD (2009).

StatLink ᵃᵐˢ᠊ http://dx.doi.org/10.1787/888932318623

the levy. The impact from relocation of manufacturing to overseas sites is also a possibility. The other possible explanation is that companies continued their SO$_x$ reduction efforts as before to cope with the levy rate which did not necessarily decline enough. Investigation into each of the above's degree of contribution is deferred to another opportunity. Ultimately during this period, the levy had an impact on SO$_x$ emissions volume because the levy was not collected from new plants and other worksites due to the halting of area designation and because the levy rate declined to some degree.

Innovation impacts of the instrument

This section uses patent-related data to survey the state of technological development regarding SO$_x$ treatment technology. Not all of those who conducted technological development necessarily patented the outcomes of their research and technological development activities, but examination of patents helps to understand the major trends in technological development involving FGD.

PATOLIS is a database that is searchable using various keywords obtained from abstracts of the technical contents drawn up from the patent gazette, in addition to basic information on all patents submitted to the Japan Patent Office (the applicant's name, date of application, etc.). The search period of 1971 to 2000 produced a sample of 5 647 patents. As shown in Figure I.4, this trend is nearly identical to the trend in sales of FGD equipment.

However, the number of patent applications related to desulphurisation technology is affected not only by the state of development of desulphurisation technology, but also by social currents concerning the patent system and intellectual property rights. Looking at trends in the ratio of desulphurisation-related patent application to total patent applications, the trend is similar to Figure I.4. However, compared to the trend in the number of applications, it is a much gentler rise. Consequently, the second peak in patent applications for technology related to desulphurisation which occurred from the mid- to late 1990s appears to have been partially resulting from a positive attitude toward patenting in general caused by changes in the system or the social environment.

Most patent applicants are plant engineering manufacturers (outside suppliers who emit hardly any SO$_x$ at their own companies). For example, the top 10 plant manufactures in terms of accumulated capacity of FGD equipment make up 54% of all patent applicants from Japan in the sample, excluding individuals. There are not a large number of patent applications from companies that emit SO$_x$, or from national or public research institutes. For example, the electric power companies, which were the largest source of SO$_x$ emissions,

Figure I.4. **FGD sales and patents**

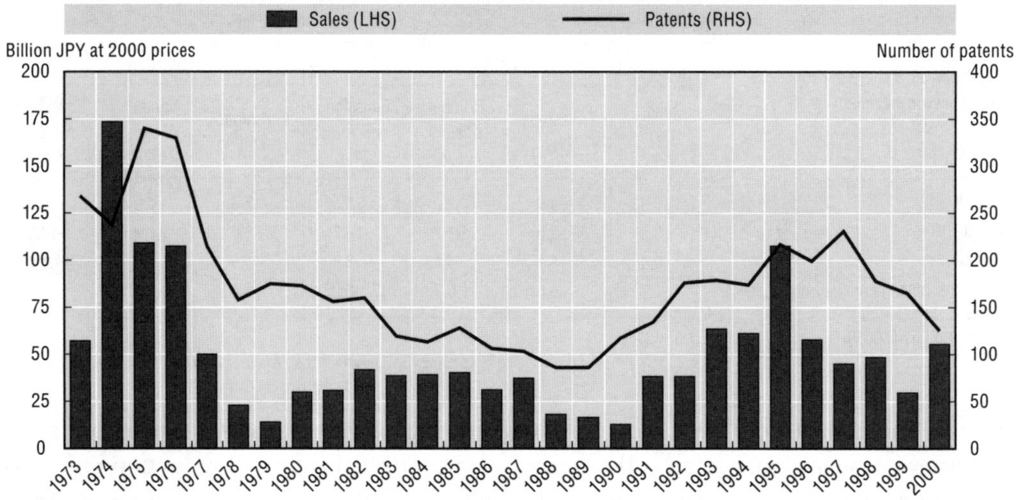

Source: OECD (2009).

StatLink http://dx.doi.org/10.1787/888932317958

set up their own pilot plants and actively undertook joint research with plant manufacturers during the development period of FGD technology from the late 1960s to the early 1970s, but not only did they not submit their own applications for patenting, they did not submit many joint applications either.

Looking again at Figure I.4, the period from the early to mid-1980s, when the CL levy rates rose sharply, was a period in which there were very few patent applications related to FGD. Conversely, applications increased again in the 1990s when recognition of new patients was halted and the CL levy rates were significantly lowered. The second peak in the number of applications occurred in the late 1990s. Consequently, it is difficult to discern evidence from patent data suggesting that the Compensation Law provided a strong incentive for technological development activities related to FGD.

Conclusion

The CL levy, the purpose of which is to ensure a source of compensation funds for victims suffering health damage due to pollution, was set at a high rate after the SO_2 environmental standards were achieved, from the late 1970s to the late 1980s, and it was a major contributor to the spread and utilisation of the above technology. It may also have had an impact on technological improvements such as the boosting of efficiency of FGD equipment, as well as the popularisation of use of LNG and natural gas.

Accompanying the halt of area designation under the Compensation Law in 1988, the CL levy affected SO_x emitters by providing an incentive to build plants and other worksites in new locations and by continuing to provide an incentive to reduce SO_x emissions at existing plants and other worksites. Increases in SO_x emissions from plants and other worksites that have no obligation to pay the levy are affected by both incentives. One can observe the impact of the latter (incentive to reduce SO_x emissions at existing plants and other worksites) in effects such as the increase in the use of natural gas by the industry.

The Compensation Law was highly successful in providing relief quickly to numerous victims of health damage due to pollution. However, the imposition of the entire burden on SO_x (as opposed to sharing the burden across all pollutants that contributed to the health problems) triggered an excessive reduction of SO_x. If resources for SO_x reduction that were used after the SO_2 environmental standards were achieved in the late 1970s had instead been directed toward measures against other types of pollution, it might have been possible to achieve a greater reduction of health damage, and thereby achieve greater cost-effectiveness.

There are two main reasons why the CL levy did not induce much technological innovation. The first reason is the failure of the CL system to address the true causes of the health damage that the system tries to compensate. When the level of compensation is exogenously determined, the rational behaviour of the unified polluters is to do no SO_x abatement in order to minimise the sum of abatement costs and levy payments. Conversely, if the revenues were not exogenously determined, but were levied on the full range of air pollutants emitted with adverse health effects, their rational unified behaviour is to abate the true cause pollutants rather than SO_x so that the sum of the abatement costs and compensation amount is minimised. In this case, even the levy on SO_x could induce technological development to abate the true cause pollutants. But what businesses did was to reduce SO_x emissions. It shows that they did not form an industry-wide cartel to not reduce SO_x or to reduce the true cause pollutants. Instead, they were united to try to change the CL system, while individually trying to reduce their share of the burden through abatement, not innovation.

Until the late 1970s, firms had to reduce SO_x emissions because of legal regulations and PCAs. The uncertainty of levy rate did not prevent businesses from investing to comply with the laws and PCAs. As the result, they reduced the emissions to the sufficient level. So, it seems reasonable that businesses started then to demand halting the certification of new patients. They managed to change the system and erased the uncertainty of the levy rate totally.

Moreover, the technologies which were developed in the 1970s because of the stringent legal regulations and PCAs in the highly accumulated industrial areas were almost sufficient enough to bring about the subsequent emissions reduction in other areas in the 1980s. The CL levy contributed more to the diffusion of the SO_x abatement technologies developed earlier than to the development of them.

For more information on Japan's levy, the full version of the case study (OECD, 2009) is available at *www.olis.oecd.org/olis/2009doc.nsf/linkto/com-env-epoc-ctpa-cfa(2009)38-final*.

Reference

OECD (2009), *The Impacts of the SO_x Charge and Related Policy Instruments on Technological Innovation in Japan*, OECD, Paris.

OECD PUBLISHING, 2, rue André-Pascal, 75775 PARIS CEDEX 16
PRINTED IN FRANCE
(23 2010 05 1 P) ISBN 978-92-64-08762-0 – No. 57481 2010